Biological Fluid–Surface Interactions in Detection and Medical Devices

RSC Detection Science Series

Editor-in-Chief:
Professor Michael Thompson, *University of Toronto, Canada*

Series Editors:
Dr Sub Reddy, *University of Central Lancashire, Preston, UK*
Professor Damien Arrigan, *Curtin University, Perth, Australia*

Titles in the Series:
1: Sensor Technology in Neuroscience
2: Detection Challenges in Clinical Diagnostics
3: Advanced Synthetic Materials in Detection Science
4: Principles and Practice of Analytical Techniques in Geosciences
5: Microfluidics in Detection Science: Lab-on-a-chip Technologies
6: Electrochemical Strategies in Detection Science
7: Peroxynitrite Detection in Biological Media: Challenges and Advances
8: Biological Fluid–Surface Interactions in Detection and Medical Devices

How to obtain future titles on publication:
A standing order plan is available for this series. A standing order will bring delivery of each new volume immediately on publication.

For further information please contact:
Book Sales Department, Royal Society of Chemistry, Thomas Graham House, Science Park, Milton Road, Cambridge, CB4 0WF, UK
Telephone: +44 (0)1223 420066, Fax: +44 (0)1223 420247
Email: booksales@rsc.org
Visit our website at www.rsc.org/books

Biological Fluid–Surface Interactions in Detection and Medical Devices

Michael Thompson
University of Toronto, Ontario, Canada
Email: mikethom@chem.utoronto.ca

Christophe Blaszykowski
University of Toronto, Ontario, Canada
Email: cblaszyk@chem.utoronto.ca

Sonia Sheikh
University of Toronto, Ontario, Canada
Email: ssheikh@chem.utoronto.ca

Cesar Rodriguez-Emmenegger
RWTH Aachen University, Germany
Email: rodriguez@dwi.rwth-aachen.de

and

Andres de los Santos Pereira
The Academy of Sciences of the Czech Republic, Prague, Czech Republic
Email: santospereira@imc.cas.cz

ROYAL SOCIETY OF CHEMISTRY

THE QUEEN'S AWARDS FOR ENTERPRISE: INTERNATIONAL TRADE 2013

RSC Detection Science Series No. 8

Print ISBN: 978-1-78262-097-6
PDF eISBN: 978-1-78262-204-8
EPUB eISBN: 978-1-78262-894-1
ISSN: 2052-3068

A catalogue record for this book is available from the British Library

Published by The Royal Society of Chemistry,
Thomas Graham House, Science Park, Milton Road,
Cambridge CB4 0WF, UK

Registered Charity Number 207890

For further information see our website at www.rsc.org

Printed in the United Kingdom by CPI Group (UK) Ltd, Croydon, CR0 4YY, UK

Preface

The quantitative detection of species of biological interest in blood, serum, urine, and cerebrospinal and interstitial fluids by biosensors has been the subject of intense research over many years. The appeal of such analytical devices resides in their promise to offer relative simplicity in their use, selectivity/sensitivity for target analytes, low cost, and significant potential for operation in both the point-of-care approach and the automated flow-through station in the clinical biochemistry laboratory. However, despite considerable effort over the years, fouling or non-specific adsorption effects associated with sensor platforms have remained something of an intractable problem. The presence of biological matrix components, such as proteins and cells, at the sample–device interface can indeed lead to serious signal interference. Although perhaps considered by some to constitute a separate area of research, analogous surface phenomena are also important in terms of the biocompatibility exhibited by medical devices, whether employed *in vivo* or in extracorporeal applications. Indeed, there is much common ground among research devoted to these topics from a surface physico-chemical point of view. In the present book, the authors have attempted to prepare a cohesive treatment of the surface chemistry involved in the fouling, and its minimization, of biosensors and biomaterials in general.

The text is modular in character, with a modest level of deliberate overlap in some chapters. We anticipate a readership originating from a disparate set of research communities. The interested reader might very well wish to zero-in on sections where there is less familiarity. For example, the work opens with a concise look at the fouling problem and deals with the role of interfacial physical chemistry in this respect (***Chapter 1***). An appraisal of relevant methods for surface characterization is included in that section.

RSC Detection Science Series No. 8
Biological Fluid–Surface Interactions in Detection and Medical Devices
By Michael Thompson, Christophe Blaszykowski, Sonia Sheikh, Cesar Rodriguez-Emmenegger and Andres de los Santos Pereira

Published by the Royal Society of Chemistry, www.rsc.org

Obviously, surface science specialists may well 'skip' this chapter all together and proceed on to other aspects of the text. On the other hand, this topic may be helpful from a learning perspective to the more biologically-inclined reader. The book continues with a molecular view of the complex physical chemistry that lies behind the interactions of proteins with surfaces, pointing out basics but also warning of the perilous exercise that is to try and establish general principles in this regard (***Chapter 2***). In addition to the kinetics and thermodynamics of such processes, the chapter also includes a look at the various models that have been proposed to explain the mechanism of protein–substrate interaction. The section is completed with an overview of the practical concerns associated with protein adsorption upon exposure of synthetic surfaces to the quintessential biofluid: blood. The ensuing chapter then proceeds to a discussion on the interaction of biological cells with surfaces, which incorporates a brief description of relevant cell types, mainly intended for the unfamiliar, surface science-focused reader (***Chapter 3***). The section then moves on to a research area, which has attracted enormous attention over recent years: that of the adhesion, growth, guidance, and proliferation of cells on surfaces; all of which are especially important in fields such as tissue engineering and regeneration. There is an emphasis in the text on extra-cellular matrix proteins employed to orchestrate and mimic natural tissue. Included here is a description of the effects of substrate morphology, rigidity, and elasticity. The chapter concludes with a concise treatment of the interaction of cells with nanoparticles, which is crucial in the field(s) of clinical diagnostics and therapy. As implied above, the interaction of biological entities with solid surfaces has ramifications for the application of a wide variety of devices and systems used in modern medicine. The next chapter deals with this subject, including topics such as blood coagulation and inflammatory response, whose effects are crucial for the successful implementation of biomedical equipment, implants, and devices such as stents and vascular grafts (***Chapter 4***). The book then covers the voluminous amount of work that has appeared in the literature on the chemical surface modification of substrates with molecular coatings in an attempt to minimize surface fouling from a variety of biofluids, whether of human or animal origin (***Chapter 5***). This includes a vast array of surface chemistries ranging from relatively simple, ultrathin self-assembled monolayers to sophisticated, thicker polymer brush entities. The last section of that chapter is dedicated to the assessment of the applicability of said antifouling surface chemistry to biosensor technology. The final chapter (***Chapter 6***) represents an attempt by the authors to coalesce and rationalize past and current mechanistic arguments regarding the antifouling properties of surfaces, especially the pivotal role played by surface hydration through the inherent, molecular-level structuration of water at interfaces. This quite naturally leads to a discussion of future surface chemistry possibilities that are consonant with the application of biosensor devices in biological fluids and substrates from a biocompatibility point of view.

Finally, we especially would like to express our deep appreciation to all past and present members of the Thompson Biosensors Group at the University of Toronto and the colleagues at the Institute for Macromolecular Science, Czech Academy of Sciences, Prague, for their many valued research contributions and unfailing support. We also wish to thank Debby Repka, Edmund Chan, and Lisa Qiu of the University of Toronto for their dedicated assistance in the preparation of some of the content of the manuscript.

Michael Thompson, Sonia Sheikh,
Christophe Blaszykowski, Cesar Rodriguez-Emmenegger
and Andres de los Santos Pereira

Foreword

The Physical Chemistry of Blood–Surface Interactions

Controlling the interactions between blood and synthetic material surfaces remains as one of the most important challenges in the field of biomaterials today. As a general rule, any foreign material surface that comes in contact with blood becomes rapidly coated by adsorbed biomolecules (mostly proteins), leading to the body's ability to somehow recognize the surface as 'non-self' and react by inducing a thrombotic response. These natural processes are, of course, in place to protect the body from vascular damage and blood loss caused over the course of life's normal daily activities. However, these same processes, if left to themselves, lead to serious and often life-threating consequences when material surfaces are intentionally placed in contact with blood as part of a medical device (*e.g.*, a cardiovascular stent) or as a diagnostic procedure (*e.g.*, an intravascular catheter or biosensor).

Of course, today's surgical practice involves the use of a large number of medical devices that are intentionally placed in contact with blood, ranging from catheters, stents, and vascular prostheses, to extracorporeal systems used during cardiopulmonary bypass surgery and renal dialysis. The clinical approach to preventing material-induced thrombosis and the generation of thromboemboli, which can lead to the failure of the device and/or severe physiological complications, is to suppress the blood coagulation system on a systemic level. This is typically done through the administration of various anti-coagulant or anti-platelet drugs that systemically inhibit blood clot formation. While widely accepted in clinical practice, the consequences of this approach are to place patients at risk for adverse bleeding events due to the suppression of one of the body's most important protective mechanisms—*i.e.*, blood coagulation. A much more desirable, but yet unattainable solution is to

RSC Detection Science Series No. 8
Biological Fluid–Surface Interactions in Detection and Medical Devices
By Michael Thompson, Christophe Blaszykowski, Sonia Sheikh, Cesar Rodriguez-Emmenegger and Andres de los Santos Pereira

Published by the Royal Society of Chemistry, www.rsc.org

prevent thrombotic events from initiating locally at the material surface of the medical device itself, while maintaining the body's ability to use its natural blood coagulation mechanisms to provide protection where otherwise needed. Achievement of this capability requires the understanding of the fundamental processes that govern the interactions of blood with material surfaces; this is the primary focus of this book.

Blood is a very complex, heterogeneous fluid composed of a broad range of macromolecules and cells within a buffered saline media. The most important components of blood with respect to the function of medical devices placed in contact with blood are the proteins in the blood (of which there are over 4500 different kinds) and platelets, which are the cellular mediators of blood coagulation. Within seconds after coming in contact with blood, material surfaces are generally covered with a tightly adsorbed layer of proteins. The characteristics of this layer of adsorbed proteins (*i.e.*, the types of proteins adsorbed and their orientation, conformation, and packing arrangement) have been shown to strongly influence the degree of stimulation of the coagulation and complement cascades, as well as platelet activation. This non-specifically adsorbed protein layer also leads to the fouling of biosensor surfaces, compromising their sensitivity and specificity. It also provides conditions that are favourable for bacterial adhesion and subsequent biofilm formation, which can lead to chronic infection. The chapters presented in this book take on the challenge of addressing important aspects of each of these processes from a physical chemistry perspective, to provide the reader with an understanding of the underlying scientific principles involved and the practical consequences for various biomedical applications.

Chapter 1 sets the stage for the book by providing an overview of the importance of the interface between synthetic materials and the biological system. It addresses basic aspects of interfacial thermodynamics, the role of surface morphology, methods used to physically and chemically characterize surfaces, and methods used to characterize protein–surface interactions. This understanding sets the stage for Chapter 2, which goes into greater depth regarding the physical chemistry of the interactions between proteins and surfaces, as well as the mathematical models used to characterize this behaviour. This chapter continues on to address the other side of the coin, namely the composition of blood and the blood coagulation system, and presents an overview of the general response of blood when it comes in contact with foreign material surfaces. The perfect word used in this chapter to epitomize these interactions is 'complexity': the complexity that involves the combined influences of protein structure, protein adsorption behaviour, and the subsequent biological consequences of these dynamic processes; all of which combine to make this an extremely challenging yet fascinating area of study.

While Chapter 2 focuses on interactions at the molecular level, Chapter 3 goes on to focus on cellular interactions with material surfaces. An overview of cellular structure and function is first presented (intended primarily for materials scientists and engineers *vs.* biologists) along with their interactions with substrates to mediate cell adhesion, growth, guidance, and proliferation.

Examples of approaches are then presented using surface chemistry, morphology, and elasticity, to control cellular responses for various types of medical devices, including stents and extracorporeal systems used for cardiopulmonary bypass and renal dialysis. This chapter concludes with sections dedicated to the important topic of nanoparticles and their use for therapeutics and diagnostics (or in their combined form as 'theranostics'). It also addresses concerns regarding nanotoxicity, which represents a unique material surface issue given the fact that nanosized particles are able to interact not only with the external surface of cells, but with intracellular processes as well.

Chapter 4 continues to build on the foundation laid by the previous chapters by going into more detail on the biological consequences of the interactions between foreign material surfaces and blood, and addresses topics including hemostasis, blood coagulation, inflammation, biofouling, and biofilm formation. As pointed out in this chapter, one of the key concepts for improving biocompatibility and biosensor sensitivity and specificity is to prevent the non-specific adsorption of proteins altogether. This topic is then squarely addressed in Chapter 5, which presents an overview of the state-of-the-art types of surface treatments and coatings that have been developed and shown to provide a high level of resistance to non-specific protein adsorption. The fundamental physical chemistry aspects governing the prevention of protein adsorption are then further addressed in Chapter 6 with a focus on the important role of water structure at the interface. Finally, Chapter 6 also provides a concluding overview of future considerations and directions toward the long-range goal of providing truly blood compatible surfaces for medical devices used for the diagnosis and treatment of disease.

As one who has spent a large part of his career studying these processes and developing methods to physically characterize protein adsorption behaviour on material surfaces, and to understand how this influences platelet response and blood compatibility, I see this book as a very welcome and valuable addition to the biomaterials literature. The content of this book is especially important for biomaterials scientists and engineers; including students, postdocs, and professionals in both academic and industrial settings, as well as for clinicians who wish to better understand the fundamental principles that scientists and engineers are trying to comprehend and control through surface design. Unfortunately, the specific answers are not yet available regarding how to achieve biomaterial surfaces for clinical applications that are completely resistant to biofouling and the generation of thrombus/thromboemboli without the systemic inhibition of coagulation processes. However, the fundamental principles presented in these chapters do provide the foundation that is needed to eventually bring the dream of developing truly blood-compatible surfaces into reality.

Robert A. Latour
Clemson University
Clemson, SC USA

Contents

RSC Detection Science Series No. 8
Biological Fluid–Surface Interactions in Detection and Medical Devices
By Michael Thompson, Christophe Blaszykowski, Sonia Sheikh, Cesar Rodriguez-Emmenegger and Andres de los Santos Pereira

Published by the Royal Society of Chemistry, www.rsc.org

CHAPTER 1

Relevant Aspects of Surface Physical Chemistry

1.1 Introduction

Advances in biosciences, the discovery of new drugs, the development of novel clinical diagnostic methods, and improvements in food safety rely heavily on modern analytical techniques. These technologies provide scientists with the ability not only to detect analytes, biomarkers, or specific pathogens, but also to quantify biochemical and biophysical interactions among biomolecules and cells. The sensitive monitoring of biomarkers allows the early detection of disease before its progression, opening the possibility of treatments with the highest probability of success.

Generally, the methods used in routine biochemical analyses are based on immunoassays and require the use of labeled antibodies as reagents. A further step, such as an enzymatic reaction, is usually necessary to produce a physical signal that can be read. However, affinity biosensors are able to quantify the analyte of interest without using additional reagents. They rely on a transducer element that can directly translate the concentration of analyte in its immediate vicinity into a physical signal. The transducer surface must be equipped with appropriate biorecognition elements that can bind the analyte. The advantages of label-free measurements include rapid analysis times, real-time monitoring of concentrations, the possibility of measuring the binding kinetics, and ease of automation and miniaturization. Automation and miniaturization are critical, as they offer the possibility of measuring a large number of analytes in parallel, with high throughput. Such detection methods rely on the conjugation of bioreceptors. Their continuous development depends both on advances in the design of

RSC Detection Science Series No. 8
Biological Fluid–Surface Interactions in Detection and Medical Devices
By Michael Thompson, Christophe Blaszykowski, Sonia Sheikh, Cesar Rodriguez-Emmenegger and Andres de los Santos Pereira

Published by the Royal Society of Chemistry, www.rsc.org

physical sensors and the implementation of novel sensing principles, and on the precise control of the interactions between these artificial devices and the biological fluids/environments in which they are meant to operate.

1.1.1 Materials and Biological Systems – The Biomaterials Interface

An effective affinity biosensor must be able to concentrate the analyte from the solution on its surface without eliciting non-specific interactions with other components present in the medium. While the transducer is able to generate a readable signal due to the presence of the analyte, it generally cannot discriminate between changes arising from the analyte concentration and those arising from the concentrations of interfering substances. The biological fluids of interest for biosensing are highly complex matrices, such as blood plasma and serum, which will be discussed in depth in later chapters. They contain an enormous and varied range of potential interferents.

The interactions of biosensors with biological fluids are critical for determining their performance; nonetheless, many applications are affected by such interactions. In fact, the whole category of biomaterials can be defined as materials (other than food and drugs) that are engineered to be used, either independently or as part of a larger system, in human or veterinary medicine for diagnostic or therapeutic procedures involving continuous contact with biological fluids. It is clear from this definition that affinity biosensor surfaces fall within this category as they can perform their function only while they are in contact with the biological matrix in which the analyte is present. Moreover, bioimplants and therapeutic devices which come into contact with biological fluids are subject to the same phenomena. Even outside the field of medical practice, biological and biophysical research is constantly faced with events that occur when artificial and biological systems come into contact.

For a biomaterial to support the interactions at the interface required for each application, the surface must be functionalized with appropriate bioreceptors. These molecules can bind a specific target in solution or, more generally, perform or promote a specific function. Examples include, but are not limited to, antibodies, enzymes, aptamers, oligonucleotides, receptor proteins, molecularly imprinted polymers (MIPs), peptide motifs, and even whole cells. However, they need to be immobilized in such a way as to ensure a sufficient quantity, homogeneous distribution, appropriate spatial orientation, and preservation of the bioreceptor structure and conformation. Particular applications may impose additional requirements, such as site specificity on the bioreceptor molecule or spatial selectivity to create patterns on the surface.

However, upon contact of artificial materials with complex biological media, components of the fluid may be spontaneously deposited on the surface. This non-specific adsorption, called fouling, can dominate the biomaterial–biological fluid interaction, ultimately impairing the performance of a device.

1.1.2 The Problem of Fouling

The biological components most frequently responsible for fouling on biomaterials are proteins. Their complex structures contain regions with varying hydrophilicity and hydrophobicity, and also charges that can mediate interactions with surfaces *via* multiple mechanisms. Chapter 2 will provide an in-depth discussion of proteins and the origins of fouling. The deposition of a layer of protein on a biomaterial surface immediately and irreversibly alters its properties. This can affect biosensors in two ways. First, the presence of the fouling layer can occlude the immobilized biorecognition elements, preventing the specific binding of the analyte from taking place. Second, label-free affinity biosensors function by directly detecting the analyte on the basis of its intrinsic properties (*e.g.*, mass, dielectric permittivity, and electrochemical redox properties), so the properties involved in fouling can themselves give rise to a detection signal. Since the attachment is non-specific and does not depend on the presence of the biorecognition elements, a signal produced by fouling can be orders of magnitude stronger than the signal related to the specific analyte. This source of interference can render the sensor inoperable, especially for media containing high concentrations of protein in solution. This is why biosensing in these media is considered particularly challenging.

A layer of deposited protein resulting from fouling has critical consequences for implantable and therapeutic devices. The proteins on the surface mediate the subsequent *in vivo* response, giving cues for an immune response. This leads to foreign body reactions and the encapsulation or rejection of implants. When these devices come into contact with blood, thrombogenesis is triggered, causing potentially deadly embolisms. Protein fouling also mediates the attachment of cells and bacteria, with an increased risk of infection.

The next section of this chapter will discuss the properties of surfaces and the mesoscopic forces that act at the nanoscale level in terms of their relevance to biosensors and biointerfaces. The last section of the chapter will describe the experimental methods used to characterize biomaterial surfaces and their interactions with proteins.

1.2 Interfaces and Their Properties

The interfaces between biological media and biomedical devices have a crucial effect on their behavior and ultimately on their performance. Accordingly, the properties that describe interfaces will now be considered. The first step is to discuss interfaces in broad terms, focusing on those composed of solid surfaces in contact with aqueous media. It should be noted that the general definitions are also applicable to the interfaces formed between two immiscible liquids, a gas and a liquid, or two solid phases. Emphasis will be placed on the qualitative rather than the quantitative aspects. For a detailed treatment, including thermodynamic derivations of the relationships, the reader is referred to specialized texts.[1–3]

Precise definitions of the main interfacial properties are imperative since they are used to characterize the behavior of materials. Understanding these properties is essential for describing the biophysical phenomena that determine the performance of biosensing and medical devices, the focus of the remaining chapters of this book.

1.2.1 Definition and Thermodynamic Aspects

The term interface refers to the boundary region between two material phases. Imagining an observer moving from one phase across the interface into the second phase makes it possible to see that the transition of properties takes places across a certain physical volume and is not infinitely sharp. The interface is defined as the region in which this transition takes place. Not only does the presence of an interface imply that the two phases coming into contact have different bulk properties, but also that new properties emerge which are associated exclusively with the boundary region. Consider the example of the interface formed when a droplet of oil is suspended in water. For certain properties such as the refractive index or viscosity, as the observer moves from inside the droplet across the interface and into the water phase, the change will be monotonic. However, this may not be the case for all properties of the bulk phases.

To illustrate the concept, consider a detergent, *e.g.*, sodium dodecyl sulfate (Figure 1.1a), dissolved in the water phase. The chemical structure of a detergent comprises two distinct regions: each molecule is composed of a polar head group capable of interacting strongly with water, and a non-polar tail with a tendency to associate with oil molecules. Its amphiphilic character causes the detergent to accumulate at the interface, each molecule

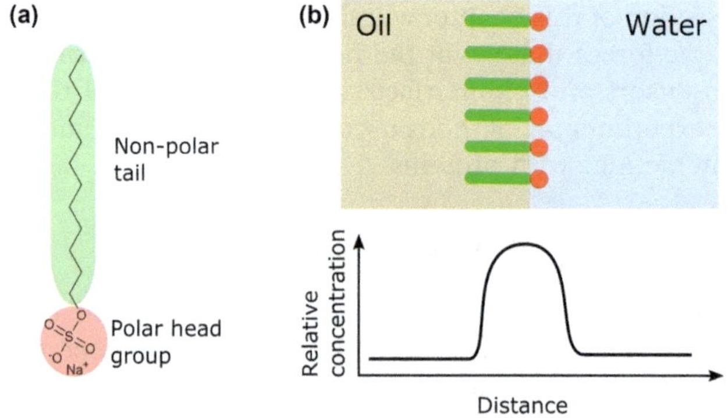

Figure 1.1 Example of a surfactant at the oil–water interface. (a) Chemical structure of sodium dodecyl sulfate and (b) schematic representation of its concentration profile at the interface.

oriented with its head group towards the water phase and its non-polar tail towards the oil phase. If the detergent concentration is the property of interest (Figure 1.1b), there will no longer be a monotonic trend leading from the bulk concentration of the first phase to the bulk concentration of the second. The concentration will have a maximum at the interface.

The many forces and interaction potentials that play a role at interfaces will be covered later in this chapter. The appearance of specific surface- and interface-related phenomena is of special importance in systems in which the ratio of surface area to volume is especially high, such as colloids and nanoparticle dispersions. The macroscopically observable properties of these systems, including their stability and rheology, are in fact determined by interfacial phenomena.

Now consider a system composed of two homogeneous immiscible phases α and β (of any type) and their interface, denoted by σ. From the above discussion, it is clear that the interface exhibits physicochemical properties that are distinct from those of either phase. Therefore, to describe the physicochemical properties of the system as a whole, the properties of the interface must be considered separately. For example, the internal energy can be expressed as

$$U = U^{\alpha} + U^{\beta} + U^{\sigma} \tag{1.1}$$

where U^{α} and U^{β} represent the internal energy of the two phases, and U^{σ} represents the internal energy of the interface region. Analogous equations can be formulated relating other extensive properties of the system to their counterparts in each of the phases and the interface. In particular, for each component i, the number of moles N_i can be expressed as

$$N_i = N_i^{\alpha} + N_i^{\beta} + N_i^{\sigma} \tag{1.2}$$

While it is obvious that the total number of moles of each component is the sum of the numbers in the different regions, it is important to note that the interface also contributes to the equation. In principle, these intuitive relations should make it possible to determine the contributions of the interface to the properties of the whole system simply by considering how the properties vary across different regions and by considering the volume of each phase V^{α} and V^{β}. In the bulk regions, the internal energy of the phases is given by

$$U^{\alpha/\beta} = V^{\alpha/\beta} u^{\alpha/\beta} \tag{1.3}$$

where $u^{\alpha/\beta}$ is the specific internal energy of each of the phases. Similarly, in a homogeneous phase, the number of moles of each component is related to the volume of the phase by the concentration c_i

$$N_i^{\alpha/\beta} = V^{\alpha/\beta} c_i^{\alpha/\beta} \tag{1.4}$$

It follows that the number of moles in the interface is given by

$$N_i^{\sigma} = N_i - V^{\alpha} c_i^{\alpha} - V^{\beta} c_i^{\beta} \tag{1.5}$$

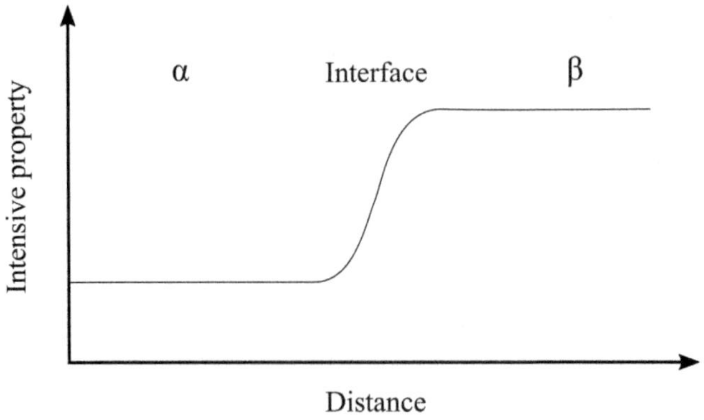

Figure 1.2 Variation of intensive material properties in the interfacial region. The property can represent density, viscosity, refractive index, *etc.*

However, as mentioned previously, the change in a property that takes place between the bulk regions of the two phases occurs over a narrow but not infinitely thin region. Accordingly, the values obtained for the quantities associated with the interface will depend on where the interface is defined to be. Even knowing that the sum of the volumes of the two phases equals the total volume does not determine exactly where one phase begins and the other ends. Thus, although the total volume V of the system is known, the individual volumes V^{α} and V^{β} of the phases are not. It is still necessary to pinpoint the location of the interface (Figure 1.2).

Following the convention introduced by Gibbs, all the properties of the interface can be assigned to an infinitely thin plane, and the boundary between the phases can be defined so that for one major component, the excess of moles at the interface is zero. For example, let the phase α be an aqueous solution, as is the case for biological fluids. Then the interface can be taken to be the dividing plane between the bulk phases such that the excess number of moles of water at the interface is zero:

$$N_w^{\sigma} = 0 \tag{1.6}$$

Knowing the concentrations c_w^{α} and c_w^{β} of water in both bulk phases, the above condition makes it possible to define the volume of the aqueous solution (phase α) using eqn (1.5). This fixes the position of the delimiting plane. Intuitively, $c_w^{\alpha} > c_w^{\beta}$ because α is an aqueous phase immiscible with β. From eqn (1.5) and (1.6), and the condition that the total volume is the sum of the volumes of α and β, it is clear that there is only one possible value of V^{α}, and therefore one value of V^{β}, since the total volume of the system is fixed and $V = V^{\alpha} + V^{\beta}$. At this stage, it is possible to calculate the contribution of the interface to the thermodynamic internal energy of the system:

$$U^{\sigma} = U - V^{\alpha}u^{\alpha} - V^{\beta}u^{\beta} \tag{1.7}$$

For the purpose of thermodynamic analysis, the interface can be considered to be an infinitely thin surface with area A, which can be determined according to the Gibbs criterion. It is useful to define the intensive interface properties, such as the specific interface internal energy:

$$u^{\sigma} = \frac{U^{\sigma}}{A} \tag{1.8}$$

1.2.2 Surface Tension and Surface Free Energy

For homogeneous systems, the enthalpy is defined as $H \equiv U + pV$, and it is a state function. For many systems and processes, especially in chemistry and biology, enthalpy is the preferred quantity for calculating energy balances because a change in enthalpy at constant pressure equals the heat input into the system plus the work output from the system other than pressure–volume expansion work. However, once an interface is introduced into the system, the mechanical work must include not only the pressure–volume work $p\Delta V$, but also the work required to enlarge or shrink the interface. This change can be associated with the difference ΔA in the surface area of the interface. The following paragraphs describe how this work can be incorporated.

The key concept that must be considered at this stage is that the creation of an interface or the extension of an existing interface requires the input of work. Therefore, there is a certain energy associated with the existence of an interface that depends on its dimensions.

Consider a liquid supported by a rectangular frame on three sides with a movable slider of length l on the fourth side, under constant pressure and temperature (Figure 1.3). Pulling the slider a distance d will increase the area

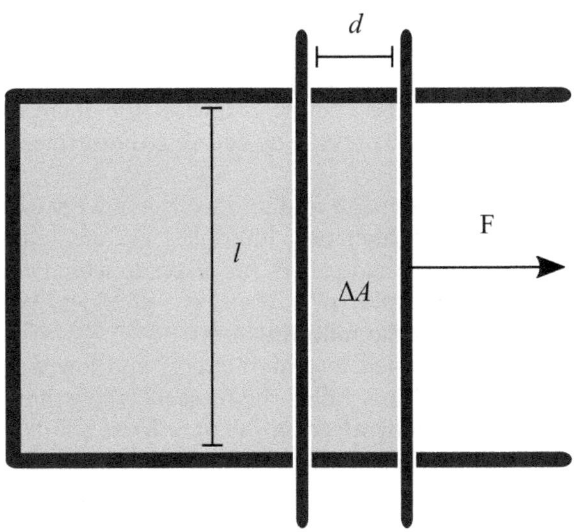

Figure 1.3 Interfacial tension. A layer of liquid is held by a frame on three sides, while the fourth side is able to slide. As it is pulled under constant force, the surface area of the layer of liquid is extended on both sides.

of the liquid interface by $\Delta A = 2ld$ The factor of 2 appears because the new interface area is created on both sides of the liquid layer. Such movement will require a force F to be applied, which is constant throughout the pulling process. The absolute value of the force exerted on the liquid to increase its area is given by

$$|\mathrm{F}| = 2l\gamma \tag{1.9}$$

The proportionality factor γ is known as the surface tension of the liquid, and it has units of force divided by distance (N m^{-1}). The work required for this process is proportional to the increase in the area of the liquid. It is given by

$$W = \gamma\,\Delta A \tag{1.10}$$

Since the process occurs under constant pressure and temperature, this work equals the increase in the Gibbs free energy of the system. As a result, γ is also the increase in the surface free energy:

$$\gamma = \left(\frac{\partial G}{\partial A}\right)_{T,P} \tag{1.11}$$

The excess free energy of a surface is the result of the loss or rupture of chemical bonds and/or attractive van der Waals and electrostatic interactions of the molecules at the interface; these losses arise because there are fewer neighboring molecules than in the bulk phase. Therefore, liquids in air tend to minimize their surface area. As an example, a droplet of water in space not subject to gravity tends to have a spherical form since this shape minimizes the surface area to volume ratio.

In a solid, the creation or enlargement of the surface area is also accompanied by an increase in free energy. Likewise, the increase in the free energy of the system can be expressed by defining the surface free energy as in eqn (1.11).

The concepts of surface tension and surface free energy are essential for the study of processes in which new interfaces are created. Wetting and adhesion are two important examples of processes in which changes in area are determined by the interplay of surface free energies. Wetting will be explained in more detail in the following section.

Typically, solids are classified into high energy and low energy solids according to the type of bond that holds the molecules together. Solids of the first type are composed of networks of covalent or ionic bonds, which cannot be broken without a large amount of energy. In contrast, in low energy solids, individual molecules are attracted to each other due to much weaker van der Waals interactions or hydrogen bonding. The energy of these bonds, which must be overcome to form a new solid surface, determines the surface free energy of a solid. In practice, because of energy dissipation and plastic deformation during the process of breaking, the energy input required to split a solid and create two new surfaces can be much greater than the energy of the bonds involved.

The surface tension of liquids follows an analogous trend. Alkanes, interacting only through van der Waals dispersion forces, have very low surface tensions. The behavior of water molecules is different; they form a network involving hydrogen bonds, which is disrupted by the presence of an interface. This carries a large free energy penalty, and consequently, the surface tension of water is high.

1.2.3 Wetting: Contact Angle and Capillarity

The presence of a surface renders a contribution, which is always positive, to the total free energy of a system. The tendency to minimize the surface free energy is evident in the wetting phenomena observed in systems consisting of multiple phases that come into contact, and of the resulting interfaces. The example of the interfaces created when a droplet of liquid comes into contact with a solid surface will be addressed in detail due to the importance of such systems in many practical applications. Similar surface energy arguments apply to systems consisting of immiscible liquids that come into contact.

As a droplet touches a solid, three interfaces are created: solid–air, liquid–air, and solid–liquid (Figure 1.4). The spreading of the droplet on the surface increases the area of the solid–liquid interface at the expense of the solid–air interfacial area. The tendency of a droplet of liquid to spread on the solid surface depends on the relative interfacial free energies of the solid–air, liquid–air, and solid–liquid interfaces. Equilibrium is reached when the interfacial areas are such that the free energy of the system is minimal. In this situation, the interfacial tensions can be analyzed at the point of contact of the three phases. At equilibrium, the sum of the components of the forces parallel to the surface must be a zero net force acting on the contact region of all the phases:

$$\gamma_{SL} + \gamma_{LA} \cdot \cos(\theta) - \gamma_{SA} = 0 \tag{1.12}$$

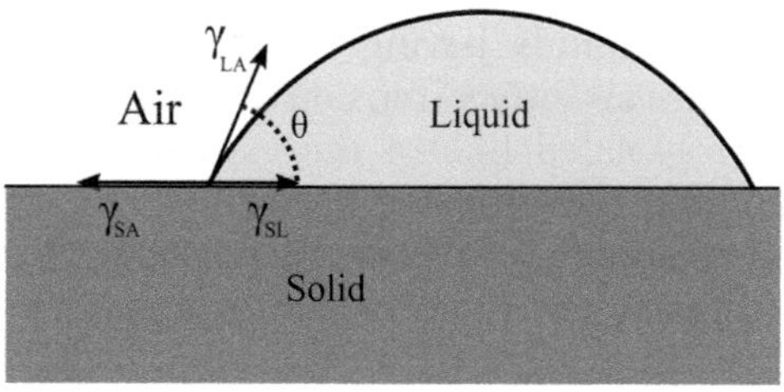

Figure 1.4 Contact angle at the interface of a solid and a liquid in air. Under equilibrium conditions, the contact angle θ formed is such that the components of the interfacial tensions parallel to the surface are balanced and Young's equation is fulfilled.

where γ_{SL}, γ_{LA}, and γ_{SA} are the interfacial tensions of the solid–liquid, liquid–air, and solid–air interfaces, respectively. The condition given by eqn (1.12), known as Young's equation for the contact angle, determines the spreading of the droplet on the solid surface. The equilibrium contact angle can be calculated from the surface free energies of the interfaces. The more energetically favorable the spreading of the droplet (*i.e.*, increasing the solid–liquid area at the cost of decreasing the solid–air interface area), the lower the contact angle and the larger $\cos(\theta)$. In the extreme case, a contact angle of zero amounts to complete spreading of the liquid on the solid surface, and the solid is said to be fully wettable by the liquid. This can occur, for example, with low energy liquids or high energy solids. At the other end of the wettability scale, very high contact angles, *e.g.*, greater than 90°, can be observed for water on smooth fluorinated surfaces such as PTFE, which has extremely low surface energy. The most important liquids for applications in the biological sciences are water and its solutions. Surfaces that are easily wettable by water, *i.e.*, those having low contact angles, are called hydrophilic; materials that repel water and have high contact angles are called hydrophobic. The affinity of a surface for water is critically important for determining its interactions with biological media and components, as will be discussed in subsequent chapters of this book. The most important physicochemical features influencing the hydrophilicity/hydrophobicity of a material surface will be addressed later in this chapter in the context of surface chemistry and surface chemical functionality.

The contact angle of water or other liquids on a surface can be a valuable source of information about the surface free energy and hence the surface chemistry.[4] However, this parameter cannot be obtained directly from Young's equation since it contains two unknown variables, γ_{SL} and γ_{SA} (the surface tension γ_{LA} of a large range of liquids at different temperatures can be found in the literature). A quantity commonly used to approximate the energy of a surface is the critical surface tension, which can be obtained by constructing a Zisman plot. The critical surface tension is defined as the maximum surface tension of a liquid that will fully wet the surface. A liquid having a surface tension higher than the critical surface tension of a solid will not spread completely and instead will form a finite contact angle. A Zisman plot, based on measurements of the contact angles between a surface and various liquids having different surface tensions, graphs $\cos(\theta)$ *vs.* γ_{LA}. The critical surface tension is obtained as the intercept at $\cos(\theta) = 1$ of the extrapolated line of the experimental points. Figure 1.5 shows the calculation of the critical surface energy for PTFE.[5]

However, several objections have been raised about the relationship of the critical surface tension to the performance of a surface in biomedical applications. To begin with, critical surface tension values are usually found for low surface tension liquids, such as alkanes and other organic solvents, to avoid the swelling of the solid by the liquid. However, it is the interactions with an aqueous environment that are critical for bioimplant and biosensing applications. The behavior of a surface in contact with air or low surface

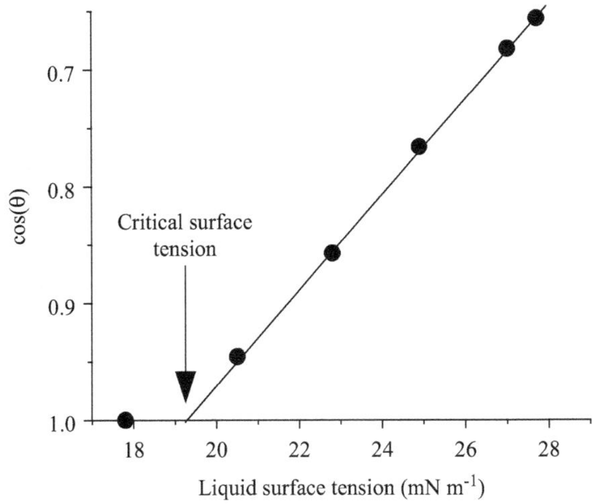

Figure 1.5 Calculation of the critical surface tension of a solid by the Zisman method. The contact angles on the surface of PTFE were measured with a series of di(*n*-alkyl)ethers. The intersect of the linear fit at $\cos(\theta) = 1$ represents the critical surface tension. (Prepared from data from Table I in ref. 5.)

tension liquids may not be a good predictor of its behavior at an interface with water.[6] The quantification of the energy of interaction with non-polar liquids will yield only the dispersive component (*i.e.*, London forces; see below) of the surface free energy. Another problem is that Zisman plots are often non-linear. In addition, while the Zisman plots used to calculate the critical tension may have a reasonably good fit for a series of liquids within the same family of compounds (*e.g.*, a series of alkanes, a series of halocarbons, or a series of alkyl ethers), the values obtained for different families can vary due to the interactions between the liquids and the surface.

The wetting of a smooth surface by a droplet of liquid is a dynamic process, and usually a different contact angle is observed for a droplet whose volume is increasing than for a droplet whose volume is decreasing. When a droplet of liquid is placed on a surface and its volume is slowly increased by the careful addition of liquid (for example, through a needle connected to a syringe pump), the contact area begins to expand, and the droplet forms a characteristic advancing contact angle. If the process is reversed and liquid is slowly drawn out of the droplet, thereby decreasing its volume, initially the contact area remains unchanged, and the droplet will be pinned while the contact angle decreases. If liquid continues to be drawn out of the droplet, reducing its volume, at some point, the contact area will start to decrease, and the droplet forms a characteristic receding contact angle. The equilibrium contact angle, as calculated from the interfacial energies and Young's equation, lies between the advancing and receding contact angles. The difference between the advancing and receding contact angles is called contact angle hysteresis, for which there are various causes. While hysteresis complicates the analysis of the ideal thermodynamic equilibrium contact angle, it can be a source of valuable information about the physical chemistry of the surface. It has been argued that contact angle hysteresis is a manifestation of the pinning of the three-phase contact line due to the activation energy required for it to start moving. The energy barrier to the advancing and receding of the contact line can be caused by an inhomogeneous surface, for example, one containing microscopic areas, each with a different surface chemistry. In the case of contact with water, the surface can comprise microscopic hydrophilic and hydrophobic patches. The contact line has to jump across these patches, creating a local energy barrier to its movement, which results in contact angle hysteresis. Swelling and relaxation of the surface also lead to contact angle hysteresis, as the physicochemical structure of the surface changes after it is wetted, lowering the interfacial free energy. This also results in the pinning of the contact line.

Furthermore, the surface morphology plays a major role, strongly influencing the equilibrium contact angles and hysteresis. For rough surfaces, there tend to be large differences between the advancing and receding contact angles. This can be visualized by picturing the contact line for a rough surface at the microscopic level. On a scale smaller than the scale of the roughness, a rough surface appears smooth but wavy, and Young's equation is satisfied locally. However, the slope of this smooth surface varies along the corrugations of the roughness. From a macroscopic perspective, the areas having different slopes behave like patches with different equilibrium contact angles. The hysteresis observed in the apparent contact angles is a result of this variation on the macroscopic scale.[7]

With regard to the equilibrium contact angle, two models have been used to explain the effect of roughness: the Wenzel model, and the Cassie–Baxter model. In the Wenzel model, a roughness parameter r is defined for the solid surface: the ratio of the actual rough surface area to the ideal geometric area that the same surface would have if it were perfectly flat, *i.e.*, the geometric area of the base of the liquid droplet. The apparent contact angle θ^* for a rough surface can be calculated by minimizing the total free energy of the system, considering the contributions from the three types of interfaces and their real surface areas, and taking the roughness into account. The equilibrium contact angle on a rough surface is related to the Young contact angle by the following equation:

$$\cos(\theta^*) = r \cdot \cos(\theta) \tag{1.13}$$

This equation shows that the apparent contact angle for a rough hydrophilic surface is lower than that for a smooth surface of the same material since the roughness ratio is always greater than 1. In contrast, for a hydrophobic surface with a contact angle greater than 90°, the effect of roughness is to increase the apparent contact angle. In extreme cases, this produces a superhydrophobic surface, *i.e.*, contact angles greater than 150°. These surfaces can trap air bubbles under the liquid droplet, effectively lowering the solid–liquid interface area, leading to a heterogeneous contact area. This behavior cannot be described by the Wenzel equation since the contact area is now heterogeneous, *i.e.*, composed of patches of solid–liquid and air–liquid interfaces. The Cassie–Baxter equation is used in this case because it takes into account the fraction of the solid surface area wetted by the liquid, as well as the roughness ratio of the wetted regions. The phenomenon of very high water contact angles resulting from air pockets on a textured surface is known as the lotus effect, since it is observed on the surface of lotus leaves. Water droplets can roll off such a surface, producing a self-cleaning effect. Artificial superhydrophobic materials with textured surfaces have been created that can replicate this behavior and are proposed for applications including self-cleaning windows and biomedical materials.

Surface tension also plays an important role in the way water rises inside a capillary tube. When a capillary (a very thin tube) is partially immersed vertically in water, the water level inside the tube rises higher than the level of the water outside the tube, a phenomenon called capillary rise. It is observed only if the surface of the tube is hydrophilic. As is the case with the wetting of planar surfaces and the formation of the contact angle, capillary rise is governed by interfacial tension: the surface tension of water pulls the contact line in such a way that the liquid can rise inside the tube until the surface tension pulling the liquid upwards is balanced by the weight of the liquid column formed. The height to which the water is able to rise inside the tube increases with the decreasing capillary inner radius. Capillary action is also responsible for the wetting of porous hydrophilic materials. For example, surface tension drives the wetting of the pore network of materials such as paper.

1.2.4 Surface Chemistry and Functionality

The chemical environment of the molecules and atoms on the outermost layer of a solid or liquid is markedly different from that in the bulk phase. The creation of an interface implies the breaking of chemical bonds that would otherwise continue into the second phase. For each atom or molecule on the surface of a material, there are fewer neighboring atoms or molecules than there would be in the bulk of the material. Microscopically, this can be seen as the origin of the surface tension of a liquid. In liquid water, a dynamic network of hydrogen bonds is formed which is responsible for the distinctive properties of water, such as its relatively high boiling point. However, at the interface between air and water, the molecules participate in hydrogen bonding only from the bulk liquid side. This loss of intermolecular bonds that are favorable in terms of potential energy gives rise to the surface free energy and therefore the surface tension of water. Furthermore, it also gives water the ability to interact at interfaces with the appropriate chemical functionality.

In the case of the surfaces of solids, typically composed of the exposed faces of crystallites or sometimes amorphous phases, the minimization of the surface free energy leads to physical and chemical relaxation and restructuring at the surface. Relaxation involves the shortening of the distance between the surface layer of atoms and the layer immediately adjacent to it because of the imbalance in the bond network that appears as a result of the interface. Moreover, the minimization of the free energy often leads to a change in the two-dimensional surface crystallographic structure of the solid. This restructuring involves changes in the distribution and periodicity of the atoms in the surface layer of a solid, which can be detected by surface sensitive diffraction techniques. These methods always require the use of ultra-high vacuum conditions (abs. pressure $<10^{-9}$ Torr) to avoid contaminating the surface.

A surface having an interface with any gas or liquid is always subject to possible interactions with the molecules, atoms, and ions present in the fluid phase. These can be the gas or solvent molecules themselves or a solute. In this way, the minimization of the free energy at the surface can lead to enrichment of the concentration of certain components in the interface, in a process termed adsorption. Adsorption is classified as either of two types, depending on the character of the interactions mediating the binding of the molecules from the fluid phase (adsorbate) and the surface: chemisorption, in which chemical bonding occurs; and physisorption, in which the binding is the result of weaker van der Waals forces.

For adsorption to occur, there must be sites on the surface to which it is energetically favorable for the adsorbate molecules to attach. The parameter typically used to describe the extent to which adsorption occurs is the surface coverage, usually denoted θ, which can be defined as the fraction of occupied adsorption sites, or the percentage of the substrate surface that is covered by a monolayer of adsorbate. The thermodynamics of the adsorption process is described using diagrams called adsorption isotherms, which express the extent of adsorption as a function of the pressure or concentration of the adsorbing molecule in equilibrium in the fluid phase at an interface with the substrate at a fixed temperature. Several models are used to describe the behavior of adsorption isotherms mathematically, based on different assumptions about the interactions between the substrate and adsorbate and among the adsorbate molecules once they are on the surface.

For chemisorption to occur, chemical groups that can bind the adsorbate must be present on the surface. A typical example, of great importance for the fabrication of biosensors, involves the attachment of molecules containing thiol groups on the surface of metallic gold. This can be exploited in combination with self-assembly phenomena (Chapter 5) to form ordered self-assembled monolayers on gold which are used to impart a desired chemical functionality, for instance, on a transducer surface.

Physisorption can take place on any solid–liquid or solid–gas interface, since no specific chemical groups are required on the surface. The adsorption of multilayers is possible since van der Waals interactions can also occur between adsorbate molecules.

Specific functional groups on the surface can also have critically important effects on the behavior of interfaces. Solid surfaces may contain sites that act as Brønsted acids or bases (proton donors or acceptors, respectively) or more generally as Lewis acids or bases (electron pair acceptors or donors, respectively). An example involving Brønsted acids is a surface on which carboxylic acid groups are present, which can be prepared *via* partial oxidation of an organic polymer. When the surface comes into contact with water, some carboxylic acid groups will dissociate and become negatively charged. This will dramatically alter the way in which the surface interacts with the aqueous phase.

Specific chemical groups, which can be created by reaction with a medium in contact with the surface, are relevant in surface-based sensing technologies since they can be exploited for the functionalization of the surface. To illustrate the principle with a common example, consider the surface of crystalline silicon.[8–10] Typically, it is covered by a native layer of about 1.5 nm of silicon oxide. This substrate can be functionalized in many ways by taking advantage of reactions on the surface. It can be activated under oxidative conditions, such as exposure to low-pressure air plasma, after which silanol groups (Si–O–H) will be present. These are often exploited for the immobilization of a further layer of functional organic silanes (Figure 1.6b).

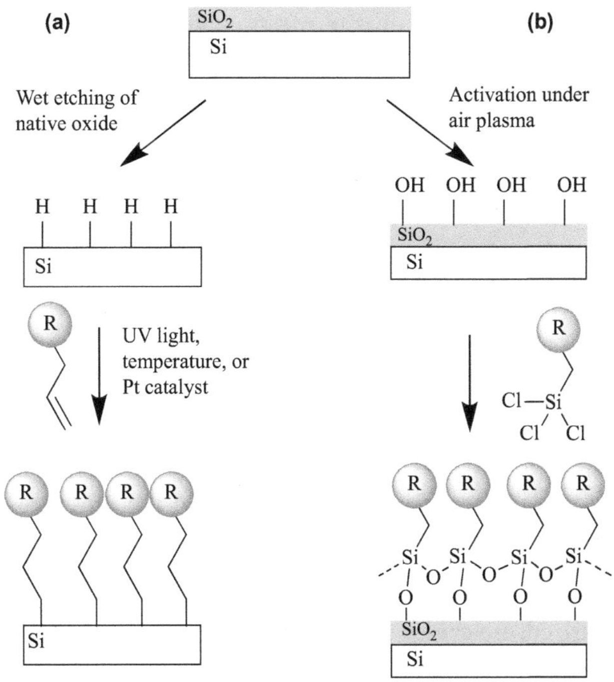

Figure 1.6 Chemical functionalization of the surface of silicon. (a) Etching of the native oxide layer followed by hydrosilylation of a functional alkene. (b) Activation to form silanol groups and subsequent silanization with a reactive trichlorosilane. The R group represents the desired chemical functionality.

A silicon surface can be also etched to remove the native oxide layer and yield Si–H groups, which under appropriate catalysis, can react with alkenes or alkynes to form a stable adlayer on the surface (Figure 1.6a).

The groups marked R represent an arbitrary chemical function that is desirable to immobilize onto the surface. Different substrates have specific chemical groups available on their surfaces and will require different chemical strategies to achieve functionalization. Therefore, knowledge and proper characterization of the available surface chemistry are essential for the design of biosensors.

1.2.5 Mesoscopic Forces Acting on Surfaces

Biological macromolecules and cells interact with each other and with interfaces through the same fundamental forces as atoms, ions, and small molecules: electromagnetic forces and gravity (typically of no importance on these scales). Van der Waals forces arise from the electrostatic interactions and the quantum-mechanical behavior of the electrons in atoms. They are also important, but they are not a fundamental force.

However, because of the larger sizes of the components of biological systems, their behavior can be understood through the appearance of forces which are better interpreted on the mesoscopic scale. These are not new forces per se, as they arise from the same fundamental forces that determine the behavior of atoms and molecules.

The familiar Coulomb force between two charged particles is also present between two charged surfaces or between a charged surface and an ion. Biological macromolecules may have a net charge, which will result in electrostatic forces. Ions and charged particles and surfaces can also participate in electrostatic interactions with neutral molecules since even in molecules with a zero net charge, the charge distribution is typically not homogeneous. While a region may contain a partial negative charge, it will be balanced by a positive charge in another region, forming permanent dipoles. The interaction potential and force, attractive or repulsive, are dependent on the orientation of the dipole with respect to the charge. If a dipole is in a liquid medium which allows it to rotate freely, there is still a net interaction with charges as they can affect the orientation of the dipole. The dipole will have a tendency to orient itself in the presence of ions and charged particles, and an attractive force will arise.

Similarly, charged particles can interact electrostatically with neutral molecules, inducing a partial charge separation referred to as polarization of the molecule. Polarized molecules are in fact temporary induced dipoles, which are also subject to electrostatic forces. Biomolecules are subject to electrostatic forces from charges by all of these mechanisms. They often possess a net charge; for example, DNA has an abundance of negatively charged phosphate groups. In the case of proteins, acidic and basic amino acid residues cause oppositely charged regions to develop in aqueous solutions. Most biomolecules also have polar groups which form permanent dipoles. Furthermore, they may experience polarization and the formation of induced dipoles.

When a solid is immersed in a liquid environment, typically aqueous, its surface may become charged by the acid–base dissociation of surface chemical groups or by the adsorption of ions from the solution. In this way, a layer of surface charge is built up; it will determine the electrostatic interactions at the solid–liquid interface. However, the presence of ions in the solution will cause a screening effect near the surface charge. Counterions, *i.e.*, ions of the opposite charge, will accumulate in a region close to the surface, attracted by the Coulomb force (Figure 1.7). Concurrently, the random motion of the ions resulting from their thermal energy tends to spread them out away from the surface in a region called the diffuse layer. An ion concentration gradient is established and, along with it, an electric potential distribution, which tends to zero with the increasing distance away from the surface into the bulk of the solution. Together, the charged surface and the diffuse counterion layer form the electrical double layer, which is of critical importance for understanding the behavior of particles and colloids. The mesoscopic forces arising from the electrostatic interactions of charged particles and surfaces can be understood in terms of the interactions caused by the overlapping of the electrical double layer regions. These are called double layer forces, and their magnitude depends on the ability of the

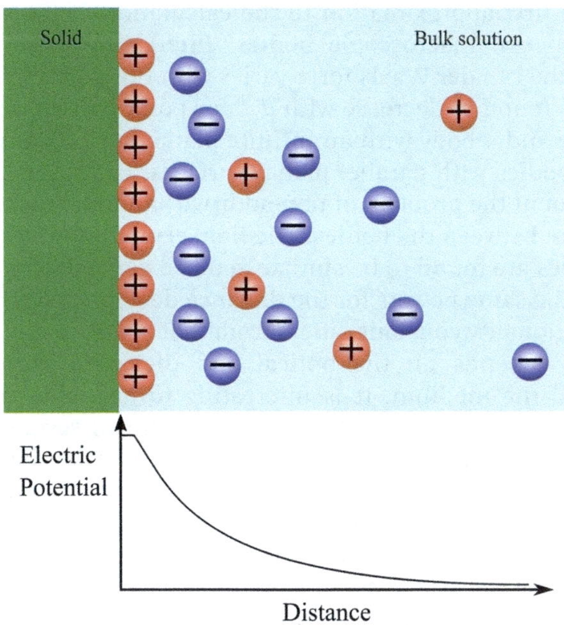

Figure 1.7 Schematic representation of the electrical double layer formed when a surface is placed in an electrolyte solution.

counterions to effectively shield the charge of the inner layers: high ion concentrations reduce the strength and range of the electrostatic repulsion force between identically charged particles. Furthermore, ions bearing higher charges will be much more effective at screening the electrostatic force arising from the surface charge.

Even without the presence of charges, neutral molecules are subject to intermolecular forces. These ubiquitous forces arise from three types of interactions: those between two permanent dipoles (the Keesom force); those with one permanent dipole inducing a dipole in a polarizable molecule (the Debye force); and those between an instantaneous dipole and an induced dipole (the London force). Collectively, they are known as van der Waals forces. Of the three, the London force dominates in terms of magnitude. In the case of the London force, the quantum nature of the molecules and the oscillations of the electrons lead to the spontaneous appearance of an instantaneous dipole. The influence of this dipole leads in turn to the polarization of a second nearby interacting molecule, forming an induced dipole. The attraction of these dipoles, which influence each other, forms the basis of the explanation of the London force.

Just like individual molecules, macroscopic bodies experience van der Waals forces. However, these forces are not linearly additive. This means that given three particles A, B, and C, the interaction between A and B is affected by the presence of C. Nevertheless, the Hamaker theory ignores this effect and adds the contributions of each molecule in a body pairwise to obtain the first approximation to the calculation of the van der Waals force between two macroscopic bodies. Interestingly, the distance dependence of the van der Waals force varies with the geometry of the bodies involved. It is found to decrease with d^{-6} for point particles; with d^{-3} for a point particle and a body with an infinite planar surface; and with d^{-2} for two infinite bodies with parallel planar surfaces. Later, the Lifshitz theory took into account the problem of non-additivity and included the influence of the medium between the bodies. The final expressions for the total force between bodies are found to be similar to those derived from the Hamaker theory, with the same powers for the distance dependence. The calculation of the proportionality constant differs from that conducted in the Hamaker approach: it depends on the optical and dielectric properties of the materials and the medium. It is interesting to note that van der Waals forces between bodies can reach distances of up to several nanometers. Furthermore, a remarkable consequence of considering the presence of a medium between the bodies is the observation that the force may be repulsive, depending on the properties of the two bodies and of the medium. However, for most relevant examples, van der Waals forces are always attractive: two bodies of the same material always attract, no matter what the medium. Repulsive van der Waals forces can occur when the medium between the two solids is attracted to one of the solids more strongly than the solids are attracted to each other.

Van der Waals interactions play an important role in molecular solids as they provide the cohesive force which holds a solid together. It is this force that needs to be overcome when the solid is split apart, creating two new surfaces.

The stability of a dispersion of colloids is determined by the balance between the attractive and repulsive forces between particles. The interplay between electrostatic interactions (repulsive or attractive) and van der Waals forces (always attractive in the case of particle dispersions) is modeled by the Derjaguin–Landau–Verwey–Overbeek (DLVO) theory. It explains the tendency of colloidal dispersions to aggregate upon the addition of an electrolyte due to the lowering of the potential barrier caused by electrostatic repulsion. Upon close approach, the particles tend to adhere due to van der Waals forces. This theory has also been used to explain the basic forces that lead to the adhesion of bacteria to surfaces, allowing them to attach.[11]

In aqueous systems, surfaces are also subject to other forces arising from the effects of the conformation of the molecules in the system and the structure of the solvent. These are often categorized as non-DLVO forces. Polar surfaces interacting strongly with water can have a hydration layer of bound water molecules, which gives rise to a short-range steric repulsion force upon the close approach of a similar surface. The range of this force is on the order of a few times the size of the water molecules. This effect has been shown to increase with the presence of certain dissolved ions, and it is believed to be enhanced by the strongly bound water molecules in the solvated shell of such ions.

Similarly, a solvation force occurs at larger distances (several molecular diameters) due to the discrete nature of the solvent. Imagining a solvent composed of hard spheres makes it possible to visualize how the ability of the solvent to fill the gap volume between two surfaces will oscillate with the separation distance. This leads to a periodic oscillation in the solvent structuring. In turn, there is an oscillation in the density of the solvent with respect to the bulk density away from the surfaces. This produces a force that exhibits periodic oscillations with respect to the distance, with a period of about the size of a solvent molecule (for water, 0.25 nm). At larger distances, the solvent density in the gap tends to approach the bulk density, and the force vanishes.

The ability of water to self-associate through a network of hydrogen bonds also has important effects on the interactions between hydrophobic surfaces in aqueous media. The hydrophobic interaction is a force of entropic origin that arises from the disruption of this network. The presence of a non-polar particle or surface, which does not participate in the hydrogen bond network, forces the water molecules at the interface to adopt an orientation with respect to the surface in order to maximize the number of hydrogen bonds among the water molecules at the interface. Thus, the surface is caged by the self-associating water molecules, forming structures called clathrates. At the same time, these water molecules lose conformational freedom, thereby decreasing the entropy of the system in proportion to the surface area of the particle. Because of this increase in free energy, the system tends to minimize the surface area by driving the particles together and joining non-polar surfaces, which do not participate in hydrogen bonding with water. The force caused by this potential is called the hydrophobic interaction, and it decays exponentially over a range of up to several nanometers, with a characteristic decay length of 1–2 nm. The hydrophobic interaction, which is much stronger than the van der Waals attraction, plays a central role in the tendency of non-polar particles to aggregate in aqueous media. In biological applications, it is thought to be one of the main factors driving the folding of proteins. The hydrophobic effect is also considered to play a major role in the non-specific adsorption of proteins and cells to surfaces, as will be discussed in Chapters 2 and 3.

Another important type of force of entropic origin arises from the presence of polymers on surfaces. A polymer chain grafted onto a surface can interact with a solvent and assume flexible conformations subject to thermal fluctuations. The approach of a second surface can induce compression of the chains, causing their fluctuations to overlap and thus restricting their conformational freedom. This produces a repulsive force called the steric entropic force. Here, the term entropic is used to avoid confusion with the purely steric energetic force that arises simply because two solids cannot overlap in space.

Finally, a perhaps less intuitive entropic force occurs upon the close approach of two particles in a liquid containing a macromolecular solute. Once the distance between the particles is smaller than the molecular dimensions of the solute, the macromolecule can no longer occupy the liquid gap between the particles. This results in a region which is depleted of solute in comparison to the bulk of the solution. Consequently, an attractive force is established between the particles, arising from the higher osmotic pressure of the solution surrounding the particles than that in the solute-depleted volume which separates them. This force, called the depletion force, leads to the destabilization of colloidal dispersions upon addition of high concentrations of a high molecular weight solute. In biology, this effect has been exploited to force the fusion of cells by the addition of poly(ethylene glycol). For example, the fusion of antibody-producing B cells and myeloma cells is performed to achieve immortalized antibody-producing cell lines for use in the preparation of monoclonal antibodies.

1.2.6 Surface Morphology

Understanding the shapes and sizes of the physical features of a surface is vital for studying its interactions with biological systems. While so far, the discussion has focused on planar geometries, *i.e.*, flat surfaces, it is worth commenting on the deviations from this idealized example.

The most extreme examples of non-flat surfaces are micro and nanoparticle systems. An immediate consequence of small particle size is that it necessitates surface curvature, resulting in an increase in the surface area to volume ratio. To visualize it simply, consider the formulae for the surface area and the volume of a sphere:

$$A = 4\pi r^2 \tag{1.14}$$

$$V = \frac{4}{3}\pi r^3 \tag{1.15}$$

As the radius decreases, the surface area of a single sphere decreases with r^2, while its volume decreases more rapidly, with an r^3 dependence. The result is that the ratio of surface area to volume is inversely proportional to the size of the particle:

$$\frac{A}{V} \propto r^{-1} \tag{1.16}$$

This phenomenon is characteristic not only of spheres, but also of any particle shape. For a given total amount of material, *i.e.*, a set mass or volume, reducing the particle size will increase the surface area, dramatically enhancing the importance of all surface-related properties and phenomena. For example, the adsorption of the components of a solution on the surface of a solid will be much more extensive when the solid is composed of finer particles. This is a direct consequence of the increase in the number of adsorption sites available on the larger surface. This phenomenon can be exploited to measure the specific surface area of a solid, *i.e.*, the area per amount of material, which depends on its morphology. The ability of a gas, usually nitrogen, to physisorb on a solid surface is measured and fitted with an adsorption isotherm model, from which the effective area can be obtained if a value for the surface density of adsorption sites is assumed.

Another morphological feature typical of solids is pores, whose presence and structure is relevant to the specific surface area. Porosity can increase the specific surface area available on solid bodies. For a porous material, it is important to characterize the pores in terms of their size distribution, volume fraction, and accessibility. These parameters determine the influence of the porosity on the overall behavior of the system in surface-related phenomena. For example, a material with a high specific surface area might not serve as an effective adsorbent if much of the surface area is located inside a pore network that is too narrow to allow the diffusion of the desired adsorbate. In this case, even if the surface chemistry is appropriate, the adsorption process might be very slow if diffusion is limited by the pore structure. However, a relatively closed pore structure can allow an enclosed substance such as a drug to be released slowly over time into the surrounding medium. Accordingly, it is necessary to characterize the morphology of the pore network.

The surfaces of solids are almost never perfectly flat as they contain bumps and valleys, which also lead to an increase in the specific surface

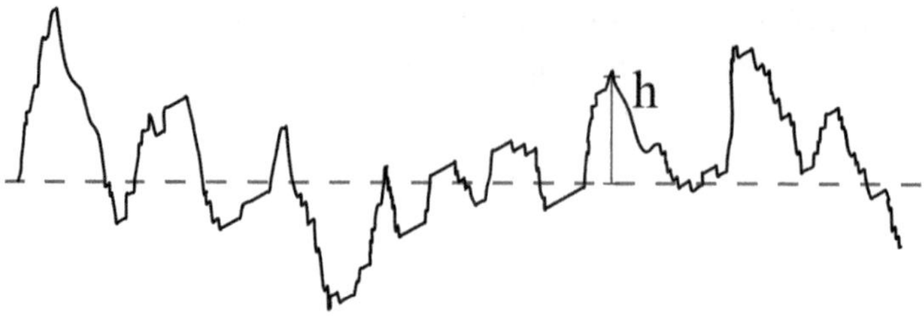

Figure 1.8 Height profile of a rough surface, obtained for example by AFM or profilometry, showing the variation of the height along a line segment. The roughness parameters are calculated by averaging the height differences between the surface topography and the line of best fit.

area. The deviation from flatness is typically quantified by the surface roughness R, which is defined as the average vertical distance h with respect to the line of best fit along a segment traced over the surface (Figure 1.8). Depending on the type of average used, different roughness expressions can be defined, the most important being the arithmetic average roughness R_a and the root mean square roughness R_q (also w or R_{RMS}):

$$R_a = \frac{1}{n}\sum_{i=1}^{n} |h_i| \tag{1.17}$$

$$R_q = \sqrt{\frac{1}{n}\sum_{i=1}^{n} h_i^2} \tag{1.18}$$

where n is the number of points sampled along the segment. The roughness value can be useful for comparing the amplitude of the height variations on surfaces with similar morphologies. Especially for surfaces with topographies that are randomly introduced by processes such as etching or sputtering, R_q can help to assess the extent of the modification and the size of the features in a direction normal to the surface.

However, roughness values give no information about the horizontal size distribution of the height variations. Depending on the material under consideration and the coating or etching techniques employed, these features can have characteristic horizontal lengths that differ by orders of magnitude, from several microns to a few tens of nanometers. Especially in biomaterial applications, this can dramatically alter the way in which a surface interacts with the medium with which it comes into contact. For the biological components sharing an interface with the material, this can change the mechanisms of interaction with respect to a flat surface, potentially leading to widely different outcomes. Features on the order of microns will appear almost flat or as having a slight curvature on the scale of individual protein molecules, whose sizes are 100–1000 times smaller. But since the sizes of such features are comparable to those of prokaryotic and eukaryotic cells, they can have a significant impact on the way cells behave upon contact with the surface. However, roughness on a scale of tens of nanometers can change the adsorption mechanisms of proteins.[12–14] The creation of nano and microscale curvatures at the edges associated with features of small size alters both the amount and the conformational state of the adsorbed proteins. Ultimately, these changes can govern the performance of a biomaterial surface. This is thought to be the reason the effects of roughness in biomaterial applications often go beyond what would be expected from calculations based on the larger surface area alone. The complex influence of surface morphology on cell attachment will be discussed in depth in Chapter 3.

For surfaces with uncontrolled or random roughness, the characteristic lateral dimensions of the roughness are often expressed in terms of the spatial frequency, which refers to the average distance between two peaks on the surface. Another parameter sometimes used to compare roughness features is the horizontal correlation length. This can be defined as the distance up to which the height of a given point influences the height of surrounding points; it is related to the average lateral size of morphological features on the surface.

An important class comprises surfaces having topographical features with precisely controlled geometry. For example, specific shapes can be introduced using lithography to induce patterns on surfaces. Generally, processing techniques that achieve uniform patterning over large surfaces (several cm^2) take advantage of self-assembly phenomena to direct the formation of topographical features. A common example is colloidal hole-mask lithography, in which a pattern of holes is obtained on the surface *via* 2D packing and self-assembly of polymer particles, typically by electrostatic interaction. After the sputtering of a metal and removal of the mask, a large-scale pattern of metal nanostructures such as nanorods remains. In the case of precisely controlled topographical features on a surface, the geometric parameters of the shapes tend to be much more important in describing the properties obtained by modification than the roughness values. These include the shape of the feature (*e.g.*, rods and cylinders, cones, pits, grooves, and creases), the height and width, the aspect ratio, the distribution on the surface and the periodicity, and the directionality.

Another important property of the surface morphology is anisotropy, which refers to the preferential orientation of a feature along an axis parallel to the surface. A typical example of an anisotropic surface is one with oriented grooves. The directional orientation arising from an anisotropic surface topography is especially important for the study of cell–surface interactions.

Surface morphological modifications are used for purposes other than moderating biological responses to the material. As mentioned in Section 1.2.3, the contact angle of a liquid droplet on a surface is strongly influenced by the roughness. The transition between wetting regimes is described by the Wenzel model or the Cassie–Baxter model, depending on the types and sizes of the surface features. As a result, a controlled surface topography is a determining factor for superhydrophobicity.

Other properties are also governed by morphology. The optical properties of the surface of a material can be tuned by the introduction of controlled topographies. Diffraction gratings have been used to achieve the coupling of incident light with surface plasmons,[15] making it possible to develop surface plasmon resonance sensors without the use of a prism. Moreover, the behavior of surface plasmons is dramatically altered by the morphology of nanostructures. This has spawned the emerging field of nanoplasmonics, which examines the interactions of controlled morphologies with electromagnetic fields. Remarkable properties can be achieved, such as the concentration of the field intensity at specific surface regions with characteristic sizes of several nanometers, resulting in the appearance of localized surface plasmons. This effect has enabled the development of novel sensing technologies.[16] Surface plasmon resonance sensors have been created to exploit localized surface plasmons using topographical surfaces with regular patterns of metallic nanostructures instead of the conventional flat metal transducers. If these nanostructures are combined with appropriate surface functionalization techniques, it is arguably possible to envision molecular detection in a microscopic volume. Another sensing technique that relies on localized surface plasmons is surface-enhanced Raman spectroscopy (SERS). Here, the characteristic Raman signal of the analyte is enhanced by its interaction with the electromagnetic field generated in the vicinity (several nm) of a nanoroughened metallic surface supporting localized surface plasmons.[17] There have been reports of enhancement factors of up to 10^{12} of the Raman signal on a surface compared to the signal obtained in solution, which is an improvement sufficient to allow the detection of single molecules.

1.3 Surface Characterization Methods

Physicochemical parameters are used to characterize a surface and ultimately determine its performance in biosensing and bioimplant applications. They can be obtained only by experiment, through techniques employed in such a way as to focus on information about the material surface rather than the bulk. The following section will briefly review some of these methods. The examples have been chosen based on their importance and prevalence, but it should be noted that a vast range of techniques has been used, and more are constantly being developed and refined, expanding the range of characterization.

The purpose of this section is to familiarize the reader with the most common surface physicochemical characterization techniques along with their basic principles of operation. The goal is to facilitate the understanding not only of the subsequent chapters of this book, but also of current scientific articles in the fields of biosensing and biomaterials, with a focus on their physicochemical properties.

1.3.1 Surface Chemical Characterization Techniques

The elemental composition and covalent chemical structure of the layers close to the surface of a solid can be investigated *via* spectrometric methods. Depending on the method and the configuration used, the surface chemical analytical techniques can be sensitive to the composition of the layers at depths ranging from a few nanometers to several microns from the surface. The emphasis will be on the spectroscopy of electrons, more precisely, on their kinetic energy, the infrared electromagnetic radiation, and the mass spectra of secondary ions.

The technique of X-ray photoelectron spectroscopy (XPS), also called electron spectroscopy for chemical analysis (ESCA), is based on the emission of electrons from the surface of a material upon the incidence of focused electromagnetic irradiation. In a standard setup, a beam of monochromatic X-rays, typically Al Kα, is focused on the region of interest on the sample surface (Figure 1.9). The incident X-rays interact with the inner shell electrons on the surface of the material. After an electron absorbs an X-ray photon, it is emitted from the surface with a kinetic energy given by

$$E_{\text{kin}} = h\nu - BE \tag{1.19}$$

where E_{kin} is the kinetic energy of the emitted electron, $h\nu$ is the energy of the incident X-rays, and BE is the binding energy of the energy level occupied by the electron in the atom. Since the energy of the X-ray photons is a known characteristic of the source, the kinetic energy of the electrons is a function of the energy level they occupied when they absorbed the radiation. After being emitted from the surface, the photoelectrons enter an energy analyzer, in which the energy spectrum is collected by counting the number of electrons having a given kinetic energy. This energy value is converted into the binding energy (eqn (1.19)). The measurement typically requires ultra-high vacuum as the electrons need to travel a relatively long distance ($\sim$1 m) to the detector. Even though the X-rays can penetrate deep into the sample, the

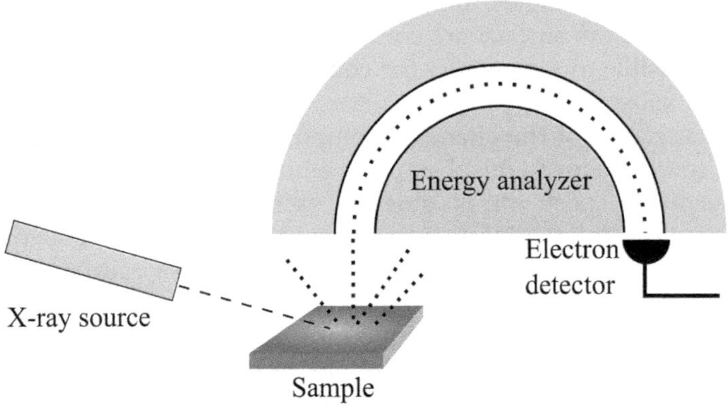

Figure 1.9 Schematic representation of the operation of an XPS spectrometer.

photoelectrons generated can travel only a very short distance inside the solid due to inelastic collisions with the sample, causing them to lose energy. Therefore, XPS is sensitive only to the surface of the material, typically to depths of up to 10 nm, as the photoelectrons emitted at greater depths do not escape the sample and so cannot reach the detector.

The exact value of the binding energy of an electron depends on the element and the atomic orbital that it occupies. Thus, the XPS spectrum gives quantitative information about the atomic composition of the sample. To obtain a precise quantification, corrections must be applied based on the different photoionization cross sections of the orbitals in each element, which produces differences in their relative sensitivities.[18] Moreover, since the photoelectrons are scattered in different ways depending on their energy, their effective attenuation lengths before leaving the sample must be calculated and taken into account. Although XPS measurements are not sensitive to the presence of hydrogen and helium, the relative composition of the other elements still provides valuable information.

The chemical state of the atoms can influence the binding energy of the orbitals. Thus, from a high resolution spectrum of the binding energy region expected for a given orbital of a certain element, it is possible to distinguish the oxidation state or to identify the bonds in which the atom is participating. For example, this idea can be applied to the 1s orbital of carbon (the relevant part of the spectrum is called the C 1s region: approximately 284–294 eV) to determine whether carbon is participating in saturated alkane chains (285.0 eV), alcohols and ethers (approx. 287.6 eV), or esters (289.0 eV), among others. This approach, which is complementary to the elemental analysis, provides a powerful tool for characterizing the chemical functionality on a surface.

A variant of the technique permits analysis of the composition profile as a function of the depth, taking advantage of the attenuation of the photoelectrons as a function of the distance traveled within the sample. In angle-resolved XPS (ARXPS), the sample is tilted so that the emitted electrons are collected as a function of the exit angle, measured with respect to the surface normal. At greater angles, only electrons being emitted from shallower depths closer to the surface are collected. With a correct model of the surface, it is possible to reconstruct the composition profile of ultra-thin layers on top of a substrate.

Characterization of the chemical composition can also be performed by applying spectroscopy to the absorption of infrared light. This portion of the electromagnetic spectrum is able to excite the vibrational modes of molecules. Conventionally, in infrared spectroscopy, the frequency of the radiation is expressed as the wavenumber in inverse centimeters (cm^{-1}), which is simply the inverse of the wavelength in vacuum or in air. The functional groups of organic compounds exhibit a number of vibration modes due to the stretching, wagging, twisting, scissoring, and rocking of their chemical bonds. To be observable by infrared spectroscopy, a vibration mode must be associated with a change in the dipole moment of the group.

Generally, the wavenumber region of interest for chemical characterization lies between 400 and 4000 cm^{-1}. Assigning the absorption bands observed in the spectrum to the vibration modes of specific functional groups makes it possible to fully characterize the chemical structure of an organic sample, except that optical isomers cannot be distinguished.

Virtually all modern instruments are based on the Michelson interferometer. They are able to measure the whole wavelength spectrum of interest simultaneously and, taking advantage of the Fourier transform, reconstruct the absorbance spectrum of the sample as a function of the wavenumber. In organic chemistry, this technique, called Fourier transform infrared spectroscopy (FTIR), is widely applied in transmission mode to bulk samples to confirm the identity of compounds. Several configurations are available that make the measurements sensitive exclusively to the composition at, or close to, the surface. These are based on the reflection of the infrared beam at an interface. The paragraphs below discuss the two basic configurations and some of their variants.

The attenuated total reflectance (ATR) configuration relies on the total internal reflection of light traveling in a medium of high refractive index at an interface with a medium with a lower refractive index. As the light beam reaches the interface, if the angle of incidence with respect to the surface normal is greater than a certain critical angle, no light is refracted into the second medium, and the beam is reflected with an angle equal to the angle of incidence. However, due to the continuity condition of the electric and magnetic fields of light, an evanescent wave appears on the other side of the interface. The intensity of this wave decays exponentially away from the surface into the second medium with a characteristic decay distance determined by the penetration depth. In ATR-FTIR spectroscopy, the beam is directed into a prism of a material with a higher refractive index than that of the sample, typically a crystal. The sample is pressed onto the prism surface from which the light is reflected, and the evanescent wave formed extends into the sample. The sample is able to absorb energy from the evanescent wave, attenuating the totally internally reflected light exiting the prism. The absorbance spectrum is obtained from the reflected beam. In some configurations, the prism is built in such a way as to allow multiple reflections on the interface, each causing attenuation due to absorption, thereby increasing the overall signal. The penetration depth of the evanescent wave depends on several factors. It is proportional to the wavelength of the light, so it decreases with the increasing wavenumber. With grazing incidence, *i.e.*, at higher angles of incidence, the penetration depth is lower than at lower angles, which are closer to the critical angle. It is also a function of the refractive indices of the crystal and the sample. The higher the former and the lower the latter, the lower the penetration depth. Typically, diamond is the material used for ATR crystals. At an angle of 45°, the penetration depth of the evanescent wave formed by an infrared beam incident on a diamond–air interface is on the order of a few microns. This deep penetration has been exploited to measure the bulk composition of liquids and solids simply by spreading or pressing them onto the ATR crystal, without the need for sample preparation. However, for surface studies, higher sensitivity to the composition of the surface layer can be achieved by employing crystal materials with a higher refractive index (*e.g.*, germanium) and a greater (grazing) angle of incidence. In this way, it is even possible to acquire the spectra of organic monolayers. Furthermore, when used to characterize ultra-thin film modifications on solids, modern instruments offering high signal-to-noise ratios make it possible to subtract the contribution from the bulk and observe only the contributions from the desired layers.

In another FTIR spectroscopy configuration, the sample is deposited as a coating on a reflective substrate, usually a polished metal, from which the beam is specularly reflected. A near-grazing angle of incidence is chosen so that the light beam travels almost parallel to the surface, thereby maximizing the optical path. This configuration is called infrared reflection-absorption spectroscopy (IRRAS). The reflection of the infrared light beam from the metal surface at near-grazing angles results in a difference in the phase shift between the p-polarized component (parallel to the plane of incidence, which includes the surface normal) and the s-polarized component (perpendicular to the plane of incidence and parallel to the surface). The reflection of each polarization is governed by the Fresnel equations, which cause the s-polarized light to be almost completely canceled at the reflecting surface, while the intensity of the p-component is nearly doubled. In consequence, IRRAS measurements are sensitive only to vibration modes oriented perpendicular to the surface, while vibrations parallel to the surface are not observed in IRRAS spectra. This gives rise to the surface selection rule. Due to the increased intensity of the p-polarized component at the surface, the measurements are more sensitive, having high signal-to-noise ratios. Furthermore, the surface selection rule makes it possible to assess the orientation of the chemical groups and molecules on the surface. Interestingly, Grunze *et al.* applied this approach to observe the conformation of poly(ethylene glycol) chains grafted on gold.[19]

A variant of the IRRAS setup, polarization modulation infrared reflection-absorption spectroscopy (PM-IRRAS), incorporates an element that is able to modulate the polarization of the infrared light beam incident on the reflective metallic surface. The main advantage of this technique is its ability to collect the spectra for both polarizations simultaneously. Both follow the same optical path but only the p-component is sensitive to the ultra-thin layers on the surface, so the s-component is used as a background signal, and the differential reflectance spectrum corresponds only to the surface species. This removes interference such as that from atmospheric water vapor and carbon dioxide, and eliminates the need to collect a background spectrum separately, while making it possible to measure the spectrum of adsorbed species or molecular monolayers with high sensitivity.

Mass spectrometry is also routinely applied to the chemical analysis of surfaces. The identification of ions of certain elements, molecules, and their fragments provides detailed information about the surface chemical structure. For the mass spectrometry analysis of a surface, ionization is typically achieved by bombardment with highly energetic ions from a primary ion gun. When they reach the surface, the resulting collision cascades fragment and ionize the molecules in the topmost layers (1–2 nm), generating secondary ions which are directed to a mass analyzer. Modern instruments generally use a time-of-flight analyzer because it can analyze a wide mass range simultaneously with high sensitivity. This technique is known as time-of-flight secondary ion mass spectrometry (ToF-SIMS); however, it damages an area of several square nanometers around the site impacted by a primary ion. To maintain excellent sensitivity to the surface molecular layer, it is operated in the static SIMS mode: the primary ion dose is kept low enough (generally below 10^{12} ions cm^{-2}) so that, statistically, a primary ion always strikes a fresh area of the substrate. In contrast, dynamic SIMS, which delivers high ion doses resulting in the breakdown of molecules, can be used to obtain a depth profile of the atomic and isotopic composition.

A significant advantage of ToF-SIMS is that the primary ion beam can be focused on extremely small spots, of sizes as small as 50 nm. Deflecting the primary ion beam makes it possible to scan larger areas with submicron resolution. Thus, the chemical composition can be represented by maps of the intensity of the different fragments identified. This leads to high resolution imaging of the chemical composition at the surface.

1.3.2 Contact Angle Techniques

The measurement of the contact angles of different liquids on a surface can provide valuable information about the surface chemistry of the outermost molecular layers. The contact angle of a surface with water is the most important for the evaluation of the potential interactions of a material in a biological environment. As mentioned in Section 1.2.3, the hysteresis in the measurement of the contact angle makes it necessary to distinguish between the advancing and receding contact angles; fortunately, this can be a source of information about the surface physical chemistry. The following paragraphs present an overview of the methods most widely used to obtain both contact angles. One of the main advantages of these measurements is the ease with which they can be performed in any laboratory, as they require very little instrumentation. While the in-depth interpretation of the data relies on profound assumptions about the behavior of the interface, the power of these approaches is that they offer easy access to important information. The hydrophilic or hydrophobic character of a surface can often provide evidence of the success of specific chemical modifications before it is necessary to apply more complex, expensive, and time-consuming spectroscopic techniques. The hydrophilic or hydrophobic character of the surface also contributes to the understanding of the surface forces that might be at play when the material is employed in biological applications.

The most straightforward technique for measuring the advancing and receding contact angles is the sessile droplet method. It involves placing a droplet of liquid on the surface under study and measuring the contact angle using a contact angle goniometer. The device consists of a horizontal sample stage, a microliter syringe that dispenses the droplet, a digital camera, and a source of illumination. The camera is connected to a computer for continuous image acquisition, and the syringe can be equipped with a motorized piston. The droplet usually has a volume on the order of several microliters, which can be dynamically increased by the motorized syringe to acquire images during the advancing of the contact line. The liquid can also be drawn back up with the syringe to reduce the droplet volume so that images of the receding droplet can be recorded. Specialized software is available that fits the droplet profiles from the acquired images, from which the contact angles on both sides of the droplet can be extracted.

A variant of the previous technique, called the captive bubble method, relies on an inverted configuration, in which the surface under study is submerged face down in a container filled with the liquid of interest. A microliter syringe with a curved needle is used to place a droplet of air on the surface. Small air bubbles are allowed to float towards the surface, and their dimensions are measured as soon as they attach to it. The value of the contact angle obtained in this way approaches that of the receding contact angle. This method has the advantage that the material is fully immersed in the liquid so that it does not dry due to contact with the air. This is beneficial in the case of volatile liquids, which would otherwise evaporate quickly, and also makes the technique suitable for studying the wetting of the surfaces of materials which absorb water (*e.g.*, hydrogels) in their swollen state. Since it is the swollen condition that is biologically relevant, this measurement is arguably more appropriate than the sessile droplet technique at the interface of the material with air. Unfortunately, it cannot determine the advancing contact angle.

The Wilhelmy plate technique examines the wetting of a surface using a balance to measure directly the force resulting from the interfacial tension of the liquid in contact with the surface. A slab of the material under study is suspended vertically from a balance and brought into contact with a liquid. The balance measures the force experienced by the slab while it is being lowered into the liquid and then pulled back up. The force F is given by

$$F = F_g - F_b + F_i \tag{1.20}$$

where F_g is the weight of the slab, F_b is the buoyancy of the immersed volume of the slab, and F_i is the force exerted at the interface due to the surface tension at the liquid–air interface. The first term is constant throughout the immersion, but the buoyancy depends on the height of the immersed portion of the plate. The force applied by the interface will depend on the contact angle θ. The terms in eqn (1.20) can be expanded to give

$$F = mg - \rho_L shg + \gamma_{LA}\cos(\theta)p \tag{1.21}$$

where m is the mass of the plate, g is the acceleration due to gravity, ρ_L is the density of the liquid, s is the cross-sectional area of the slab, h is the height of the immersed portion of the plate, γ_{LA} is the surface tension of the liquid, and p is the perimeter of the slab. Clearly, due to the hysteresis of the contact angle, the force recorded during immersion is different from the force measured while the slab is being raised. During immersion, the advancing contact angle appears at the surface of the liquid, while the receding contact angle appears during the ascending stage of the measurement. Since the density and surface tension of liquids can be found in the literature, the contact angle can be calculated from eqn (1.21). An advantage of the Wilhelmy plate method is that the contact angle is averaged over the whole perimeter of the plate, yielding high accuracy and reproducibility. However, both sides of the slab must be prepared in the same way to ensure that both exhibit the type of surface which is to be characterized. This can cause practical difficulties, especially when surface modifications are performed on one side of a substrate, for example, by deposition processes.

1.3.3 Surface Structure and Morphology

Typically, microscopy is used to observe the morphology and structure of a surface. The following paragraphs present an overview of the most common techniques, as well as methods often employed to study layer structures and obtain depth profiles.

Scanning electron microscopy (SEM) is designed to perform the direct observation of surface topographical features, using a beam of high energy electrons to image a sample. High energy electrons have very short wavelengths, on the order of tens of picometers, much smaller than those of the visible light used in optical microscopy. The electron beam can be focused to an extremely small spot size, which makes it possible to obtain image resolutions of less than a nanometer. The beam is generated by an electron gun and accelerated by a high voltage to achieve a kinetic energy of between 1 and 30 keV. The beam is then focused on a small spot and used for raster scanning a selected area of the sample. As the electrons reach the surface, they interact with the material, producing different types of signals that can be observed by specialized detectors in the sample compartment. The signal is mapped to the position of the spot being scanned to generate an image. The most common type of signal is low-energy secondary electrons released from the surface by inelastic scattering, which can be captured by a detector placed near the sample. Since secondary electrons are far more prone to scattering before they leave the sample, only those generated close to the surface (within a few nanometers) can escape and be detected, so the technique is sensitive exclusively to the surface. The surface topography determines the intensity of the secondary electron signal. When the electron beam hits the surface at a shallow angle, the secondary electrons are generated closer to the surface, allowing them to escape more easily, in a manner analogous to the behavior of low-angle beams in angle-resolved X-ray photoelectron spectroscopy. Thus, when the beam hits a topographical feature containing faces or surfaces that form a steep angle with respect to the plane of the surface, the intensity of the secondary electrons will be stronger than when the beam is incident on a surface from a normal direction. This difference in signal strength is responsible for the topographical contrast in SEM with secondary electron detection, which makes it well suited to morphological characterization. However, since electrons are also scattered in air, SEM measurements are typically performed under high vacuum. This can be problematic for some types of samples, notably biological tissue, and may require special sample preparation. Since the sample is constantly bombarded by an electron current during measurement, it needs to be grounded to avoid the buildup of charge, which could lead to artifacts. Non-conductive samples need to be coated with a thin film (with a thickness of tens of nanometers) of a conductive metal, typically gold, to allow neutralization of the charge. In this case, the morphology observed is that of the metallic film, which conforms to the shape of the underlying sample. Nevertheless, at very high magnification, this can lead to artifacts.

A backscattered electron detector measures the intensity of the electrons which are backscattered due to the elastic scattering of the primary electron beam by the sample. The scattering intensity depends on the composition of the sample: heavier atoms scatter more strongly. Thus, the image obtained from the signal reflects differences in sample composition. However, backscattered electrons have a much higher energy, so they are scattered by the sample much more weakly before reaching the detector. For this reason, they can be detected even if they originated at a greater depth within the sample. Thus, the signal is not sensitive exclusively to the surface region.

The third type of detector often installed in scanning electron microscopes is the X-ray detector, which is able to convert the energy of the photons into a signal. Such detectors are used to analyze the elemental chemical composition of the region being scanned *via* energy-dispersive spectroscopy (EDS). When the high energy electron beam interacts with the scanned region of the sample, it causes the release of electrons from the inner shells of the atoms. When the hole created is filled with an electron coming from a higher energy orbital, a photon is released in the form of X-rays of characteristic energies, which depend on the electronic structure of the atom. The energies of the X-rays captured by the detector are assigned to the specific emission lines of each element, and in this way, the atomic composition of the sample can be obtained. Since the information is limited to the region scanned, which is also visualized using SEM, EDS can provide valuable insight into differences in composition over regions of a material, making it possible, for example, to visualize phase segregation in metals. However, due to the deep penetration of the electron beam into the sample and the very weak scattering of the emitted X-rays (as opposed to the strong scattering of secondary electrons or photoelectrons in XPS), the analysis depth of the technique is much greater, on the order of a micron.

The imaging of surface topography can also be performed with a family of techniques called scanning probe microscopy (SPM). This rapidly developing field encompasses a number of methods that are able to access different features beyond the topography, including mechanical, chemical, and electrical properties, and their distribution on the surface. The basic operating principle involves a probe mounted on a cantilever that interacts with the surface of a sample. The sample and the probe are moved relative to each other by piezoelectric actuators. Such devices are capable of achieving extreme precision of movement, on the order of fractions of nanometers. In these techniques, the probe is scanned over the sample surface on a grid while the strength of the probe–surface interactions is recorded. The high resolution in the movements achievable by piezoelectric scanners allows even single atoms to be resolved by some of these techniques. The type of interaction or property probed on the surface depends on the technique. The first SPM method, scanning tunneling microscopy (STM), detects the quantum tunneling current that occurs when the sample is subjected to a bias potential and the probe is in close proximity. Nowadays, the most common SPM implementation is atomic force microscopy (AFM), which measures the deflection in the cantilever caused by the mesoscopic interaction forces between the sample surface and a very sharp tip mounted on the cantilever, which acts as the probe. It can be operated in several modes, depending on how the interactions of the tip and the sample surface are quantified. The simplest, called contact mode, involves keeping the tip in contact with the sample while scanning it over the surface. The tip follows the surface topography, and its height can be recorded by measuring the height of the cantilever, usually by optical means. When AFM is operated in tapping mode, the cantilever is made to oscillate near its resonant frequency as it scans the sample surface, coming into contact with the sample during oscillation. A feedback mechanism maintains a constant oscillation amplitude. The main advantage of this mode is that it minimizes damage to the sample surface since the tip is not dragged over it. This is especially important for the measurement of soft samples, such as biological materials or some polymers. Non-contact mode is similar in principle to tapping mode. In this case, the cantilever oscillations do not come into direct contact (repulsion) with the surface; instead the tip interacts with the sample *via* attractive van der Waals forces. In all modes, the lateral resolution obtainable by AFM is limited by the sharpness of the tip. Since at the nanoscale level accessible by AFM, the tip cannot be perfectly sharp, the topography that is recorded is the convolution of the real surface morphology and the curvature of the tip. For certain morphologies, for example, those exhibiting very steep vertical surfaces or sharp edges, this gives rise to imaging artifacts. The main parameter used to characterize the sharpness of the tip is the radius of curvature. From the topographical image obtained using AFM, it is possible to calculate the roughness parameters R_a and R_q of the surface. Even though AFM provides excellent lateral and vertical resolution for the characterization of surface morphology, it suffers from low throughput due to the slow scanning speed of most systems, usually on the order of pixels per second. The development of high-speed AFM is currently an active area of research, with the goal of reaching video-like frame rates. Further limitations of AFM are the maximum size of the areas that can be scanned (on the order of 100 μm wide) and the vertical dynamic range (a few tens of microns). A family of techniques known as profilometry is used to characterize surface topography on larger scales, both vertically and horizontally; the X- and Y-ranges are on the order of several centimeters. Profilometry is also commonly applied to measure the roughness of a coating or surface finish.

With AFM, it is possible to obtain information about the sample other than the topography. When it is operated in tapping mode, the phase of the oscillation with respect to the driving voltage of the cantilever can be recorded. Even on samples that have no important topographical features (*i.e.*, height differences), the phase image can reveal differences arising from factors such as adhesion of the tip to the sample and local mechanical or viscoelastic properties. This makes it possible to detect features such as the segregation of phases on apparently homogeneous surfaces. In addition, contact mode AFM can be employed to record the lateral deflection of the cantilever as it is scanned over a sample. The lateral deflection, which is caused by the friction force between the tip and the sample, can reveal differences due to local variations of the coefficient of friction even in the absence of topographical features.[20] This version of AFM is called friction force microscopy (FFM) or lateral force microscopy (LFM).

However, none of the SPM techniques is able to directly access structural or chemical features arranged perpendicular to the surface, *i.e.*, varying with the sample depth. Of particular interest is the presence of coatings on the surface of a bulk material, consisting of one or multiple layers with varying properties. Such coatings are commonly used to tune the surface properties of a device without modifying the bulk properties of the material. The thickness of these layers can be measured using two approaches. If a particular region of the surface can be left uncoated (*e.g.*, by patterning the coating) or the coating can be completely removed from that area (*e.g.*, by intentionally scratching the sample), the jump in step height that appears where the coated area begins represents the thickness of the layer; it can be measured by profilometry or AFM.

The layer structure of thin films is often examined using optical techniques if the sample is reflective or transparent at the appropriate wavelengths, or if the modification or layer of interest can be applied on an appropriate model surface that is reflective. A routine method for characterizing thin films and multilayers is ellipsometry, which involves measuring the changes in the polarization state of a beam of light after specular reflection from the sample, or transmission through the sample if the substrate is transparent. The data obtained is expressed in terms of the complex reflectance ratio between the s- and p-polarized components of the reflected light. The reflectance ratio is a complex number given by an amplitude Ψ and a phase difference Δ. The power of ellipsometry is that it measures the amplitude ratio and phase difference between the two polarizations, so no referencing is necessary. Furthermore, the setups available for performing the measurements allow extremely precise determination of the ellipsometric parameters. With appropriate treatment of the data, this technique can achieve sensitivity to changes in single atomic layers. The parameters are influenced by the (complex) refractive indices of the substrate and each of the layers that form a stack on the surface, as well as the thickness of each layer and its structure, including interpenetration and roughness. To obtain the relevant information about the sample surface, the data must be fitted with a model of the surface which includes all the above-mentioned parameters. The model is based on the Fresnel equations of reflection and is implemented in ellipsometry software. To maximize the amount of data available and achieve more reliable fitting of the model, measurements can be performed at multiple wavelengths; spectroscopic ellipsometers are able to simultaneously acquire data between 200 and 1000 nm as well as at multiple angles of incidence. The size of the area analyzed is typically several millimeters wide, but focusing optics can be installed in the beam path, allowing measurements on areas less than 100 μm wide. This makes it possible to map the thickness or layer structure along the surface of a sample by scanning multiple points while moving the sample on a translator stage. Mapping can also be performed using imaging ellipsometry, in which instead of a single photodetector or spectrometer, the detector is equipped with a CCD digital camera that records the intensity of the reflected light on an area of the sample instantaneously. One of the main advantages of ellipsometry is that it can be performed *in situ*, so the data can be acquired while the sample is immersed in a liquid environment that can be manipulated to simulate conditions mimicking the intended application of a device. In this way, the behavior of the surface, including swelling, adsorption, and dissolution, can be observed under relevant conditions.

Techniques related to ellipsometry are X-ray and neutron reflectometry (XRR and NR, respectively): the reflectivity of a beam of either monochromatic X-rays or neutrons with a defined wavelength is measured as a function of the angle of incidence in the specular reflectance configuration at grazing angles. The techniques are based on the same principle but are complementary because they use different types of probes. While X-rays are electromagnetic radiation, sensitive to differences in electron densities, neutrons detect the contrast caused by the differences in the scattering lengths of the atoms in the different layers, each having a different scattering length density. Due to constructive and destructive interference, the reflections of the incident X-rays or neutrons at the interfaces between each pair of layers on the surface cause oscillations of the reflected intensity as a function of the angle of incidence, called Kiessig fringes. The period of these intensity modulations depends on the thickness and properties of the layer structure. It is important to note that the observation of the fringes is highly sensitive to the roughness of the layers and the interfaces. Information about the layer structure is obtained by fitting the reflectivity curve with a model of the surface. While X-ray sources for XRR are easily accessible, the use of high energy electromagnetic radiation can damage the sample, especially if it is an organic material. Furthermore, since X-rays are absorbed by water, *in situ* measurements are challenging, requiring the use of high energy and high intensity synchrotron radiation. However, NR can be readily performed *in situ*. Since the source of the contrast is the scattering length density, the contrast can be readily tuned by selectively changing the isotopic composition of some of the layers or of the contacting medium, typically by replacing hydrogen with deuterium, which has a much greater scattering length. For example, an experiment can be performed in which the surface is in contact with water and the neutron reflectivity profile is recorded. Subsequently, the medium is replaced by a mixture of water and deuterium oxide with the same scattering length density as the outermost layer. This renders the interface between the outer layer and the medium invisible to the neutrons, which makes it possible to obtain information about the other layers in the stack independently. However, NR experiments require access to a neutron source, usually a nuclear reactor, which is prohibitively expensive. Moreover, the samples become radioactive after prolonged exposure to neutrons. Nevertheless, NR is a powerful tool for resolving the density profiles and layer structures of thin films on the subnanometer scale, which is important because they are thought to play a major role in determining the behavior of solid interfaces in contact with biological media, as will be elaborated in later chapters.

Another important issue is characterizing the variation in the chemical composition as it depends on the depth in a surface. As mentioned earlier, angle-resolved XPS can uncover differences in the range of up to 10 nm. To obtain the chemical composition profile at greater depths, XPS and ToF-SIMS can be coupled with ion milling. In this technique, chemical surface analysis alternates with cycles of surface etching performed by bombardment with ions under vacuum inside the analysis chamber. Many state-of-the-art XPS and ToF-SIMS spectrometers include an ion gun for the specific purpose of ablating the sample surface by bombardment with ions or ion clusters; typically Ar is used, but other ions can also be selected, depending on the nature of the sample. Repeating the analysis and etching cycles makes it possible to obtain the composition profiles through the sample depth. The depth range analyzed depends on the intensity and duration of ion milling, but typically ranges from tens of nanometers to microns.

1.3.4 Surface–Biomolecule Interactions

The complex interactions between biological macromolecules and artificial materials and surfaces will be addressed in the next chapter. The comprehensive characterization of the structure and physicochemical properties of a surface is the first step in understanding the phenomena involved in these interactions. Direct examination of the adsorption of biomolecules onto the biomaterial interface requires techniques capable of quantifying the amount adsorbed. Since adsorption is a dynamic process, an ideal method for observing surface–biomolecule interactions should also provide information about the kinetics of adsorption. This section will review the experimental techniques commonly applied to quantify the adsorption of biomolecules onto surfaces.

Since proteins account for most of the non-specific adsorption from biological media and mediate the biological response to materials, the emphasis will be placed on measurement techniques for protein adsorption. Direct quantification of protein adsorption involves bringing the surface into contact with the biological medium of interest and then analyzing the amount of material deposited. While the principle is simple, there are trade-offs among the various methods proposed for bringing the materials into contact and those for quantifying adsorption. Simple model solutions consisting of a single protein dissolved in a buffer can be brought into contact with the surface. However, even if the protein concentration mimics that found in natural biological fluids, these solutions generally fail to replicate the complexity of the adsorption phenomena. Alternatively, a natural biological fluid such as whole human blood, blood plasma, or serum can be studied. The fluid chosen should be relevant to the intended application of the device, *e.g.*, blood serum would be used to test a biosensor surface intended for medical diagnostics. While this method approximates the conditions in real applications much more closely, the complexity of the fluid limits the techniques available for quantification of the amount adsorbed.

Whatever the solution or fluid used for adsorption tests, special care must be taken to avoid contact between the surface and air during measurement, as dewetting and drying of the solution leads to measurement artifacts caused by the deposition of proteins. Prior to testing, the surface should be protected by immersing it in an appropriate buffer solution. If complex biological media of natural origins are used, it is essential to take into account the variability of the biological samples. It has been shown that apparently similar biofluids obtained from different sources or donors can lead to dramatically different results for protein adsorption on the same surface.[21,22] For example, samples of blood plasma from different donors showed variation of up to an order of magnitude. Even using pooled blood plasma to test the antifouling properties of polymeric coatings only partially alleviated the problem as plasmas from different suppliers also exhibited significant variation.

The amounts of proteins that adsorb on surfaces vary widely, depending on the type of surface tested. Ultra-low fouling and non-fouling surfaces are typically defined to be those whose adsorption of proteins is below the detection limit of the most sensitive techniques available (on the order of 0.1 ng cm^{-2}). At the other extreme, hydrophobic polymeric materials often adsorb levels of up to μg cm^{-2} of proteins on their surfaces. Nevertheless, for the direct quantification of adsorption, for specimens having surface areas on the order of square centimeters, it is not a viable option simply to weigh the sample on an analytical balance before and after contact. Solution-depletion methods measure the concentration of protein in the solution before and after contact with the sample. The protein concentration is typically measured by UV spectrophotometry. The amount of adsorbed protein is then obtained by a simple mass balance calculation. However, the limit of detection for the amount of adsorbed protein depends on the sensitivity of the measurements of protein concentration. Since the overall mass of the adsorbed protein is extremely small, especially in the case of samples with low specific surface area, this method generally has very low sensitivity, especially for macroscopic specimens and flat surfaces. However, solution-depletion has sometimes been used in applications involving nanoparticles.

Nevertheless, there are various instruments and protocols that make possible the direct quantification of the amount of protein adsorbed on a surface.[23] Labeled assays are mainly based on fluorescence or radioactivity measurements. A label is attached to the proteins before contact with the sample surface for a specified time, and it emits a signal that can be readily detected in spite of the extremely small mass of the adsorbed protein. The tag is usually a fluorescent molecule or a radioactive isotope, most commonly ^{125}I. The limits of detection for both methods are low, on the order of 1 ng cm^{-2}. Regardless of the type of tag, the sample must be thoroughly rinsed and dried to remove any remaining unbound protein, which would produce a signal indistinguishable from that emitted by the adsorbed protein. Furthermore, while fluorescence assays are easy to implement, fluorescent molecules are expensive, and labeling with radioactive isotopes requires careful handling of the reagents due to their potential toxicity. However, the main limitation of both approaches is that the label itself can change the properties of the protein, potentially altering the adsorption behavior of the labeled proteins relative to the proteins in their natural state. Usually, it is implicitly assumed that such effects can be neglected, but this should be verified independently using other methods.[24] Other drawbacks of label-based assays are that they can be used only with single-protein model solutions, and generally they cannot evaluate adsorption of proteins from complex biological media. In addition, the need for rinsing and drying steps prevents these methods from providing results in real time, limiting the possibilities for studying adsorption kinetics.

There are yet other instrumental techniques for the direct quantification of adsorbed proteins, based on the physicochemical characterization techniques already discussed in previous sections. They measure the amount of adsorbed proteins by comparing surfaces before and after adsorption. For instance, XPS is able to detect the changes in chemical composition caused by the deposition of a layer of protein on the surface. In FTIR, the adsorption of proteins is characterized by the appearance of the amide I and II bands, which absorb in the spectral region between 1500 and 1700 cm^{-1}. These bands arise from the numerous peptide bonds present along the chains of the amino acids of which proteins are composed. The shape of the pattern of these spectral bands, which depends on the conformation of the protein, can be detected even in the adsorbed state. Comparing the spectra of proteins before and after adsorption makes it possible to infer the loss or preservation of α helices and β sheets and maybe to observe their denaturation.[25] It is also possible to detect the presence of an adsorbed layer of proteins using ToF-SIMS. This technique has the advantage that it is extremely sensitive to the composition of the surface, though only at the outermost 1–2 nm. However, interpreting the data can be challenging because the penetration depth of the analysis is less than typical protein dimensions, so only fragments of small mass are detected. Since only the amino acid residues pointing out from the surface are sampled, ToF-SIMS spectra are also sensitive to the conformation and orientation of the proteins on the surface.

Alternatively, the amount of adsorbed protein can be quantified by measuring the change in thickness of the layers after contact with a protein solution. For full surface coverage with a protein monolayer, the expected increase in thickness is on the order of several nanometers. This is well within the sensitivity of ellipsometry, which can achieve a limit of detection of 0.1 nm, so it has been used to study protein adsorption.[26] However, as with all ellipsometry measurements, it can be applied only to flat reflective samples, and the modeling of the data can be challenging.

A possible limitation of the techniques mentioned above is that they require the sample to be rinsed and dried after contact with the protein solution and before analysis, and in the case of XPS and ToF-SIMS, measurements are performed under high vacuum. It is of particular interest to study the adsorption of proteins *in situ*. If it is possible to observe not only the amount of protein deposited but also the structure of the proteins on the surface, then it is possible to distinguish different types of adsorption. The technique of choice for the *in situ* observation of protein–surface interactions is neutron reflectometry, which gives access to a wide range of structural parameters by fitting the data with an appropriate model. As previously explained, the contrast can be tuned through the use of isotopic substitution (deuteration) in one of the layers or in the solvent. Choosing the scattering length density of the medium to match exactly the scattering length density of the surface layer onto which proteins are adsorbed eliminates the contrast between the coating and the medium, so the reflectivity plot will correspond to the protein layer exclusively. In this way, the density profile of the adsorbed protein layer can be found independently of the other layers. However, even though the measurements are performed *in situ*, NR cannot be considered a real-time technique for performing kinetics experiments due to the long acquisition time (several hours) required to obtain a reflectivity curve. This is because of the relatively low contrast between the neutrons reflected by the different interfaces. Moreover, the sample surfaces need to be prepared on perfectly flat, large areas (tens of square centimeters) to obtain sufficiently strong signals. However, the main disadvantages of NR are the considerable expense involved in performing the measurements and the need for a neutron facility. *In situ* ellipsometry has also been applied to study the structure in surface–protein interactions,[27] but it requires careful selection of the model used to interpret the data.

Only a few techniques are able to track *in situ* the adsorption of proteins on flat surfaces in real time. Each of these measurements is based on the change in the physical signal generated by the attachment of a protein on the surface of a transducer. In reflectometric interference spectroscopy, the increase in the thickness of a layer on the transducer due to the attachment of proteins is measured by optical means. White light is directed to an optical layer that is brought into contact with the solution of interest. The path difference and the phase difference of the light reflected from the interface between the substrate and the optical layer and from the interface between the optical layer and the solution create an interference pattern. Attachment of proteins to the surface increases the optical thickness of the layer, changing the path difference between the two types of reflections. The signal recorded is the shift in the interference pattern, which is directly proportional to the increase in the layer thickness.

In contrast, surface plasmon resonance (SPR) is based on the measurement of the refractive index in the immediate vicinity of the surface; the index is sensitive to the deposition of mass, for example by protein adsorption.[28] Surface plasmons, the collective oscillations of the electrons of a metal at the interface with a dielectric, form an evanescent wave that extends into the dielectric. They are characterized by a complex propagation constant that is highly sensitive to the refractive index of the medium in contact with the metal surface in the region immediately adjacent to the interface. The metal film is almost always made of gold because its advantageous optical properties support surface plasmons and because it is chemically inert. In the Kretschmann geometry, surface plasmons on a thin (50 nm) metal film on top of a prism are excited by light reflecting on the back side of the metal (Figure 1.10). After total internal reflection in the prism, the light forms an evanescent wave that extends into the metal film and stimulates surface plasmons if the condition for resonance is met. The light that fulfills the matching conditions is extinguished and not reflected out of the prism. Coupling can occur only for light for which the component of its wave vector parallel to the surface matches the real part of the surface plasmon propagation constant. This means that monochromatic light of a given wavelength can excite the surface plasmons only at a particular angle of incidence. Alternatively, when white light is directed at the prism at a particular angle of incidence, only one specific wavelength can excite the surface plasmons.

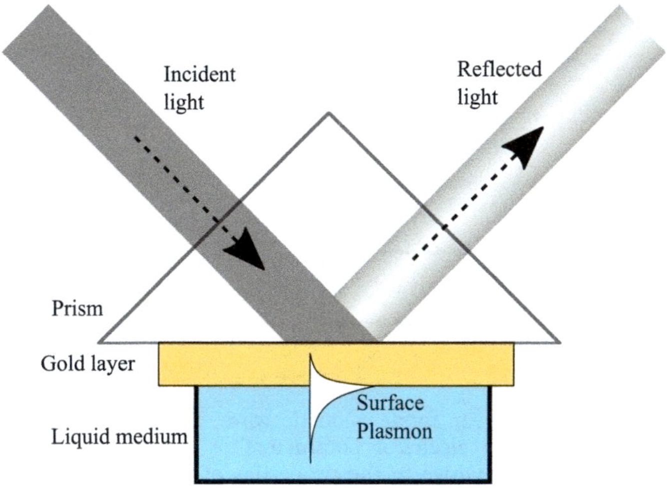

Figure 1.10 Surface plasmon resonance in the Kretschmann geometry. The changes in refractive index in the vicinity of the gold layer, arising from changes in the attached mass, are observed by monitoring the intensity of the reflected light as a function of the wavelength or the angle (see text).

Monitoring the reflected light emerging from the prism makes it possible to identify the signal either from the wavelength or the angle of incidence at which the light intensity is attenuated. Fixing the angle and measuring the wavelength or fixing the wavelength and measuring the angle are the two most common detection methods, but in either case, the signal corresponds to a change in the refractive index on the surface. The field intensity of the surface plasmons decreases into the dielectric as an exponential function of the distance to the metal surface. The region in which the refractive index is measured extends a few hundreds of nanometers into the medium, but the sensitivity is greatest for changes immediately on the surface. After suitable calibration, it is possible to convert the signal, expressed in refractive index units (RIUs), into a measure of the surface coverage by proteins (ng cm^{-2}).

Acoustic wave sensors operate by measuring the changes in the resonant oscillation of a crystal caused by the interactions of biomolecules on the surface. There are two types of devices: those based on the propagation of surface acoustic waves (SAW), and those based on bulk acoustic waves (BAW).[29–31] Acoustic oscillations, having frequencies between 4 MHz and several hundred MHz, are produced by applying a voltage to a piezoelectric crystal, typically quartz cut in the AT direction. In the most common BAW configuration, gold electrodes deposited on both sides of the quartz crystal are used to excite a transverse shear acoustic oscillation mode at the resonance frequency of the crystal or at higher order harmonics (only odd order harmonics can be excited); this is the basis of a transverse shear-mode (TSM) device (Figure 1.11). The resonator is placed in a flow cell so that it comes into contact with the solution of interest, which is made to flow using a

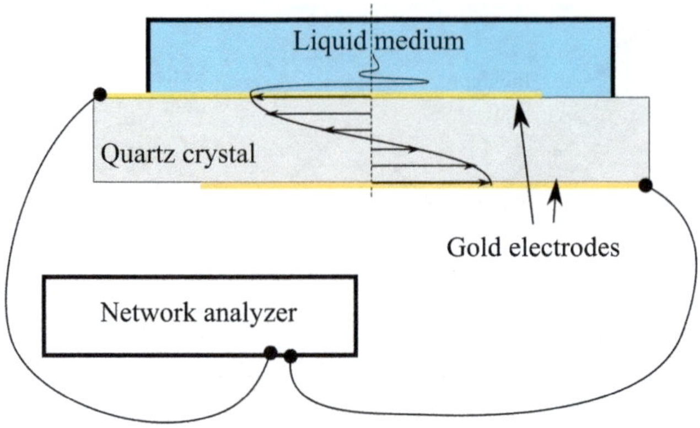

Figure 1.11 Transverse-shear mode acoustic wave device. The network analyzer monitors the electrical parameters of the quartz-crystal resonator, which are sensitive to the properties of the layers and the liquid on top of the gold electrode.

syringe pump or a peristaltic pump. The electrodes are connected to an amplifier, which is used to excite the oscillations, and the resonance frequency is recorded. As a layer of biomolecules from the solution attaches itself to the surface, the oscillation frequency is affected by the mass loading and by the viscoelastic properties of the layer on the gold electrodes. Under the assumption that the deposited material behaves like a rigid solid firmly attached to the electrode, the dependence of the shift in the resonance frequency with an increase in mass is given by the Sauerbrey equation:

$$\Delta f = -2f_0^2(c\rho)^{-1/2}\frac{\Delta m}{A} \tag{1.22}$$

where f_0 is the resonant frequency of the unloaded crystal, c is the stiffness of the crystal and ρ is its density, A is the area of the resonator, Δf is the observed frequency shift, and Δm is the increase in mass. This equation is valid only for frequency shifts of up to 2% of f_0. While it was derived for mass deposition under the assumption that the device operates in a vacuum (or gas) environment, it is commonly used for devices operating in liquid media. Since, subject to these restrictive assumptions, the device can measure mass differences, it is often referred to as a quartz crystal microbalance (QCM). However, for sensing applications in a liquid environment, this is inappropriate because factors other than mass strongly influence the frequency, so the device cannot be considered a balance. The Sauerbrey model explicitly assumes that the newly deposited mass forms a rigid layer attached to the sensor surface. The oscillatory mode of the crystal extends into this layer as if it were a continuation of the sensor. However, in real examples involving the attachment of biological macromolecules, a significant proportion of the solvent is bound in the deposited layer, and the resulting viscoelastic behavior increases the energy dissipation. In the network analysis method, the impedance of the system is measured, and the resonator is modeled by an equivalent electrical circuit using the Butterworth–Van Dyke model. The properties of the deposited layers, *i.e.*, both the mass and viscoelastic contributions, are inferred from the observed electrical parameters. The modeling of the changes in these parameters and how they relate to the impedance of the equivalent elements in the network analysis method can yield valuable information about the layer, beyond its thickness or mass. Another mode of operation for transverse shear-mode resonators is based on driving the oscillation at resonance frequency and stopping the voltage to observe the decay in the oscillation amplitude over time. Operating at several cycles per second makes it possible to directly record the resonance frequency and dissipation factor in real time without building an electrical model for the resonator. This mode is referred to as a quartz crystal microbalance with dissipation (QCM-D) and, in recent years, it has been increasingly applied in the study of biomolecule interactions. The viscoelastic modeling of the adlayer is comparatively simple because it is performed by specialized software, based on the frequency and dissipation data. The device is not a balance *per se*, but the parameters that it can obtain include the adlayer thickness, density, and shear modulus (both the loss and storage components, attributed to viscosity and elasticity, respectively).

Considerable interest has been focused on the attachment of proteins on nanoparticles, but because of their high specific surface area, solution depletion methods can be used. Certain techniques not applicable to flat surfaces are also available. One example is isothermal titration calorimetry, in which a suspension of particles is mixed with a protein-containing solution, and the heat released is monitored. Once the enthalpy of mixing is taken into account, the data can provide insight into the thermodynamic parameters driving the adsorption equilibrium. Another example is dynamic light scattering, which measures the nanoparticle size distribution. The changes in particle radius before and after contact with a protein solution are indicative of adsorption.

When protein is adsorbed onto a surface after it comes into contact with a complex protein mixture, it is important not only to quantify the adsorption, but also to identify which proteins are present in the deposit. After exposure to, for example, blood plasma or serum, the surfaces are rinsed extensively to remove non-bound proteins, and then identification is carried out using immunoblotting and liquid chromatography with tandem mass spectrometry detection (LC-MS/MS).[32–34] In one approach, the proteins are desorbed from the surface by washing with a sodium dodecylsulfate (SDS) solution; they are then separated *via* gel electrophoresis according to their molecular weight and transferred onto a membrane. Subsequently, the membranes are exposed to labeled antibodies that can be detected by an appropriate method, *e.g.*, radioactivity, fluorescence, or enzymatic reactions. However, this method has the significant limitation that only proteins for which an antibody is available can be detected. Considering that there are over 4500 types of proteins in blood plasma, the fraction accessible *via* immunoblotting is small. In the LC-MS/MS method, trypsin is used to partially digest the proteins in the deposit into smaller peptides, which are identified by liquid chromatography coupled to tandem mass spectrometry. The peptide signals detected are compared to a peptide spectral database and assigned to the proteins in the original deposit. Mass spectrometry makes possible the relative quantification of the identified proteins.

The preservation of the structure of proteins upon attachment to a surface can have dramatic effects on the interactions of materials in biological environments. Typically, AFM has been applied to observe the changes in the dimensions of adsorbed proteins on smooth surfaces. Taking repeated high magnification scans in tapping mode over the same area makes it possible to observe single protein molecules in an aqueous environment, and the evolution of their height and spreading on the surface can be tracked and correlated with the surface properties.[35] In addition, spectroscopic methods have been used extensively to study the secondary structure of proteins.[36] In circular dichroism (CD) spectroscopy, the interaction of the surface-bound proteins with circularly polarized light is an indicator of the conformation.[37] The principle underlying the measurement is the differential absorption by proteins of left- and right-handed circularly polarized light in the UV region, depending on the presence of structural elements such as α-helices and β-sheets. The CD spectrum is acquired by measuring the difference in absorbance of the two polarizations of light in the 190–250 nm range. Since different conformations give characteristic CD spectra, the spectra recorded for the proteins of interest attached on a surface can be fitted to determine the fractions of these secondary structural features.

Because of the dynamic nature of the interactions between surfaces and proteins and the large number of factors that influence them, it is vital to understand the interaction forces by measuring them directly. The forces between two interacting proteins, a single protein and a surface, or a ligand and receptor pair can be obtained by AFM operating in the single-molecule force spectroscopy (SMFS) mode.[38,39] The same technique has been used to quantify the strength of single chemical bonds. The tip of the AFM cantilever is functionalized with the protein of interest, or more generally one of the components of the interacting pair, and the cantilever is made to approach the surface until contact is established. The measurement is performed by pulling the cantilever back up while recording the deflection as a function of the distance pulled. Since the cantilever deflection is proportional to the total force between the tip and the sample surface, any sudden drop in the deflection corresponds to the disruption of an individual bond. The force–distance curve is constructed by multiplying the recorded deflections by the cantilever spring constant, which can be calculated from the cantilever geometry and material properties, or measured experimentally. The jagged appearance of the separation curve is due to the overlapping disruptions of multiple interactions. The same technique has been applied to examine the adhesion of individual bacteria to different surfaces, as well as different individual cells to each other. These measurements are performed by selectively immobilizing a single cell on the AFM probe, so the technique is called single-cell force spectroscopy (SCFS).[40,41] Integrating the area under the force–distance curve makes it possible to calculate the work necessary for detachment to occur, which corresponds to the energy of adhesion.

References

1. H.-J. Butt, K. Graf and M. Kappl, *Physics and Chemistry of Interfaces*, Wiley-VCH GmbH & Co. KGaA, Weinheim, 2003.
2. A. Baszkin and W. Norde, *Physical Chemistry of Biological Interfaces*, Marcel Dekker, Inc., New York, 1999.
3. J. N. Israelachvili, *Intermolecular and Surface Forces*, Academic Press, Amsterdam, 3rd edn, 2011.
4. J. D. Andrade, L. M. Smith, and D. E. Gregonis, in *Surface and Interfacial Aspects of Biomedical Polymers*, ed. J. D. Andrade, Springer US, Boston, MA, 1985, pp. 249–292.
5. H. W. Fox and W. A. Zisman, *J. Colloid Sci.*, 1950, **5**, 514–531.
6. E. A. Vogler, *Adv. Colloid Interface Sci.*, 1998, **74**, 69–117.
7. M. Bracke, F. Bisschop and P. Joos, *Trends in Colloid and Interface Science II*, 1988, vol. 259, pp. 251–259.
8. N. S. Bhairamadgi, S. P. Pujari, F. G. Trovela, A. Debrassi, A. A. Khamis, J. M. Alonso, A. A. Al Zahrani, T. Wennekes, H. A. Al-Turaif, C. van Rijn, Y. A. Alhamed and H. Zuilhof, *Langmuir*, 2014, **30**, 5829–5839.
9. Q.-Y. Sun, L. C. P. M. de Smet, B. van Lagen, M. Giesbers, P. C. Thüne, J. van Engelenburg, F. A. de Wolf, H. Zuilhof and E. J. R. Sudhölter, *J. Am. Chem. Soc.*, 2005, **127**, 2514–2523.
10. M. E. McGovern, K. M. R. Kallury and M. Thompson, *Langmuir*, 1994, **10**, 3607–3614.
11. R. Bos, H. C. van der Mei and H. J. Busscher, *FEMS Microbiol. Rev.*, 1999, **23**, 179–229.
12. P. Roach, D. Farrar and C. C. Perry, *J. Am. Chem. Soc.*, 2006, **128**, 3939–3945.
13. F. Variola, J. B. Brunski, G. Orsini, P. T de Oliveira, R. Wazen and A. Nanci, *Nanoscale*, 2011, **3**, 335–353.
14. F. A. Denis, P. Hanarp, D. S. Sutherland, J. Gold, C. Mustin, P. G. Rouxhet and Y. F. Dufrêne, *Langmuir*, 2002, **18**, 819–828.
15. J. Homola, I. Koudela and S. S. Yee, *Sens. Actuators, B*, 1999, **54**, 16–24.
16. K. M. Mayer, J. H. Hafner and A. À. Antigen, *Chem. Rev.*, 2011, **111**, 3828–3857.
17. E. C. Le Ru, E. Blackie, M. Meyer and P. G. Etchegoin, *J. Phys. Chem. B*, 2007, **111**, 13794–13803.
18. J. H. Scofield, *J. Electron Spectros. Relat. Phenom.*, 1976, **8**, 129–137.
19. S. Tokumitsu, A. Liebich, S. Herrwerth, W. Eck, M. Himmelhaus and M. Grunze, *Langmuir*, 2002, **18**, 8862–8870.
20. S. A. Ahmad, A. Hucknall, A. Chilkoti and G. J. Leggett, *Langmuir*, 2010, **26**, 9937–9942.

21. C. Rodriguez-Emmenegger, E. Brynda, T. Riedel, Z. Sedlakova, M. Houska and A. B. Alles, *Langmuir*, 2009, **25**, 6328–6333.
22. A. de los Santos Pereira, C. Rodriguez-Emmenegger, F. Surman, T. Riedel, A. B. Alles and E. Brynda, *RSC Adv.*, 2014, **4**, 2318–2321.
23. Q. Wei, T. Becherer, S. Angioletti-Uberti, J. Dzubiella, C. Wischke, A. T. Neffe, A. Lendlein, M. Ballauff and R. Haag, *Angew. Chem., Int. Ed.*, 2014, **53**, 8004–8031.
24. E. A. Vogler, *Biomaterials*, 2012, **33**, 1201–1237.
25. P. Schwinté, V. Ball, B. Szalontai, Y. Haikel, J. C. Voegel and P. Schaaf, *Biomacromolecules*, 2002, **3**, 1135–1143.
26. H. Elwing, *Biomaterials*, 1998, **19**, 397–406.
27. S. Reichelt, K. J. Eichhorn, D. Aulich, K. Hinrichs, N. Jain, D. Appelhans and B. Voit, *Colloids Surf., B*, 2009, **69**, 169–177.
28. J. Homola, *Chem. Rev.*, 2008, **108**, 462–493.
29. M. I. Rocha-Gaso, C. March-Iborra, Á. Montoya-Baides and A. Arnau-Vives, *Sensors*, 2009, **9**, 5740–5769.
30. B. A. Čavić, M. Thompson and G. L. Hayward, *Analyst*, 1999, **124**, 1405–1420.
31. G. N. M. Ferreira, A.-C. da-Silva and B. Tomé, *Trends Biotechnol.*, 2009, **27**, 689–697.
32. L. D. Unsworth, H. Sheardown and J. L. Brash, *Biomaterials*, 2005, **26**, 5927–5933.
33. T. Riedel, Z. Riedelová-Reicheltová, P. Májek, C. Rodriguez-Emmenegger, M. Houska, J. E. Dyr and E. Brynda, *Langmuir*, 2013, **29**, 3388–3397.
34. G. Gunkel and W. T. S. Huck, *J. Am. Chem. Soc.*, 2013, **135**, 7047–7052.
35. A. Agnihotri and C. A. Siedlecki, *Langmuir*, 2004, **20**, 8846–8852.
36. Y. Wei, A. A. Thyparambil and R. A. Latour, *Biochim. Biophys. Acta, Proteins Proteomics*, 2014, **1844**, 2331–2337.
37. K. Takasu, K. Kushiro, K. Hayashi, Y. Iwasaki, S. Inoue, E. Tamechika and M. Takai, *Sens. Actuators, B*, 2015, **216**, 428–433.
38. E. M. Puchner and H. E. Gaub, *Curr. Opin. Struct. Biol.*, 2009, **19**, 605–614.
39. P. Hinterdorfer and Y. F. Dufrêne, *Nat. Methods*, 2006, **3**, 347–355.
40. C. Rodriguez-Emmenegger, S. Janel, A. de los Santos Pereira, M. Bruns and F. Lafont, *Polym. Chem.*, 2015, **6**, 5740–5751.
41. A. Beaussart, S. El-Kirat-Chatel, R. M. A. Sullan, D. Alsteens, P. Herman, S. Derclaye and Y. F. Dufrêne, *Nat. Protoc.*, 2014, **9**, 1049–1055.

CHAPTER 2

Protein Adsorption on Surfaces: Understanding the Complex Nature of a Common Phenomenon

2.1 Protein Adsorption on Surfaces: a Brief Overview of a Critical Issue

Contact of artificial materials with biofluids (*e.g.* blood, urine) is very common in many aspects of modern medicine in procedures/applications aiming to replace diseased or damaged body parts, assist in healing, improve function, correct abnormalities, or aid in treatment. Whether applied *in vivo* or *in vitro*, the range of equipment, implants, and devices, which millions of people resort to, is wide and includes, for instance: the circuitry and pump systems (and oxygenator/thermoregulator apparatuses) used to circulate blood extracorporeally during bypass procedures such as in hemodialysis and cardiopulmonary surgery, indwelling catheters for intravenous fluid injection, coronary stents for expansion of constricted arteries, cardiac valves and pacemakers, restorative prostheses (hip/knee joints, dental implants, intraocular lenses), plates and screws to hold fractured bones during reconstructive osteogenesis, as well as biosensor devices for clinical diagnostics/monitoring (*e.g.* glucose meters), to name a few. However, this multi-billion US$ industry continues to suffer from such worrisome issues as the bio-incompatibility of the synthetic materials used to manufacture the biomedical implements (which may lead to potentially severe, short- or

RSC Detection Science Series No. 8
Biological Fluid–Surface Interactions in Detection and Medical Devices
By Michael Thompson, Christophe Blaszykowski, Sonia Sheikh, Cesar Rodriguez-Emmenegger and Andres de los Santos Pereira

Published by the Royal Society of Chemistry, www.rsc.org

long-term complications caused by tissue inflammation and the generation of thromboemboli), and the degradation of the analytical performance due to signal interference by biological matrices with respect to the sensor technology.[1–3] One major recurrent issue with foreign materials contacting biofluids is their tendency to accumulate biological species on their surface (most notably proteins) – a natural adsorption phenomenon referred to as *fouling* that negatively affects the functionality, longevity and integrity of biomedical/bioanalytical equipment, implants, and devices. In fact, when artificial surfaces are exposed to biological milieux, proteins – within mere seconds – start to spontaneously adsorb into a conditioning layer that mediates subsequent cellular adhesion and the aforementioned biological responses/processes orchestrated by the immune and coagulation systems.[4,5] With their central role established, tremendous research efforts have been dedicated over many years to try and understand the fundamentals of protein adsorption onto synthetic materials. The situation today is such that the phenomenon is still not fully understood, due to multiple key factors (such as the nature, strength, and amount, along with the dynamics of adsorption) that quickly turn an already complex mechanism on an individual basis into an indecipherable puzzle when dealing with the multi-component nature of biological media. The present chapter focuses on the current knowledge and understanding of the phenomenon of protein adsorption, addressing such critical questions as: Why do proteins adsorb on artificial surfaces? How does this process proceed mechanistically? Are there any general principles governing adsorption? To conclude, this chapter offers an overview of the real, potential consequences and risks – in both biomedical and bioanalytical realms – of protein adsorption on synthetic materials, taking as an example the quintessential proteinaceous biofluid: blood.

2.2 Structural Complexity and Diversity of Proteins

Prior to discussing the phenomenon of protein adsorption on surfaces, a molecular-level understanding of protein makeup and structure is needed. Proteins are intricate nanoscale biopolymers, which are constructed from a repertoire of various amino-acid building block monomers that, due to the unique nature of their side-chain, present varied properties in terms of hydrophobicity/hydrophilicity, polarity, acidity/basicity, and charge (Figure 2.1).[1,6–11] At any given time, the dynamic surface of a protein may be regarded as a mosaic simultaneously exposing hydrophobic, hydrophilic, cationic and/or anionic areas, whose magnitude and distribution can be altered *via* environmental conditions such as pH, ionic strength, and temperature. The amino-acid sequence, and three-dimensional arrangement, are critically important in determining protein function (*e.g.* metabolic, signaling, architectural) and interaction with the external environment.

Structurally, proteins can be described through four levels of increasing complexity (Figure 2.2). The first and simplest is the *primary structure*, which

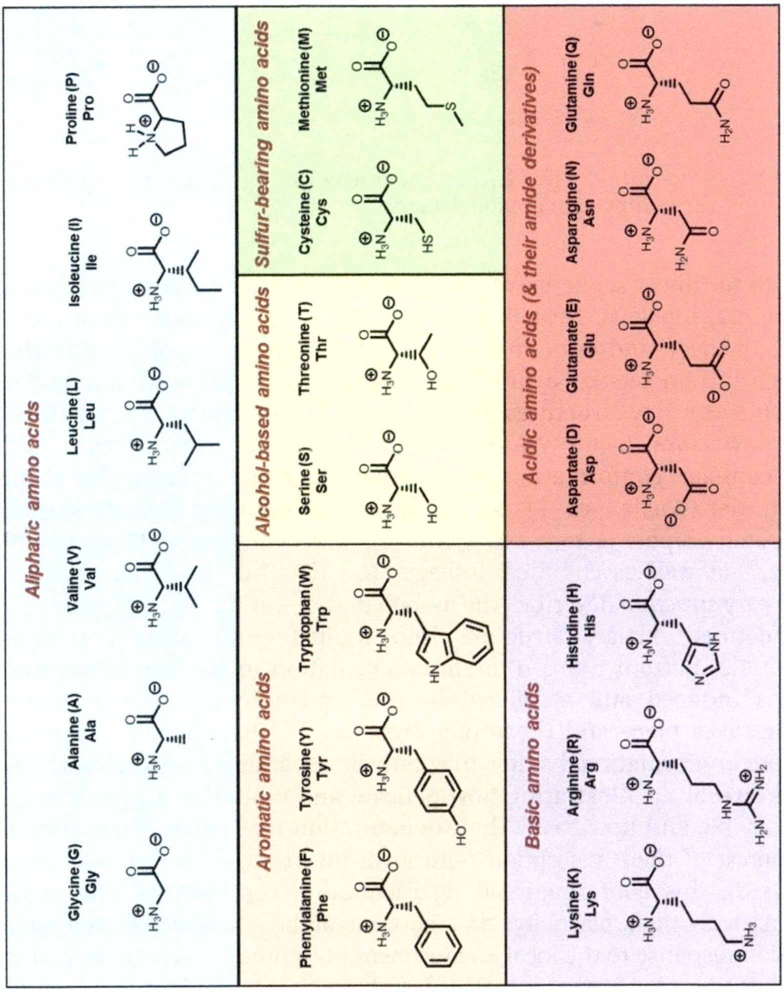

Figure 2.1 Chemical structure of amino-acids in their natural L-configuration.

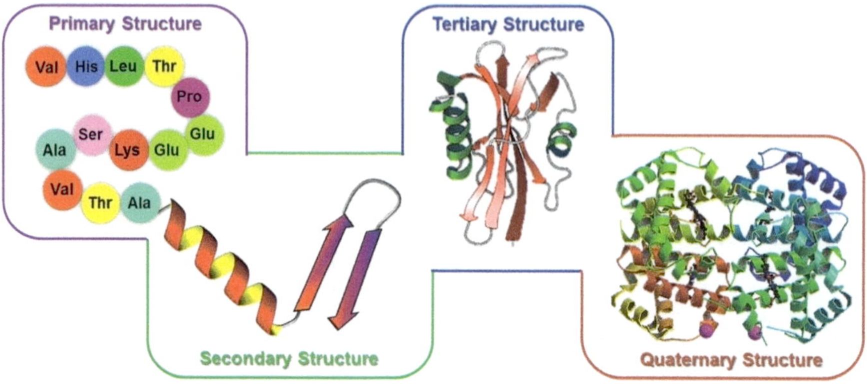

Figure 2.2 From primary (far left) to quaternary (far right) structure: proteins are three-dimensional biopolymers.

refers to the linear sequence of amino-acids linked *via* covalent amide bonds (Figure 2.2, top left). Next is the *secondary structure*, which describes the α-helix, β-sheet, and random loop structures formed by polypeptide chains through intramolecular hydrogen bonding (Figure 2.2, bottom left). These defined secondary structures coordinate with one another to establish the *tertiary structure* (Figure 2.2, top right), wherein proteins adopt a folded, more compact conformation stabilized through non-covalent interactions of van der Waals (<5 kJ mol^{-1}), hydrogen bonding (<5–40 kJ mol^{-1}), electrostatic/dipole (5 to >190 kJ mol^{-1}), and hydrophobic (5–40 kJ mol^{-1}) nature,[12] as well as chemical linkages (*i.e.* disulfide bridges). Finally, the *quaternary structure* describes the association of multiple polypeptide chains with defined tertiary structure into multi-subunit protein complexes (Figure 2.2, bottom right). Protein conformation in the lowest free-energy state is induced and stabilized by the aqueous environment in which folding takes place and the protein resides.[7,9,13] It is pivotal to note that different conformational states may only be separated by energies as low as 2–14 kcal mol^{-1},[7] illustrating how delicate and malleable a protein's structure may be, and how easily their denaturation may occur. Translated into the context of their adsorption onto artificial materials (which will be later discussed), this conformational restructuration capability is what endows proteins with their plasticity; that is, their ability to adopt altered shapes/unfold in response to the local environment presented by a contacting surface.

Proteins can be further described with respect to their morphology and classified into one of the three following main categories: membrane, fibrous, and globular. Membrane proteins, as the name suggests, are proteins associated with the membrane of cells or organelles. 'Integral membrane proteins' that are found permanently attached to/embedded in phospholipid bilayers would be an example (Figure 2.3, top). Fibrous proteins, such as collagen and keratin, are characterized by asymmetrical and elongated rod-like features (Figure 2.3, bottom left), and are typically insoluble in water.

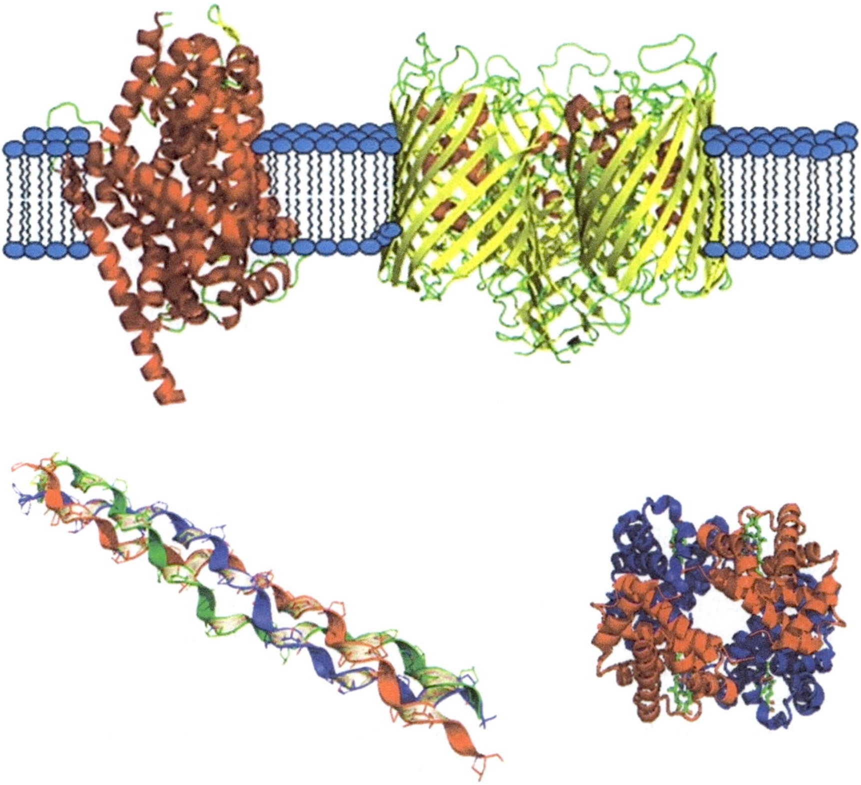

Figure 2.3 Examples of the different shapes proteins can adopt: α-helical bundle and β-barrel *membrane* proteins (top); cylindrical, *fibrous* collagen (bottom left); and spherical, *globular* hemoglobin (bottom right). *Note:* these structures are not to scale.

In contrast, globular proteins, such as albumin and hemoglobin, present more of a roughly spherical shape (Figure 2.3, bottom right). These proteins generally fold in such a way that hydrophilic/charged amino-acid residues are exposed on the outer surface, while hydrophobic ones are buried in the inner core of the protein held through hydrogen bonding, weak acid–base and/or hydrophobic interactions. The aim of such concealment is to minimize the energetic cost associated with less-favorable non-polar/polar interactions between hydrophobic protein domains and the surrounding aqueous environment.[9] Still, proteins typically display on their surface an assortment of hydrophilic and hydrophobic regions giving rise to their amphipathic nature.[8,14] As a result, globular proteins present variable solubility in aqueous media. At last, it is also possible to group proteins into families according to their functional role (*e.g.* metabolic, signaling, architectural).

As per the virtually countless amino-acid combinations and sequences, which can be arranged from the available repertoire of 20 – to form polypeptides of unlimited length folding into intricate three-dimensional structures with unique plasticity – it has become quite clear that trying to rationalize on an individual basis how proteins will adsorb on surfaces (which incidentally also possess their own unique physicochemical characteristics) is understandably a daunting task, which further gains complexity as all-interacting species in multi-component systems are considered. Nonetheless, intense research has been, and continues to be, conducted in this field (and the intimately linked field of protein repellence), which has materialized into substantial, and at times contradictory, literature. The following presents the salient features of the generally accepted understanding of protein adsorption behavior onto artificial surfaces.

2.3 Protein–Surface Interaction

2.3.1 A Glimpse into Protein Adsorption

Proteins are amphipathic, surface-active macromolecules with a heterogeneous, adaptable surface that can, as discussed earlier, display high affinity for virtually any type of man-made material.[14–16] While capable of moving freely in their carrier solution, proteins experience a decline in their rotational degree of freedom as they start to adsorb and adopt certain orientations at the surface/bulk solution interface.[4] For irregularly shaped proteins, two types of orientation can be discerned depending on whether adsorption occurs 'end-on' or 'side-on' – that is, parallel to the protein's long or short axis, respectively (Figure 2.4).[4] Understandably, the nature of the orientation affects both the thickness and saturation state of the final protein layer, end-on adsorption producing thicker and denser protein layers compared to its side-on counterpart.[4]

The ability of proteins to adopt new conformations upon adsorption is linked to their structural 'softness' or 'hardness'.[17,18] This classification

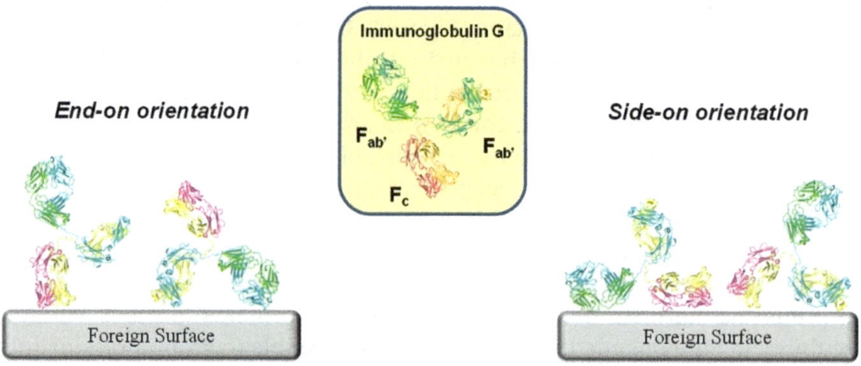

Figure 2.4 End-on (left) *vs.* side-on (right) surface orientation of asymmetrical proteins (illustrated with immunoglobulin G). *Note:* not represented here is the situation where the protein would rest flat on the surface, occupying as such the largest surface area.

arranges proteins based on their thermodynamic stability when in contact with a surface; 'soft' proteins, such as albumin and fibrinogen, are more prone to undergo conformational rearrangement (*i.e.* unfold) than 'hard' ones like lysozyme, a situation that energetically corresponds to a gain in entropy favoring the process of adsorption (see upcoming Table 2.4).[4,17–20] Predicting, with some confidence, whether a given protein will likely adsorb entails also taking into consideration the hydrophobic *vs.* hydrophilic nature of the contacting surface, as well as its electrostatic potential together with that of the protein (Table 2.1).[19]

At this stage, one can already begin to sense the complexity of the mechanism of protein adsorption on foreign surfaces. Nevertheless, fortunately there exists a number of exploratory techniques to study the various aspects of this phenomenon (Table 2.2).[21,22] Several comprehensive reviews/book

Table 2.1 Whether adsorption may be a favorable process ('Yes' *vs.* 'No') depends on the 'hard' *vs.* 'soft' nature of proteins and their positive *vs.* negative overall charge, as well as on the hydrophobic *vs.* hydrophilic and positive *vs.* negative electrostatic potential properties of the contacting surfaces. *Adapted from ref.* 19.

	Type of surface			
	Hydrophobic		Hydrophilic	
Type of protein	(+) Charge	(−) Charge	(+) Charge	(−) Charge
Hard, (+) charge	Yes	Yes	No	Yes
Hard, (−) charge	Yes	Yes	Yes	No
Soft, (+) charge	Yes	Yes	Yes	Yes
Soft, (−) charge	Yes	Yes	Yes	Yes

Table 2.2 Examples of experimental and computational techniques available to study protein adsorption behavior.

Analysis technique	Quantitative information	Structure – conformation	Kinetic information
Ellipsometry	Yes	Yes	Yes
Neutron reflectometry (NR)	Yes	Yes	Possible
Surface plasmon resonance (SPR)	Yes	Possible	Yes
Thickness shear mode (TSM)	Yes	Yes	Yes
Optical waveguide light-mode spectroscopy (OWLS)	Yes	Yes	Yes
Total internal reflection fluorescence (TIRF)	Yes	Yes	Yes
X-ray photoelectron spectroscopy (XPS)	Semi	Possible	Possible
Atomic force microscopy (AFM)	Yes	Yes	Possible
Attenuated total reflectance Fourier transform infrared spectroscopy (ATR-FTIR)	Yes	Yes	Yes
Circular dichroism (CD)	No	Yes	No
Time-of-flight secondary ion mass spectrometry (ToF-SIMS)	Possible	Yes	Possible
Molecular dynamics (MD) simulation	Yes	Possible	Possible

chapters, which interested readers are encouraged to acquire, have been published over the years to cover the topic of protein–material interaction from both a theoretical and experimental stance.[4,8–10,23–27] Despite the accessibility to numerous, evermore powerful (sensitive) analysis techniques, the truth of the matter is that the phenomenon, to this day, is still not fully understood. The reason for this is simple and lies in the fact that there is more to protein adsorption than their mere deposition on a surface; the process being indeed far more complex, potentially featuring such highly involved phenomena as conformational restructuration and surface aggregation.[4,18] The following subsection looks at the current understanding of protein adsorption on artificial materials from a molecular viewpoint.

2.3.2 Mechanism of Protein Adsorption: a Molecular Perspective

When the surface of an exogenous (artificial) material is exposed to a biological milieu, a large number of interactions take place.[4,8,28] One of the first events, which is often overlooked or simply ignored, is actually the spontaneous adsorption (on the nanosecond timescale)[29] of water molecules and ions into a water/electrical double layer that is physically distinct from the local bulk environment,[30] with a thickness defined by the Debye length.[9,31] It is only shortly thereafter (within seconds)[7,32,33] that proteins migrate from the bulk aqueous medium (where they reside solvated) towards the surface *via* convection/diffusion, whereupon they adsorb through the combination of multiple van der Waals, hydrogen bonding, electrostatic, and hydrophobic interactions (Figure 2.5).[9,23,29,31,32,34–38] It has been postulated that proteins initially adsorb to surfaces in their native state[9,10,39] and that, for adsorption to take place, both protein and material surfaces must undergo dehydration (see *Subsection 2.3.3*).[8,9,40] In any event, it is likely the case, as discussed earlier, that flexible (*i.e.* 'soft' *vs.* 'hard'), adsorbed proteins will – in order to optimize the energy of interaction,[8,41] and in response to their environment – eventually assume altered conformational states and interact with surfaces in multiple ways of varied strength, some of which are depicted in Figure 2.5.[9,26,34,35,42] The adhesion of the much larger, less motile entities that are biological cells happens last, through the adsorbed aqueous layer of ions and proteins.[30,40] Whether bioactive receptor sites will be (made) available for subsequent cellular binding is dictated by a combination of the type, amount, orientation, conformation, and packing density of proteins present in the adsorbed layer. One motif in particular, which has been shown to mediate cellular attachment, is the Arg-Gly-Asp (RGD) tripeptide sequence found notably in fibronectin as well as other cell-adhesive proteins.[40,41,43,44]

It is safe to say that protein adsorption on artificial surfaces is a complex process. This process is dynamic in nature and hinges around the following time-dependent key stages: (i) migration from the bulk phase towards the surface; (ii) random surface adsorption–desorption (with distinct

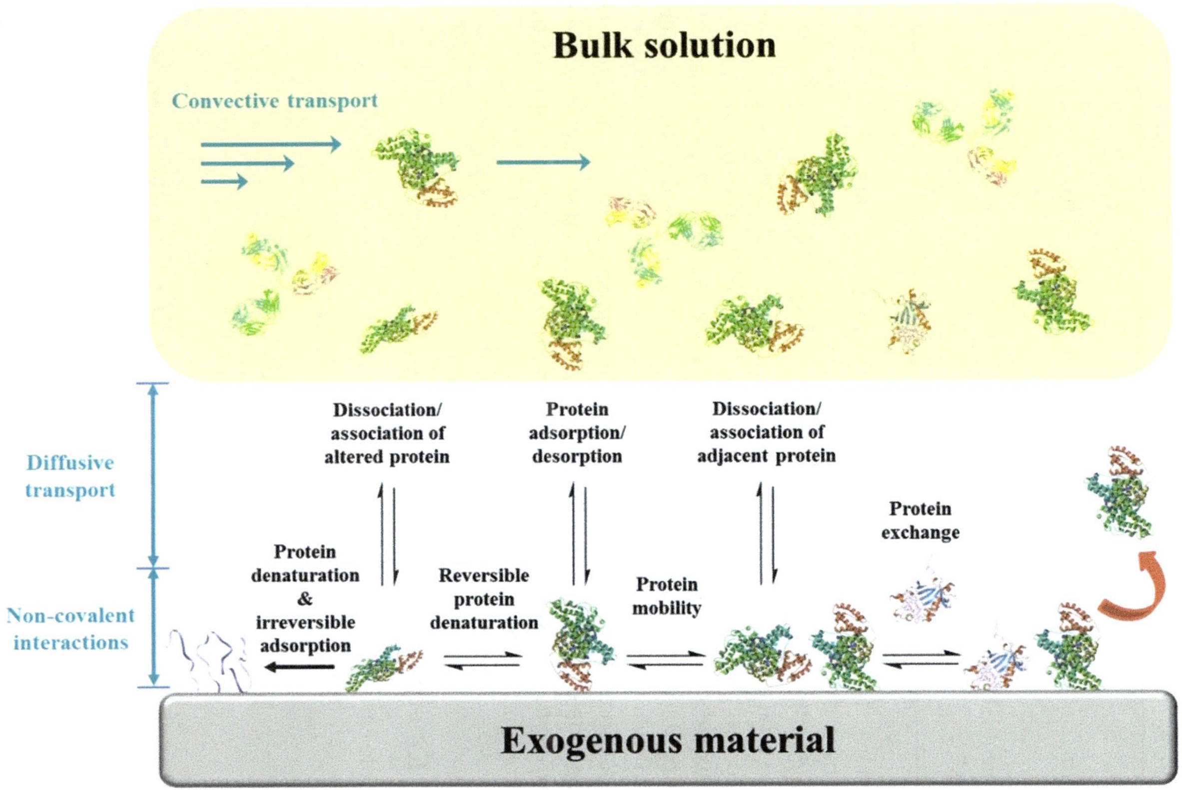

Figure 2.5 A molecular-level look into the mechanism of protein adsorption onto artificial surfaces. Non-exhaustive, schematic representation of the various possible scenarios that can occur upon exposure of exogenous materials to proteinaceous media.

Table 2.3 Molecular weight, concentration, and diffusion coefficient of some of the main proteins of blood plasma. 'IgG', 'HDL', 'IgA', and 'LDL' respectively stand for 'Immunoglobulin G', 'High-density lipoprotein', 'Immunoglobulin A', and 'Low-density lipoprotein'. *Adapted from ref.* 20.

Plasma protein	Molecular weight (kDa)	Plasma concentration (mg mL^{-1})	Diffusion coefficient (10^7 cm^2 s^{-1})
Albumin	66	40	6.1
IgG	150	8–17	4.0
Transferrin	77	2.3	5.0
HDL	170	18	4.6
IgA	150	1–4	4.0
Complement C3	180	1.6	4.5
Fibrinogen	340	2–3	2.0
LDL	2000	2.0	2.0

orientation); (iii) spreading through conformational alteration (accompanied by a decrease in the rate of desorption); and (iv) exchange or irreversible adhesion and denaturation[15,45] (Figure 2.5).[7,10,24,35,37,39,46,47] Besides surface affinity, it is thus clear that the final composition of an adsorbed layer of proteins will also be greatly influenced by the various kinetic aspects of adsorption, which are unique to each (type of) protein competing for a surface site (a phenomenon later discussed as the 'Vroman effect' in *Subsection 2.3.5.2*). One important parameter that describes the ability of proteins to transport from the bulk phase towards the interface (and depends on the protein size as well as the bulk concentration) is the *diffusion coefficient*, values of which are given in Table 2.3 for some of the main proteins found in blood plasma.[20]

2.3.3 Kinetics, Thermodynamics, and Dehydration

During the early, diffusion-controlled[6,23] stage of protein adsorption, surface coverage can be expressed mathematically as eqn (2.1):

$$\theta = 2C_0\sqrt{\frac{D \times t}{\pi}} \tag{2.1}$$

where θ is the surface coverage, D is the diffusion coefficient, C_0 is the protein bulk concentration, and t is time.[9,10]

However, early arrival to a surface does not necessarily ensure a place within the eventual protein adlayer. In fact, as time progresses, already adsorbed proteins are displaced by newly incoming ones that display higher affinity for the surface. This sequential phenomenon, known as the '*Vroman effect*', will be discussed in greater detail in *Subsection 2.3.5.2*.[16,26,40,46,48,49] In this time-dependent exchange process, the surface concentration and composition of proteins do not necessarily match those found in the bulk solution.[2,8,10,29,47,50]

Surveying the literature, it seems as though there still exists a lack of consensus regarding the exact mechanistic nature of protein adsorption, in spite of tremendous research efforts over the last few decades, and a thorough theory is still pending.[51,52] Nevertheless, it has been generally accepted that the phenomenon can be described through the following general thermodynamic expression (eqn (2.2)):[6,10,25,32,53]

$$\Delta G_{\text{ads}} = \Delta H_{\text{ads}} - T\Delta S_{\text{ads}} \tag{2.2}$$

where ΔG_{ads}, ΔH_{ads}, and ΔS_{ads} respectively are the changes in Gibbs free energy, enthalpy, and entropy of the surface–protein–water system upon protein adsorption, and T is the temperature.

Protein adsorption is energetically favored when $\Delta G_{\text{ads}} < 0$, a situation which would actually be the favorable net result of the summation of a series of enthalpic and entropic individual terms associated with a variety of favored or disfavored interactions/subprocesses occurring during protein adsorption (Table 2.4).[13,25,54,55]

In a similar manner, it was recently proposed that the term ΔG_{ads} could be divided into the following three components (eqn (2.3)):[31]

$$\Delta G_{\text{ads}} = \Delta G_{\text{hydrophobic}} + \Delta G_{\text{dehydration}} + \Delta G_{\text{interaction}} \tag{2.3}$$

where $\Delta G_{\text{hydrophobic}}$ would be the free energy gain resulting from the expulsion of 'hydrophobic' protein entities from the aqueous solution upon adsorption; $\Delta G_{\text{dehydration}}$ would be the free energy cost due to the resulting displacement of interfacial water molecules; and $\Delta G_{\text{interaction}}$ would take into account protein–surface and protein–protein interactions (a free energy gain).

For adsorption to take place, it would thus once again appear that both surface and proteins must undergo dehydration (see eqn (2.3) and previous Table 2.4),[13,31,40] a chaotropic process transferring into the bulk aqueous solution the molecules of water that were once coordinated within the surface's and proteins' layer/shell of hydration (Figure 2.6).[9,10,13,25,31,32,40,54,56] The structure of water molecules covering the surface and surrounding proteins plays a crucial role on the interactivity between the surface and the proteins, the possibility for proteins to adsorb being linked to the ability for

Table 2.4 A glimpse into the multi-component nature of the thermodynamics of protein adsorption. 'VdW' stands for 'van der Waals'. *Adapted from ref. 55.*

Energy term	+/− Sign	Interaction/process	Effect on protein adsorption
ΔH_{ads}	−	VdW interaction (short range)	Favoring
ΔH_{ads}	−	Ion–ion interaction (long range)	Favoring
ΔH_{ads}	+	Dehydration (surface & protein)	Opposing
ΔH_{ads}	+	Protein unfolding	Opposing
ΔS_{ads}	+	Water molecule desorption	Favoring
ΔS_{ads}	+	Counter-ion release	Favoring
ΔS_{ads}	+	Protein unfolding	Favoring
ΔS_{ads}	−	Protein adsorption	Opposing

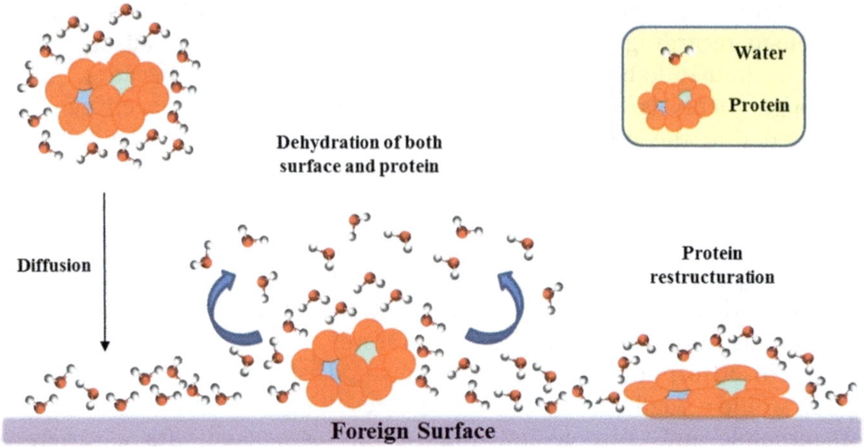

Figure 2.6 Protein adsorption onto artificial surfaces: illustration of the key involvement of both surface and protein dehydration.

such interfacial water network(s) to be disrupted and displaced through the breaking of hydrogen bonds (Figure 2.6).[54] The associated cost in energy would sharply rise with the substrate hydrophilicity and need to be overcome in order to promote protein adsorption.[31]

In summary, protein adsorption features a number of energetically favorable situations (*i.e.* $\Delta H_{ads} < 0$ and $\Delta S_{ads} > 0$) such as those related to non-covalent[8,46] (and covalent) bond formation, and protein (*i.e.* unfolding) and surface (*e.g.* interfacial reorganization of charged groups) structural alteration,[25] as well as solvation effects.[51,57] Conversely, enthalpically/entropically unfavorable processes also exist, this notably being the case when hydrogen bonds are cleaved during surface and protein dehydration, or when proteins leave the bulk solution for the surface. Although all these factors may seem somewhat simple to identify, it is in actuality tremendously challenging to account for each and every enthalpic and entropic term (and gauge their relative contribution) involved in the overall change in free energy, another reason why it is difficult to fully comprehend the phenomenon of protein adsorption.[32]

2.3.4 General Governing Principles: a Tale of Caution

The variety of influential factors is undoubtedly a characteristic of protein adsorption. With that being said, it is possible to identify for each interacting part, and their surroundings, a relatively limited number of key parameters. These relate to: (i) the nature of proteins with respect to their morphology and flexibility, size and molecular weight, and charge/polarity degree; (ii) the physicochemical properties of the contacting surface (*e.g.* topography, electrostatic potential, surface energy); and (iii) the environmental conditions (*e.g.* pH, ionic strength, temperature).[2,8,16,58–60]

Some empirical 'principles' have even emerged in the literature.[9,31] With respect to their dimension, for instance, it seems as though the larger a protein, the more likely it is to possess multiple adhesion sites and hence, readily adsorb.[16,50] The same would hold true for proteins near their isoelectric point (that is, the pH value at which a protein is electrically neutral).[50] With regards to their structure, the less stable so-called 'soft' proteins (see *Subsection 2.3.1*) have a greater tendency to unfold and fit underlying surfaces upon adsorption. Vis-à-vis substrate properties, proteins tend to display higher affinity for 'hydrophobic' *vs.* 'hydrophilic' surfaces.[4,8,40,41,57,61,62] It is reasonable that this observation may be linked to the ease with which the latter undergo dehydration (a phenomenon discussed earlier in *Subsection 2.3.3*, and depicted in Figure 2.6). The physicochemical properties of surfaces have also been investigated and, generally speaking, it would appear that stronger protein adhesion occurs on high- *vs.* low-surface energy substrates, and that adsorption is preferred on charged *vs.* neutral surfaces.[4] Furthermore, it has been shown that surface roughness/nanostructure[60] also impacts protein adsorption, since greater amounts of proteins adsorb on rougher surfaces.[63] Lastly, regarding the environmental conditions, it would appear that the higher the temperature, the faster proteins diffuse from the solution to the surface, resulting in a greater amount of adsorbed proteins.[4,64]

The complexity of the mechanism of protein adsorption, as well as the multiplicity of influencing parameters (and their interconnection) discussed earlier, call for these 'generalizations' and others to be handled with extreme caution, as there always exists the possibility of finding equally compelling counter-examples. An illustrative case would be how the literature is composed of opposing views as to whether proteins adsorb in a mono- or multi-layer fashion.[31,45] A reason for this apparent disagreement is that small proteins, with which most adsorption studies are performed, have a tendency to form monolayers (or near-monolayers), whereas less investigated, larger proteins have been shown experimentally to produce multilayers.[31] Furthermore, to add another level of complexity, it is important to note that blood proteins from different mammalian species may display different adsorption behavior.[31] Finally, one must also consider the possibility that a given protein–surface system may not always yield the same final organization/distribution in terms of adsorbed protein orientation, conformation, and packing,[32] owing to discrepancies (however subtle) in the multivariate, complex mechanism of protein adsorption.[8]

2.3.5 Protein Adsorption Models

2.3.5.1 Single-protein Solutions

Protein adsorption studies, which date back to the early 1900s, have gained popularity since the 1930s.[10,39] By allowing sufficient time for a protein–surface system to interact at constant temperature, it can be assumed that an equilibrium will be established between the amount of protein adsorbed on the surface and that of protein remaining in the bulk solution.[65] This equilibrium relationship is commonly referred to as an *adsorption isotherm*, which can be supplemented with other empirical data to help understand protein–surface interactions.[65] In general, adsorption isotherms are represented graphically by plotting the amount of adsorbed material as a function of non-adsorbed material.[65] The resulting plot can then be fitted with a suitable adsorption model to provide information regarding the thermodynamics of adsorption along with other factors related to the process.[66] Several different adsorption isotherm models, each of which is based on different assumptions and conditions, exist to describe the process of protein adsorption (Figure 2.7).[66]

For instance, the model developed by Freundlich describes monolayer protein adsorption on heterogeneous surfaces displaying binding sites that vary with respect to the rate and energy of protein adsorption.[66,67] The Brunauer–Emmett–Teller (BET) model, which depicts multilayer protein adsorption, is another example.[66,68] However, one of the simplest and most frequently used and reported[66] models to address protein adsorption is the Langmuir model,[69] even though it was originally developed to explain

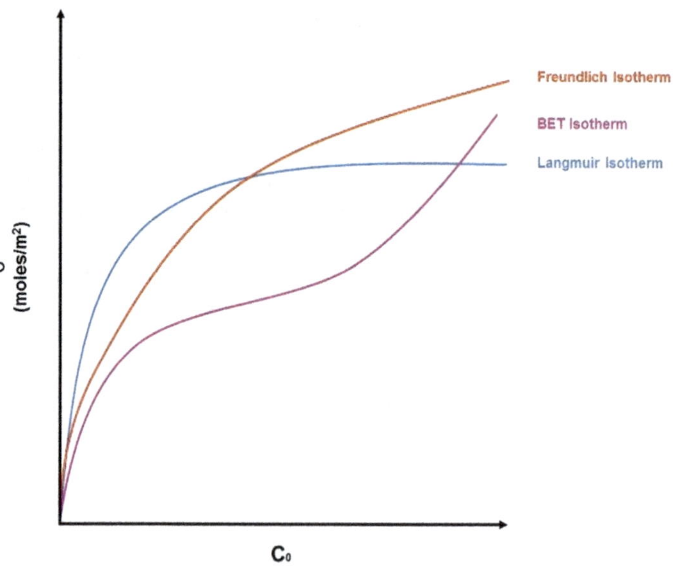

Figure 2.7 General graphical shape of Freundlich (red), Brunauer–Emmett–Teller (purple), and Langmuir (blue) adsorption isotherm models plotting surface coverage of adsorbed proteins as a function of protein bulk concentration.
Adapted from ref. 66 with kind permission of John Wiley and Sons. © 2015 Wiley Periodicals, Inc.

gas-to-solid phase adsorption.[4,9,38,65] The *Langmuir isotherm* expression can be derived by treating the adsorption–desorption of a material (the 'adsorbate') on/from a surface (the 'adsorbent') as a state of equilibrium. In the case of protein adsorption from a solution onto a solid surface, this can be represented as follows (eqn (2.4)):

$$\mathrm{P} + \mathrm{S} \underset{\text{desorption}}{\overset{\text{adsorption}}{\rightleftharpoons}} \mathrm{PS} \tag{2.4}$$

where P is the adsorbing protein, S is the surface (*i.e.* available binding sites), and PS represents the surface-bound protein.[66]

The adsorption equilibrium constant, K_{ads}, can then be defined as (eqn (2.5)):

$$K_{\mathrm{ads}} = \frac{[\mathrm{PS}]}{[\mathrm{P}][\mathrm{S}]} \tag{2.5}$$

Considering that the amount of protein bound to the surface ([PS]) is proportional to the surface coverage of adsorbed proteins (θ), that [S] is proportional to the number of vacant binding sites ($1 - \theta$), and that [P] is proportional to the protein bulk concentration (C_0), another equilibrium constant – the adsorption constant (A), which measures the adsorption capability – can next be defined through the following expression (eqn (2.6)):

$$A = \frac{\theta}{C_0(1 - \theta)} \tag{2.6}$$

Rearrangement of eqn (2.6) isolating for surface coverage provides the familiar mathematical form of the Langmuir adsorption isotherm (eqn (2.7)):

$$\theta = \frac{AC_0}{1 + AC_0} \tag{2.7}$$

Graphically, when the surface coverage (θ) is plotted as a function of the bulk concentration of protein (C_0), as shown in Figure 2.8, a distinct steep linear segment is first observed, followed by a plateau at saturation. At this stage, the number of proteins adsorbing to the surface equals that of desorbing proteins. Surface saturation typically occurs between 0.1–0.5 $\mu\mathrm{g\,cm^{-2}}$ (depending on the protein dimension and orientation) – a value characteristic of a close-packed, monolayer-like coverage.[38]

The Langmuir isotherm model makes several assumptions: (i) there is only one type of molecule adsorbing from solution; (ii) the solution is dilute; (iii) there are a finite number of available binding sites; (iv) there is single occupancy per site (*i.e.* monolayer coverage); (v) the surface is energetically uniform; (vi) adsorption sites are independent of each other (*i.e.* lateral interactions between adsorbed/adsorbing molecules are insignificant); and (vii) adsorption is reversible.[9,66] An adsorption system that fulfils these criteria is expected to produce the profile of a Langmuir isotherm depicted in

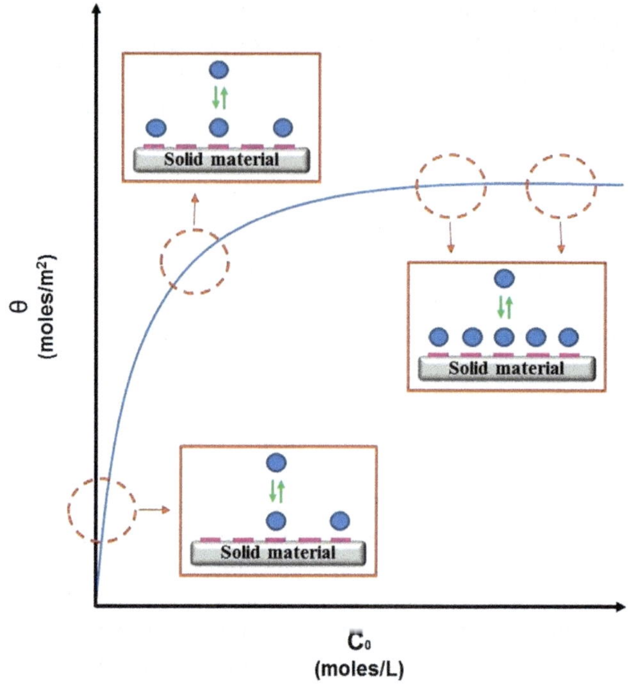

Figure 2.8 General graphical form of Langmuir isotherm (blue curve), and schematic representation of the process of protein surface adsorption–desorption on available binding sites until saturation (insets). Adapted from ref. 66 with kind permission of John Wiley and Sons. © 2015 Wiley Periodicals, Inc.

Figure 2.8. In practice, however, most of these conditions are usually violated.[4,66] For instance, the manner an incoming protein adsorbs onto a surface surely is dictated by its interaction with pre-adsorbed proteins, whether the effect is cooperative or not.[4] Not only that, but also something Langmuir himself recognized is the fact that surfaces are not energetically homogeneous.[4] Moreover, very large proteins can occupy more than one binding site.[16] This undoubtedly simple model still continues to be employed to fit experimental adsorption data despite the fact that the aforementioned requirements associated with it may not be satisfied.[66] The issue and risk with inappropriately fitting an experimental isotherm to the Langmuir model because of its mere resemblance are to misinterpret the protein adsorption behavior, as illustrated in Figure 2.9 (*vs.* Figure 2.8), and to extract erroneous information.[8,66]

More sophisticated variants have been developed to account for the more complex reality of the process of protein adsorption, including parameters such as adsorption irreversibility,[9,70] protein–protein interactions,[66] adsorbed protein conformational changes,[62] as well as protein rearrangement over time[39] resulting in a completely irreversible protein layer.[9,26]

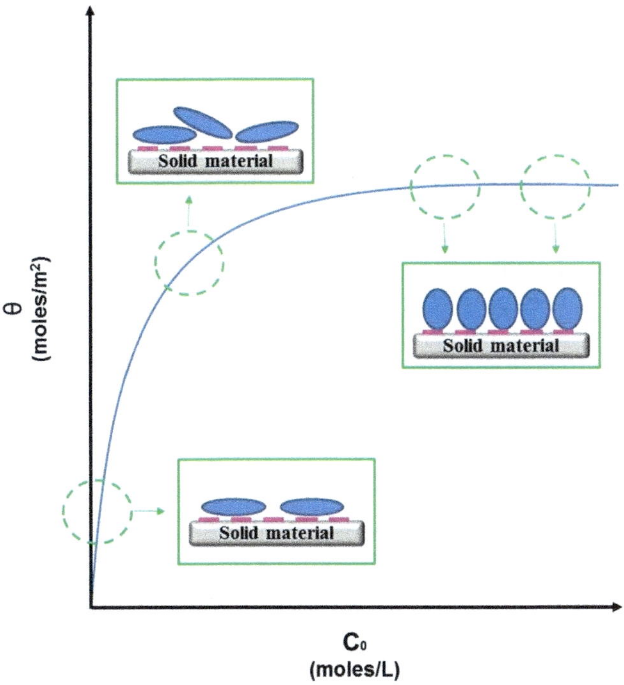

Figure 2.9 An example illustrating the possibility for an adsorption isotherm profile to resemble that of the Langmuir model, yet the process of protein adsorption greatly deviates from the assumptions made by such a model. Adapted from ref. 66 with kind permission of John Wiley and Sons. © 2015 Wiley Periodicals, Inc.

2.3.5.2 Multi-protein Solutions

Adsorption studies and models have traditionally focused on simple single-protein solutions. This chapter would not be complete, however, without mentioning the attempts that have been made to model the more complex competitive adsorption of two or more types of protein. One example is that of the two-protein model developed by Bessinger and Leonard with albumin and IgG.[62] Unfortunately, this model is highly impractical to predict protein adsorption behavior as it involves 12 variable parameters.[9,63] Not surprisingly, modeling and prediction of protein adsorption behavior grow more complex as the number of proteins included in the model/system increases. Empirically however, it has been observed that mixtures of proteins interact with artificial surfaces in a sequential manner. As discussed earlier in *Subsection 2.3.2*, the various types of protein present in a mixture initially diffuse towards the surface at a rate depending on their diffusion coefficient and bulk concentration.[9] A temporal adsorption pattern thus takes place, wherein higher mobility, more abundant proteins adsorb first transiently before being gradually replaced by less motile, scarcer ones with higher surface affinity (Figure 2.10).[3,8,49,57] This general, well-established phenomenon of sequential protein adsorption was first observed in the late 1960s by Vroman and Adams through ellipsometry studies, and is generally termed the 'Vroman effect'.[16,26,40,46,48,49,71–73] The initial thought was that exchange was specific to fibrinogen; however, later studies confirmed that other plasma proteins such as albumin and IgG also undergo this process.[9] What was eventually found is that blood plasma proteins adsorb (onto glass and some metal oxide surfaces) in the following order: albumin, γ-globulin, fibrinogen, fibronectin, clotting Factor XII, and high-molecular weight kininogen.[9,74] It is important to note that several other factors also influence the Vroman effect, such as the surface chemistry of the substrate, the solution properties (*e.g.* ionic strength, pH), temperature, and shear strain rate.[8]

A final comment deserves to be made regarding the fact that diffusion from the bulk phase towards the surface entails an environmental change, which may induce proteins to conformationally rearrange. The consequence is the possibility for hydrophobic domains, once buried within the protein and hidden from the aqueous medium, to be revealed and adsorb on the sorbent material. Further conformational restructuration, now surface-induced, may ensue and lead to the exposure of key cryptic bioactive sites capable of initiating biological processes/responses upon binding with cell

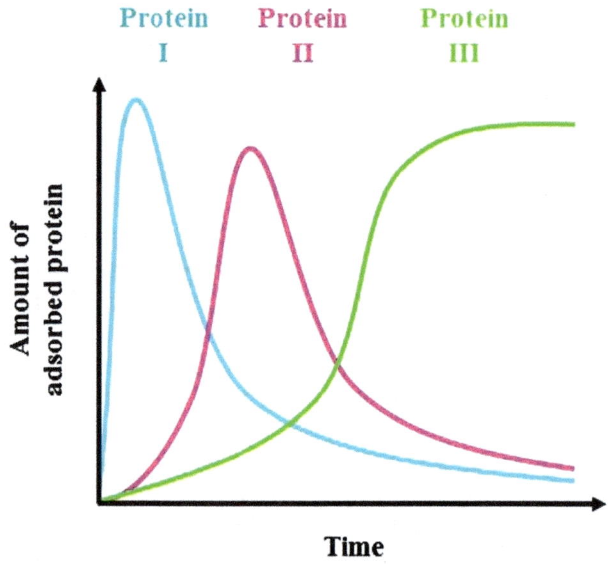

Figure 2.10 Graphical representation of the dynamic 'Vroman effect' for a three-protein mixture. Early surface adsorption is determined by the mobility and concentration of proteins (*i.e.* their ability to migrate/diffuse from the bulk solution towards the surface). However, as time progresses, proteins sequentially exchange with respect to their affinity for the surface.

integrin receptors (for instance).[8,13,75,76] [Indeed, it would be through cells interrogating the first-adsorbed layer of proteins – and not the substrate material itself – that surface-induced cellular activation may be triggered.[8]] It has been recognized that the presence of fibrinogen in an adsorbed layer of proteins stimulates platelet and leukocyte adhesion, and plays a role in the surface-induced activation of the coagulation and immune responses.[77–82] Similarly, it has also been reported that high thrombogenicity can be triggered by the presence of adsorbed γ-globulins, which enhance platelet[83] and leukocyte[84] adhesion.[62] Interestingly, platelets have even been shown to be capable of adhering to adsorbed albumin – a protein that yet lacks any known platelet binding receptor – when the latter conformationally unfolds beyond a critical point and reveals binding sites.[85,86] Research has made it quite clear that (host) biological responses to artificial material exposure are mediated by surface-activated, adsorbed proteins.[75,82,87,88]

Deciphering the mechanism of protein adsorption onto foreign material surfaces down to its most fundamental level would pave the way towards a better understanding of the consequences (*i.e.* surface-induced activation of biological processes/responses, sensor signal interference) this phenomenon may have in practice in the biomedical and bioanalytical fields. The objective is to devise rational means of preventing the potential occurrence of related risks (*i.e.* bio-incompatibility and misdiagnosis). The final section of this chapter provides a concise look into the practical concerns associated with protein adsorption in the case of blood, the quintessential proteinaceous biofluid for which a succinct but solid description (in terms of composition, role, and importance) will be given prior to discussing the issues at stake.

2.4 The Case of Blood and its Plasma/Serum Derivatives

2.4.1 Blood: the Proteinaceous Biofluid of Life

In the realm of Biotechnology, blood constitutes arguably one of the most complex and challenging biofluids. A physiological medium in which bathe both cellular and non-cellular species, blood is essential in maintaining a human organism healthy and functional through the supply of vital nutrients and gases to cells (or, conversely, the disposal of metabolic waste products), the regulation of the internal environment (*e.g.* pH, osmosis), the defense of the body against foreign invaders (*via* the action of the complement and immune systems), or the prevention of hemorrhages (*via* hemostasis and coagulation).[89–91] On average, the volume of blood that permeates a healthy adult individual is about 5 liters, or about 7% of the body weight. Normally, blood circulates throughout the body in a forward, laminar fashion *via* a closed-loop system of channels known as *blood vessels*, an illustration of which is provided in Figure 2.11.

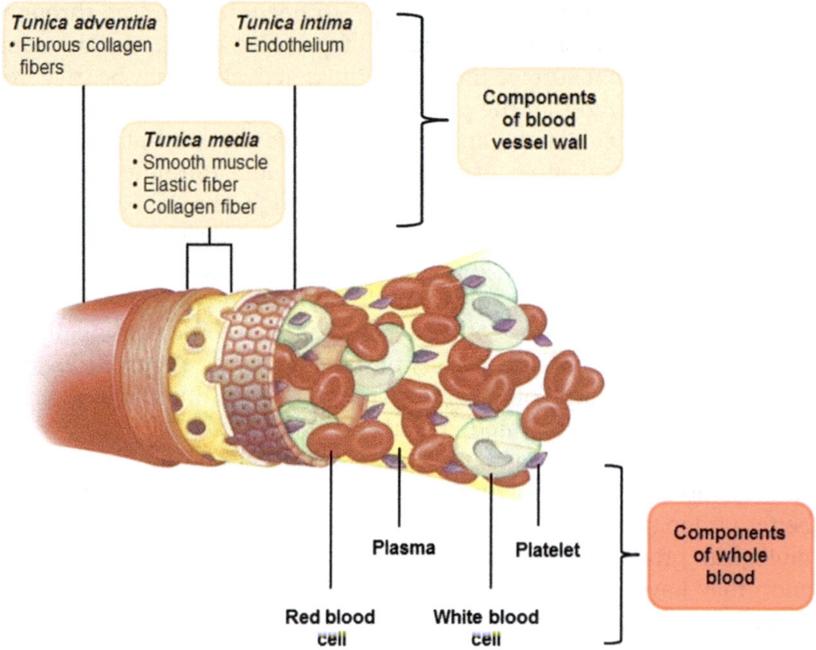

Figure 2.11 Blood vessel structure (showing the various concentric layers constituting the vascular wall), and whole blood components (plasma, platelets, and white and red blood cells).

In terms of composition, whole blood essentially consists of three parts – *plasma*, *buffy coat*, and *red blood cells* – which easily separate upon centrifugation (Figure 2.12).

The least dense fraction is plasma, a viscous aqueous liquid that accounts for approximately 54% of the volume of blood. The second, thinner phase is called the 'buffy coat' (~1% in volume) and contains platelets (or '*thrombocytes*') and white blood cells (or '*leukocytes*'). Lastly, red blood cells (or '*erythrocytes*') make up the densest fraction, which accounts for the remaining ~45% of the blood volume.[89–91] Together, the buffy coat and red blood cells constitute the cellular portion of blood (or '*formed elements*'). On the other hand, plasma gathers the non-cellular components of blood, and is mainly composed of water (90%) – the remaining material consisting of such elements as ions, lipids, vitamins, and a genuine myriad of thousands of proteins[92] present at vastly different concentrations across a range that spans over 10 orders of magnitude (Figure 2.13).[20,93] The most abundant (10^6–10^{10} pg mL^{-1}) are the classical plasma proteins (*e.g.* albumin, complement and coagulation factors). Next, in the ~10^3 to 10^6 pg mL^{-1} range, come the marker proteins that are released as a result of cell damage/death (*e.g.* creatine kinase, cardiac troponins). Finally, signaling cytokines rank last in terms of concentration ($\leq 10^2$ pg mL^{-1}).

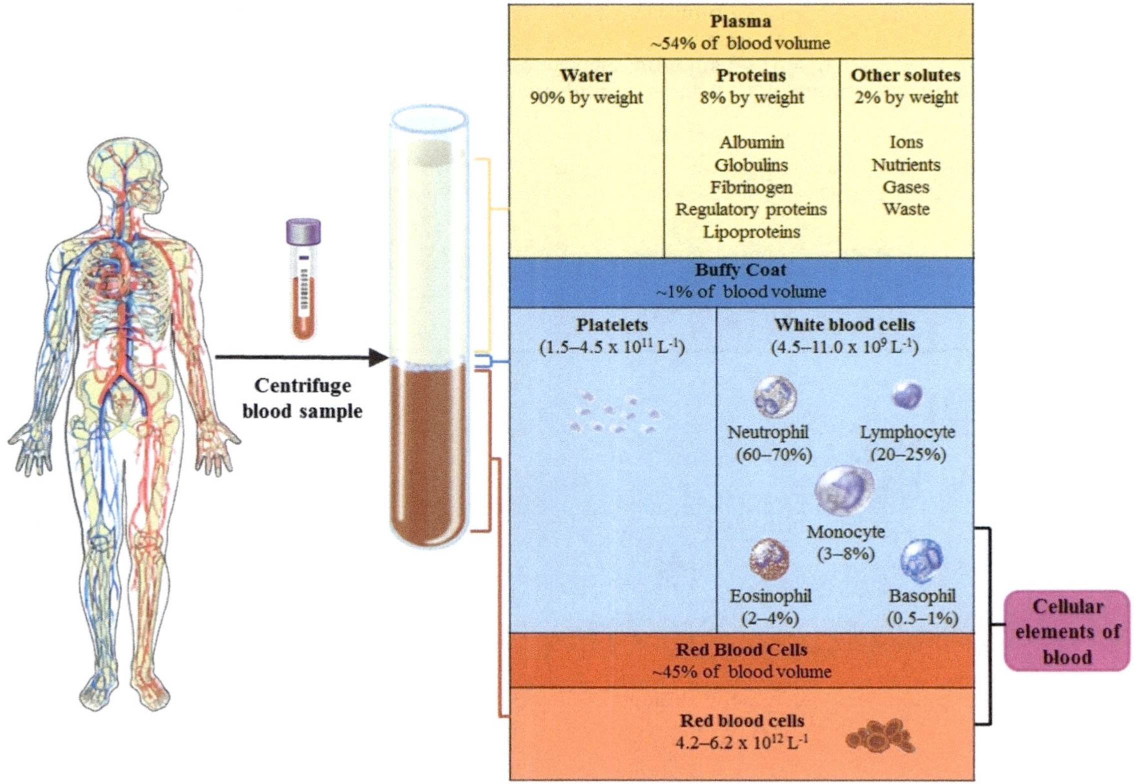

Figure 2.12 The composition of blood. Upon centrifugation, blood separates into three fractions (from top to bottom): (i) plasma, which gathers the non-cellular components of blood; (ii) the buffy coat, made of platelets and white blood cells; and (iii) red blood cells. *Note:* buffy coat and red blood cells collectively constitute the cellular or 'formed elements' of blood.

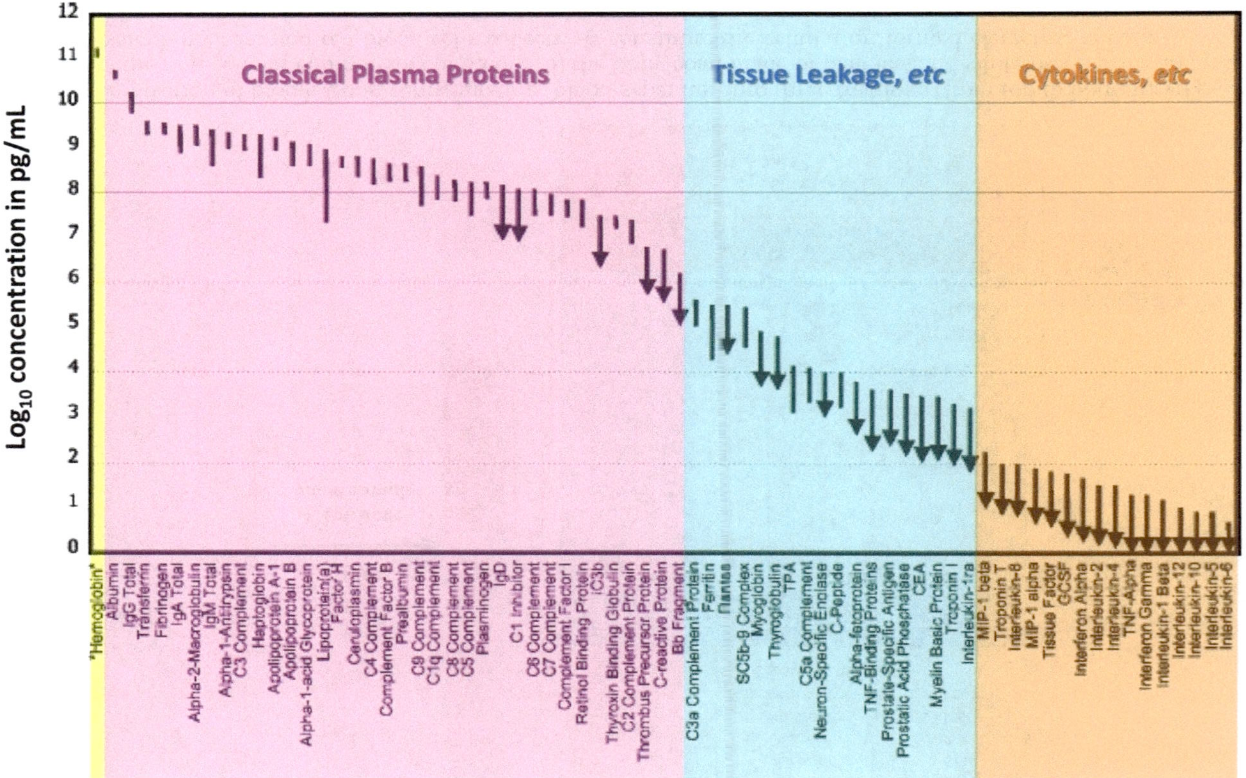

Figure 2.13 Concentration interval for some blood plasma proteins plotted along a logarithmic scale. From the 'classical plasma proteins' (the most abundant at 10^6–10^{10} pg mL^{-1}) to 'tissue leakage' proteins ($\sim 10^3$–10^6 pg mL^{-1}) to cytokines (the least abundant with $\leq 10^2$ pg mL^{-1}), the concentration of blood plasma proteins greatly varies across a range that spans over 10 orders of magnitude. *Note:* hemoglobin (highlighted in yellow far left) is found in red blood cells ($\sim$29 pg per cell).[89] As well, arrows indicate that only an upper limit was available.
Adapted from ref. 93 with permission. © 2002 The American Society for Biochemistry and Molecular Biology.

A closer look at blood plasma reveals that protein composition and distribution do not coincide. Indeed, the most abundant protein – albumin (30–50 mg mL^{-1}) – alone accounts for ~55% of the total amount of blood plasma proteins.[20,89] It is an osmotic regulator that also functions as a carrier for various metals, ions, fatty acids, amino-acids, enzymes and drugs.[89] Immunoglobulins – which are characterized by great structural diversity[20,89] – are the next most abundant plasma proteins, followed by fibrinogen with an approximate blood concentration of 1.5–4 mg mL^{-1},[20,89] which is essential for blood clot formation. When fibrinogen and other clotting factors are removed from plasma (upon coagulation then centrifugation of whole blood), the residual biofluid is known as blood *serum*.[89]

2.4.2 Deleterious Consequences of Blood-artificial Surface Interaction in the Biomedical and Bioanalytical Fields: an Overview

2.4.2.1 Biological Process/Response Activation: the Pivotal Role of Protein Adsorption

Most biomedical/bioanalytical equipment, implants, and devices are fabricated from materials foreign to the human body such as titanium, stainless steel, ceramic, or plastic.[1,2,8] A commonality among these exogenous materials is the invariable, spontaneous deposition of a layer of adhesive proteins on their surface upon their exposure to blood (whether *in vivo* or *in vitro*).[7,33,94,95] As shown in Figure 2.14, this early phenomenon of protein adsorption (discussed at length in the previous sections) may then be followed by as many interconnected events as platelet and other blood cell adhesion/aggregation, and the activation of the complement, coagulation, and other (regulatory) systems.[40,43,71,96–100]

Even though there is subtly more to it than meets the eye,[86,95] it is indeed widely acknowledged that the biological responses/processes with potentially deleterious outcomes orchestrated by the immune and coagulation systems are triggered by fouling[101] – an ubiquitous adsorption phenomenon, where (in this case) exogenous material surfaces become irreversibly covered with the matrix components of blood, most notably proteins.[2,33,94,95,102] In the perspective of biomedical/bioanalytical applications, being able to reliably prevent this sequence of events from occurring is paramount to devise biocompatible materials (*i.e.* biomaterials capable of performing an intended function without inducing adverse effects),[81,103] as well as biosensor platforms with full functionality and extended life expectancy.[104]

Paradoxically, the normally desired activation of the host's immune and coagulation protective mechanisms may become problematic when blood is purposely exposed to exogenous, foreign materials for biomedical/bioanalytical reasons. When ensuing harmful effects manifest themselves, they may affect both foreign and host bodies – as will be discussed next.

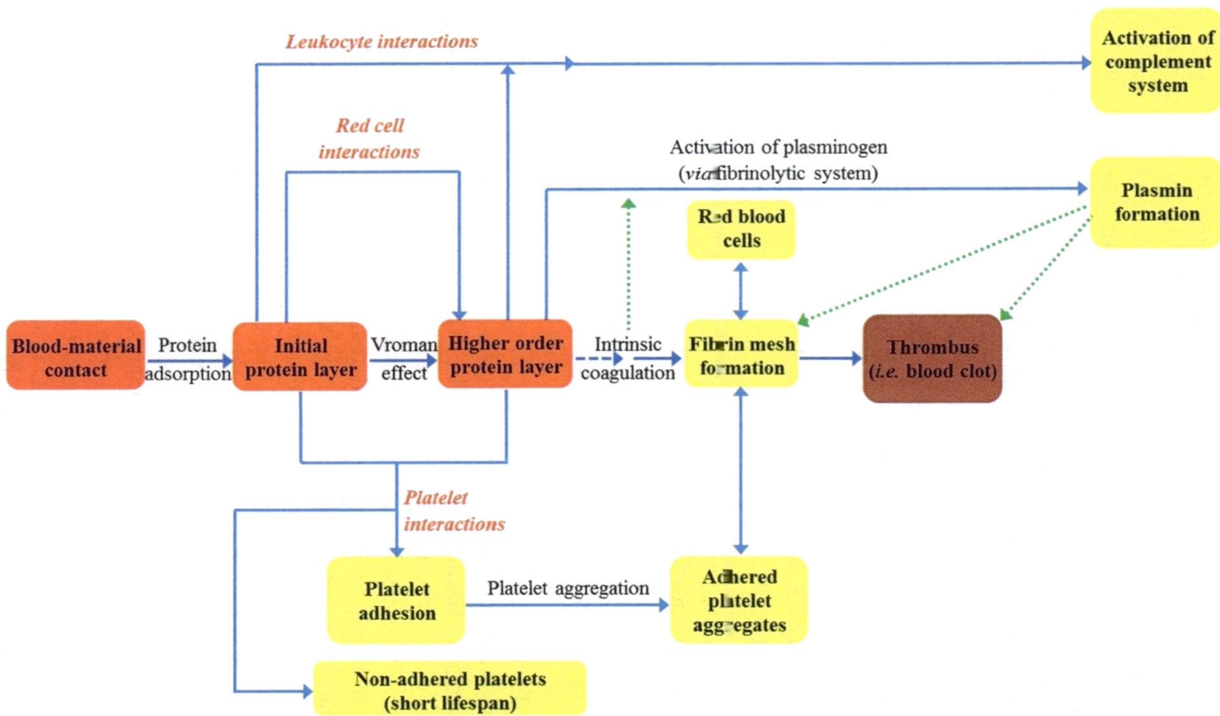

Figure 2.14 Schematic representation of the highly involved nature of the biological processes/responses that may be triggered by the contact of blood with foreign materials. Of note is the pivotal, mediating role played by the early-adhered layer of proteins (orange boxes).
Adapted from ref. 97 with kind permission of John Wiley and Sons. © 2006, John Wiley and Sons.

2.4.2.2 *Practical Complications of Blood Protein Adsorption*

2.4.2.2.1 Foreign Body Viewpoint. Serious complications can arise when an exogenous, foreign material surface in biomedical/bioanalytical applications comes into contact with blood, and biological processes/responses[105] are elicited. A first example would be the rejection of bio-incompatible implants through the *'foreign body reaction'* – a host protective mechanism that overall typically lasts a few weeks (and accompanies normal wound-healing following implant insertion), and eventually leads to the physical fibrous encapsulation of the implanted object along with consequences related to its functionality and fate as such a confining environment is also highly degradative (Figure 2.15, top).[5,88,106]

As will be discussed in greater detail in Chapter 5, concerns are also recurrently raised in sensor technology with respect to signal interference caused by the *'non-specific adsorption'* (fouling) of blood plasma/serum proteins onto detection platforms (Figure 2.15, bottom).[3,94,104,107] In fact, even with these

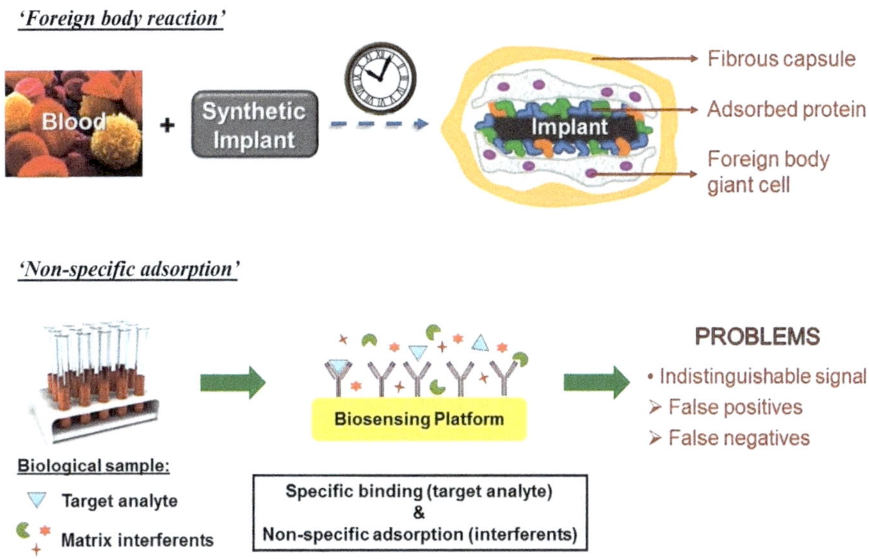

Figure 2.15 Risks incurred by exogenous materials in biomedical and bioanalytical applications upon their interaction with blood, or its cell-cleared plasma/serum derivatives. Top: the 'foreign body reaction', wherein artificial surfaces become covered *in vivo* with successive layers of proteins, 'foreign body giant cells' (fused macrophages), and a final fibrous capsule made of collagen. The end result is the function- and fate-altering isolation of implanted objects from the surrounding host biological environment. Bottom: the 'non-specific adsorption' of interfering species (*i.e.* proteins) on detection platforms in sensor technology, which causes signal interference and leads to erroneous readings. Adapted from ref. 3 with permission from the Royal Society of Chemistry.

Table 2.5 List of blood clotting factors with their normal plasma concentration.

Blood clotting factors		
Roman numeral	Synonymous names	Normal plasma concentration (μg mL^{-1})
I	Fibrinogen	2500
II	Prothrombin	100
III	Tissue factor, tissue thromboplastin	0
IV	Ionized calcium (Ca^{2+})	45–56
V	Proaccelerin, labile factor, Ac-globulin (Ac-G)	7
VII	Serum prothrombin conversion accelerator (SPCA), proconvertin, stable factor	0.5
VIII[a]	Antihemophilic factor (AHF), antihemophilic globulin (AHG), antihemophilic factor A	0.1
IX	Plasma thromboplastin component (PTC), Christmas factor, antihemophilic factor B	6
X	Stuart-Prower factor	10
XI	Plasma thromboplastin antecedent (PTA), antihemophilic factor C	4
XII	Hageman factor	30
XIII	Fibrin-stabilizing factor	15
Prekallikrein	Fletcher factor	50
High-molecular weight kininogen	HMWK, Fitzgerald factor	70
von Willebrand factor (platelet adhesion cofactor)[a]	VWF	7
Platelet/tissue phospholipid	—	—

[a]In plasma, Factor VIII and von Willebrand factor are complexed.

cell-cleared bioanalysis media derivatives of blood, the total concentration of proteins ranges between 60–80 mg mL^{-1},[108] which is up to nine orders of magnitude greater than that of the target analytes.[3,94,107] As shown in Figure 2.15 (bottom), the clinical danger is the misdiagnosis of diseases/conditions due to the possible occurrence of 'false positive' or 'false negative' results.[3,107,109] Additionally, implanted sensors – for continuous *in vivo* monitoring (*e.g.* of glycemia) – not only face this problem of signal interference, but also that of device bio-incompatibility (as mentioned above).[110–112]

2.4.2.2.2 Host Body Viewpoint. Another concern – for the well-being of patients this time – is the foreign surface-induced formation of blood clots (or '*thrombi*') *via* the aforementioned activation of the blood coagulation cascade (Figure 2.14).[75,113] Historically, the study of the influence of interfacial adsorption on blood clotting can be dated back to 1929.[114]

A number of theories were proposed in the following decades to explain the formation of blood clots, but it was not until 1964 that the cascade[115] or 'waterfall'[116] models emerged to rationalize the sequential role played (*via* their chain-reaction activation) by various so-called protein '*coagulation factors*'.[82,89,90,117] Among the clotting factors involved in blood coagulation (Table 2.5), at least a dozen proteins normally circulate in plasma throughout the body in their inactive form.[117] To note, the Roman numeral labeling system[115] used for most clotting factors (Table 2.5) correlates with the order in which these were discovered, not the order in which they are actually involved in the coagulation cascade.[117] Their active form is denoted by the addition of the suffix 'a' next to their Roman numeral. For example, the notation for activated Factor X is 'Xa'.

Two pathways exist for the clotting mechanism (Figure 2.16) that can be initiated simultaneously or separately.[43,82,89,90] On the one hand, the *intrinsic pathway* is activated when: (i) blood is exposed to collagen following

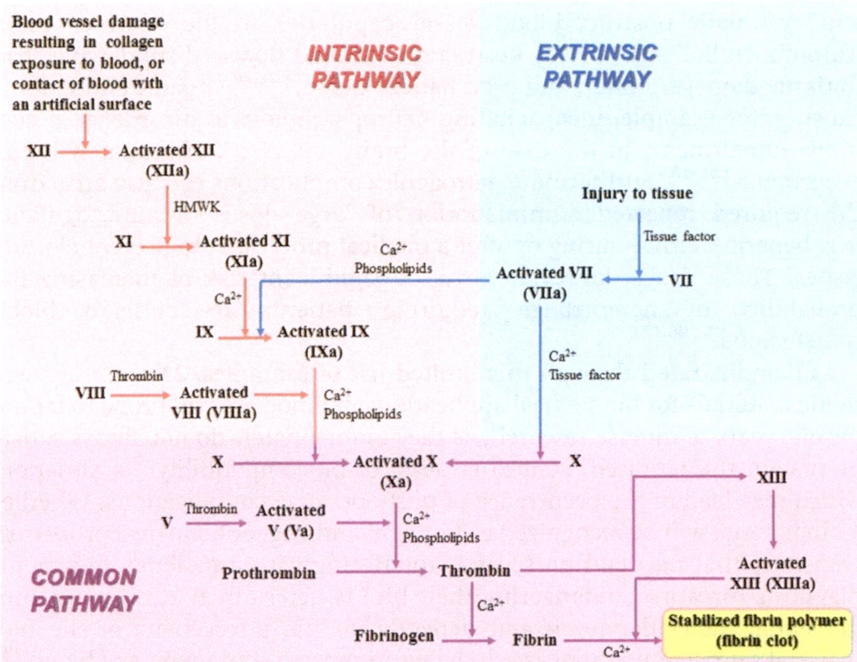

Figure 2.16 Simplified representation of the intricate blood coagulation cascade, which features two differently activated pathways: the *intrinsic pathway* (pink panel, top left), and the *extrinsic pathway* (blue panel, top right). Both eventually merge into a *common pathway* (purple panel, bottom) at the end of which a stabilized fibrin clot is formed.
Adapted from M. B. Gorbet and M. V. Sefton, Biomaterial-associated thrombosis: Roles of coagulation factors, complement, platelets and leukocytes, *Biomaterials*, **25**, 5681, Copyright 2004 with permission from Elsevier.

the lesion of a blood vessel; (ii) substances within blood are damaged; or (iii) blood contacts a negatively-charged, foreign surface.[82,91,117] On the other hand, the *extrinsic pathway* is activated by the rupture of the vascular wall and exposure of blood to damaged tissue.[2,82,90,117] Both processes involve phospholipids ('platelet phospholipids' for the intrinsic pathway, 'tissue phospholipids' for the extrinsic pathway) with which certain activated coagulation factors complex, bridged by Ca^{2+} – a factor necessary for clot formation.[2,82,90] In fact, with the exception of factor activation through contact, Ca^{2+} is required in most steps of blood coagulation, as can be seen in Figure 2.16. Finally, although the intrinsic and extrinsic pathways are initiated by different means, both merge into a *common pathway* that eventually leads to the formation of a stabilized fibrin clot (Figure 2.16).[2,82,90,117]

For the recipients of an implant (*e.g.* heart valve, indwelling catheter) or for those undergoing a surgical/medical procedure (*e.g.* bypass surgery, renal dialysis), the main risk with surface-induced clot formation is to suffer from such vascular accidents as myocardial infarction and stroke – should thrombi detach from the surface, circulate in the bloodstream, and eventually obstruct blood vessels/capillaries in the heart or brain (thromboembolism).[43,118,119] Restriction of blood flow and supply may lead to tissue damage/infarct, and even patient death.[1,2,120] Organ dysfunction – causing, for example, incapacitating neuropsychological disorder (*e.g.* cognitive impairment) in the case of the brain – is also unfortunately a real occurrence.[121–123] Furthermore, iatrogenic complications can also arise from the required repeated administration of large doses of anticoagulants (*e.g.* heparin), either during or after a medical procedure, to prevent clotting issues. These drugs, however, carry the significant risk of increasing the probability of hemorrhage requiring patients to undergo blood transfusions.[87,96,124]

As demonstrated through this limited list of examples, 21st century synthetic materials for biomedical applications in blood are still prone to failure despite years of intense research, as they unfortunately do not always satisfy or sustain the required, contextual level of biocompatibility – a situation, which may lead to the occurrence of post-operative complications. Whether a biomaterial will be recognized as 'foreign' and trigger biological processes/responses that may end up severely interfering with a patient's treatment/diagnosis (or worse endangering their life) is determined at the most fundamental, molecular level, and depends on the interactivity of the biomaterial's exogenous surface with the multi-protein components of blood.[125]

2.5 Conclusion

If one word were to describe protein adsorption onto artificial surfaces, it would undoubtedly be 'complex'. Firstly, 'complex' because proteins themselves are elaborate biopolymers – a virtually countless number of which exists – that fold into intricate three-dimensional structures, each with unique plasticity and the ability to interact with synthetic materials through multiple (types of) interactions. 'Complex' again since, even on an individual basis, there exist multiple factors influencing protein adsorption (those linked to the different protein characteristics but also to the various physicochemical properties of the contacting surface and various environmental conditions); a process that further gains complexity with all-interacting multi-component biological media. Finally, 'complex' due to the multi-step nature of the adsorption mechanism, which is dynamic and features highly involved conformational restructuration phenomena. These incidentally play a determining role in the surface-induced activation of the coagulation and immune systems by revealing concealed bioactive sites, when blood contacts a foreign surface. The issues in practice with protein adsorption onto exogenous materials are those associated with the occurrence of potentially severe complications. In the biomedical world, bio-incompatibility is of particular concern with such risks as the inflammatory rejection of implants and thromboembolism-triggered vascular accidents. In bioanalysis, the problem is that of signal interference from proteinaceous biological matrices, and the dangers associated with misdiagnosis. Despite all this complexity, which makes modeling and the establishment of general governing principles perilous, the overall phenomenon of protein adsorption today can be considered reasonably well understood. Nevertheless, there still remains some important questions to be answered, for instance: Do proteins co-adsorb onto surfaces?[126,127] If so, how prevalent would that phenomenon be? As well, what is the identity of protein-escorting molecules?[126,127] Lastly, how influential is the interaction in terms of structure and potential bioactivity of adsorbed proteins?[128] Complete and thorough understanding of the phenomenon of protein adsorption is the goal of many researchers, as such an accomplishment would allow for antifouling/biocompatible materials for bioanalytical and biomedical applications to be most rationally engineered. Each day that passes brings us closer to achieving these goals.

References

1. B. D. Ratner, A. S. Hoffman, F. J. Schoen and J. E. Lemons, Introduction – Biomaterials Science: A Multidisciplinary Endeavor/A History of Biomaterials, in *Biomaterials Science: An Introduction to Materials in Medicine*, ed. B. D. Ratner, A. S. Hoffman, F. J. Schoen and J. E. Lemons, Elsevier Academic Press, USA, 2nd edn, 2004, pp. 1–19.
2. D. G. Castner and B. D. Ratner, *Surf. Sci.*, 2002, **500**, 28.
3. M. Thompson, S. Sheikh, C. Blaszykowski and A. Romaschin, Biosensor Technology and the Clinical Biochemistry Laboratory – Issue of Signal Interference from the Biological Matrix, in *Detection Challenges in Clinical Diagnostics*, ed. P. Vadgama and S. Peteu, RSC Detection Science Series No. 2, RSC Publishing, UK, 2013, pp. 1–34.
4. M. Rabe, D. Verdes and S. Seeger, *Adv. Colloid Interface Sci.*, 2011, **162**, 87.

5. S. Franz, S. Rammelt, D. Scharnweber and J. C. Simon, *Biomaterials*, 2011, **32**, 6692.
6. T. A. Horbett, B. D. Ratner, J. M. Schakenraad and F. J. Schoen, Some Background Concepts, in *Biomaterials Science: An Introduction to Materials in Medicine*, ed. B. D. Ratner, A. S. Hoffman, F. J. Schoen and J. E. Lemons, Academic Press, USA, 1996, 1st edn, pp. 133–141.
7. B. D. Ratner, D. G. Castner, T. A. Horbett, T. J. Lenk, K. B. Lewis and R. J. Rapoza, *J. Vac. Sci. Technol., A*, 1990, **8**, 2306.
8. R. A. Latour Jr., Biomaterials: Protein-Surface Interactions, in *Encyclopedia of Biomaterials and Biomedical Engineering*, ed. G. E. Wnek and G. L. Bowlin, Taylor & Francis, USA, 2005, pp. 1–15.
9. J. D. Andrade and V. Hlady, *Adv. Polym. Sci.*, 1986, **79**, 1.
10. W. Norde, *Adv. Colloid Interface Sci.*, 1986, **25**, 267.
11. H. R. Horton, L. A. Moran, R. S. Ochs, J. D. Rawn and K. G. Scrimgeour, Amino Acids and the Primary Structures of Proteins, in *Principles of Biochemistry*, Prentice Hall, USA, 3rd edn, 2002, pp. 51–80.
12. R. Shenhar, T. B. Norsten, V. M. Rotello, Self-Assembly and Self-Organization, in *Introduction to Nanoscale Science and Technology*, ed. M. Di Ventra, S. Evoy and J. R. Heflin Jr., Springer, USA, 2004, pp. 41–74.
13. W. Norde and J. Lyklema, *J. Biomater. Sci., Polym. Ed.*, 1991, **2**, 183.
14. P. Gong and D. W. Grainger, Nonfouling Surfaces: A Review of Principles and Applications for Microarray Capture Assay Designs, in *Microarrays Volume 1 Synthesis Methods*, ed. J. B. Rampal, Methods in Molecular Biology, Humana Press Inc., USA, 2nd edn, 2007, vol. 381, pp. 59–92.
15. W. Norde, Driving Forces for Protein Adsorption at Solid Surfaces, in *Biopolymers at Interfaces*, ed. M. Malmsten, Surfactant Science Series, Marcel Dekker Inc., USA, 2nd edn, 2003, vol. 110, pp. 21–43.
16. T. A. Horbett and J. L. Brash, Proteins at Interfaces: Current Issues and Future Prospects, in *Proteins at Interfaces – Physicochemical and Biochemical Studies*, ed. J. L. Brash and T. A. Horbett, ACS Symposium Series, American Chemical Society, USA, 1987, pp. 1–33.
17. D. R. Schmidt, H. Waldeck and W. J. Kao, Protein Adsorption to Biomaterials, in *Biological Interactions on Materials Surfaces – Understanding and Controlling Protein, Cell, and Tissue Responses*, ed. D. A. Puleo and R. Bizios, Springer, USA, 2009, pp. 2–17.
18. K. Nakanishi, T. Sakiyama and K. Imamura, *J. Biosci. Bioeng.*, 2001, **91**, 233.
19. W. Norde and J. Lyklema, *Adv. Colloid Interface Sci.*, 2012, **179–182**, 5.
20. J. D. Andrade and V. Hlady, *Ann. N. Y. Acad. Sci.*, 1987, **516**, 158.
21. Y. F. Yano, *J. Phys.: Condens. Matter*, 2012, **24**, 503101.
22. M. Jasieniak, D. Graham, P. Kingshott, L. J. Gamble and H. J. Griesser, Surface Analysis of Biomaterials, in *Handbook of Surface and Interface Analysis Methods for Problem-Solving*, ed. J. C. Rivière and S. Myhra, Taylor and Francis, USA, 2nd edn, 2009, pp. 549–564.
23. T. A. Horbett, *Cardiovasc. Pathol.*, 1993, **2**, 137S.

24. J. J. Gray, *Curr. Opin. Struct. Biol.*, 2004, **14**, 110.
25. C. A. Haynes and W. Norde, *Colloids Surf., B*, 1994, **2**, 517.
26. J.-H. Kim and J.-Y. Yoon, Protein Adsorption on Polymer Particles, in *Encyclopedia of Surface and Colloid Science*, ed. A. T. Hubbard, Marcel Dekker Inc., USA, 2002, vol. 1, pp. 4373–4381.
27. E. P. K. Currie, W. Norde and M. A. Cohen Stuart, *Adv. Colloid Interface Sci.*, 2003, **100–102**, 205.
28. S. I. Jeon, J. H. Lee, J. D. Andrade and P.-G. de Gennes, *J. Colloid Interface Sci.*, 1991, **142**, 149.
29. P. Roach, D. Eglin, K. Rohde and C. C. Perry, *J. Mater. Sci.: Mater. Med.*, 2007, **18**, 1263.
30. B. Kasemo and J. Gold, *Adv. Dent. Res.*, 1999, **13**, 8.
31. E. A. Vogler, *Biomaterials*, 2012, **33**, 1201.
32. T. S. Tsapikouni and Y. F. Missirlis, *Mater. Sci. Eng., B*, 2008, **152**, 2.
33. J. L. Brash, *J. Biomater. Sci., Polym. Ed.*, 2000, **11**, 1135.
34. P. Lin, C.-W. Lin, R. Mansour and F. Gu, *Biosens. Bioelectron.*, 2013, **47**, 451.
35. M. Mrksich and G. M. Whitesides, *Annu. Rev. Biophys. Biomol. Struct.*, 1996, **25**, 55.
36. D. Leckband and J. Israelachvili, *Q. Rev. Biophys.*, 2001, **34**, 105.
37. M. Wahlgren and T. Arnebrant, *Trends Biotechnol.*, 1991, **9**, 201.
38. S. G. Eskin, T. A. Horbett, L. V. McIntire, R. N. Mitchell, B. D. Ratner, F. J. Schoen and A. Yee, Some Background Concepts, in *Biomaterials Science: An Introduction to Materials in Medicine*, ed. B. D. Ratner, A. S. Hoffman, F. J. Schoen and J. E. Lemons, Elsevier Academic Press, USA, 2nd edn, 2004, pp. 237–246.
39. M. E. Soderquist and A. G. Walton, *J. Colloid Interface Sci.*, 1980, **75**, 386.
40. C. J. Wilson, R. E. Clegg, D. I. Leavesley and M. J. Pearcy, *Tissue Eng.*, 2005, **11**, 1.
41. D. L. Elbert and J. A. Hubbell, *Annu. Rev. Mater. Sci.*, 1996, **26**, 365.
42. J. J. Ramsden, *Chem. Soc. Rev.*, 1995, **24**, 73.
43. X. Liu, L. Yuan, D. Li, Z. Tang, Y. Wang, G. Chen, H. Chen and J. L. Brash, *J. Mater. Chem. B*, 2014, **2**, 5718.
44. J. M. Grunkemeier, W. B. Tsai, C. D. McFarland and T. A. Horbett, *Biomaterials*, 2000, **21**, 2243.
45. B. R. Young, W. G. Pitt and S. L. Cooper, *J. Colloid Interface Sci.*, 1988, **125**, 246.
46. P. M. Claesson, E. Blomberg, J. C. Fröberg, T. Nylander and T. Arnebrant, *Adv. Colloid Interface Sci.*, 1995, **57**, 161.
47. R. J. Rapoza and T. A. Horbett, *J. Biomed. Mater. Res.*, 1990, **24**, 1263.
48. A. Sadana, *Chem. Rev.*, 1992, **92**, 1799.
49. L. Vroman and A. L. Adams, *Surf. Sci.*, 1969, **16**, 438.
50. K. C. Dee, D. A. Puleo and R. Bizios, Protein-Surface Interactions, in *An Introduction to Tissue-Biomaterial Interactions*, John Wiley & Sons Inc., USA, 2002, pp. 37–51.
51. V. V. Hlady and J. Buijs, *Curr. Opin. Biotechnol.*, 1996, **7**, 72.

52. W. Norde, Adsorption of Biopolymers, with Special Emphasis on Globular Proteins, in *Soil Mineral Microbe-Organic Interactions – Theories and Applications*, ed. Q. Huang, P. M. Huang and A. Violante, Springer-Verlag, Germany, 2008, pp. 99–114.
53. R. A. Latour, *J. Biomed. Mater. Res., Part A*, 2006, **78**, 843.
54. J. H. Lee, T. Li and K. Park, Solvation Interactions for Protein Adsorption to Biomaterial Surfaces, in *Water in Biomaterials Surface Science*, ed. M. Morra, John Wiley & Sons, UK, 2001, pp. 127–135.
55. A. S. Hoffman, *J. Biomater. Sci., Polym. Ed.*, 1999, **10**, 1011.
56. C. A. Haynes and W. Norde, *J. Colloid Interface Sci.*, 1995, **169**, 313.
57. M. Malmsten, *J. Colloid Interface Sci.*, 1998, **207**, 186.
58. S. Y. Jung, S. M. Lim, F. Albertorio, G. Kim, M. C. Gurau, R. D. Yang, M. A. Holden and P. S. Cremer, *J. Am. Chem. Soc.*, 2003, **125**, 12782.
59. S. Pasche, J. Vörös, H. J. Griesser, N. D. Spencer and M. Textor, *J. Phys. Chem. B*, 2005, **109**, 17545.
60. P. Koegler, A. Clayton, H. Thissen, G. N. C. Santos and P. Kingshott, *Adv. Drug Delivery Rev.*, 2012, **64**, 1820.
61. K. L. Prime and G. M. Whitesides, *Science*, 1991, **252**, 1164.
62. R. L. Beissinger and E. F. Leonard, *J. Colloid Interface Sci.*, 1982, **85**, 521.
63. K. Rechendorff, M. B. Hovgaard, M. Foss, V. P. Zhdanov and F. Besenbacher, *Langmuir*, 2006, **22**, 10885.
64. P. G. Koutsoukos, W. Norde and J. Lyklema, *J. Colloid Interface Sci.*, 1983, **95**, 385.
65. K. Y. Foo and B. H. Hameed, *Chem. Eng. J.*, 2010, **156**, 2.
66. R. A. Latour, *J. Biomed. Mater. Res., Part A*, 2015, **103**, 949.
67. H. M. F. Freundlich, *Z. Physiol. Chem.*, 1906, **57**, 385.
68. S. Brunauer, P. H. Emmett and E. Teller, *J. Am. Chem. Soc.*, 1938, **60**, 309.
69. I. Langmuir, *J. Am. Chem. Soc.*, 1916, **38**, 2221.
70. I. Lundström, *Prog. Colloid Polym. Sci.*, 1985, **70**, 76.
71. L. Vroman, A. L. Adams, M. Klings, G. C. Fischer, P. C. Munoz and R. P. Solensky, *Ann. N. Y. Acad. Sci.*, 1977, **283**, 65.
72. A. Krishnan, C. A. Siedlecki and E. A. Vogler, *Langmuir*, 2004, **20**, 5071.
73. C. A. LeDuc, L. Vroman and E. F. Leonard, *Ind. Eng. Chem. Res.*, 1995, **34**, 3488.
74. L. Vroman and A. L. Adams, *J. Colloid Interface Sci.*, 1986, **111**, 391.
75. P. Thevenot, W. Hu and L. Tang, *Curr. Top. Med. Chem.*, 2008, **8**, 270.
76. T. Ballet, L. Boulange, Y. Bréchet, F. Bruckert and M. Weidenhaupt, *Bull. Pol. Acad. Sci.: Tech. Sci.*, 2010, **58**, 303.
77. M. Shen and T. A. Horbett, *J. Biomed. Mater. Res.*, 2001, **57**, 336.
78. W.-J. Hu, J. W. Eaton and L. Tang, *Blood*, 2001, **98**, 1231.
79. W. G. Brodbeck, E. Colton and J. M. Anderson, *J. Mater. Sci. Mater. Med.*, 2003, **14**, 671.
80. B. Sivaraman and R. A. Latour, *Biomaterials*, 2010, **31**, 832.
81. L. Vroman, *Bull. N. Y. Acad. Med.*, 1988, **64**, 352.
82. M. B. Gorbet and M. V. Sefton, *Biomaterials*, 2004, **25**, 5681.

83. G. Evans and J. F. Mustard, *Surgery*, 1968, **64**, 273.
84. A. L. Adams, G. C. Fischer and L. Vroman, *J. Colloid Interface Sci.*, 1978, **65**, 468.
85. B. Sivaranam and R. A. Latour, *Biomaterials*, 2010, **31**, 1036.
86. B. Sivaranam and R. A. Latour, *Langmuir*, 2011, **28**, 2745.
87. J. M. Courtney and C. D. Forbes, *Br. Med. Bull.*, 1994, **50**, 966.
88. J. M. Anderson, A. Rodriguez and D. T. Chang, *Semin. Immunol.*, 2008, **20**, 86.
89. W. F. Ganong, Circulating Body Fluids, in *Review of Medical Physiology*, Lange Medical Books/McGraw-Hill (Medical Publishing Division), USA, 22nd edn, 2005, pp. 515–546.
90. A. C. Guyton and J. E. Hall, in *Textbook of Medical Physiology*, Elsevier Saunders, USA, 11th edn, 2006, pp. 419–468.
91. J. F. Dailey, in *Dailey's Notes on Blood*, Medical Consulting Group, USA, 3rd edn, 1996.
92. Y. Shen, J. Kim, E. F. Strittmatter, J. M. Jacobs, D. G. Camp II, R. Fang, N. Tolié, R. J. Moore and R. D. Smith, *Proteomics*, 2005, **5**, 4034.
93. N. L. Anderson and N. G. Anderson, *Mol. Cell. Proteomics*, 2002, **1**, 845.
94. C. Blaszykowski, S. Sheikh and M. Thompson, *Chem. Soc. Rev.*, 2012, **41**, 5599.
95. C. Blaszykowski, S. Sheikh and M. Thompson, *Trends Biotechnol.*, 2014, **32**, 61.
96. M. Torn, W. L. E. M. Bollen, F. J. M. van der Meer, E. E. van der Wall and F. R. Rosendaal, *Arch. Intern. Med.*, 2005, **165**, 1527.
97. J. L. Brash, *Ann. N. Y. Acad. Sci.*, 1987, **516**, 206.
98. B. D. Ratner, *J. Biomed. Mater. Res.*, 1993, **27**, 837.
99. A. S. Hoffman, *Ann. N. Y. Acad. Sci.*, 1987, **516**, 96.
100. J. M. Courtney, N. M. K. Lamba, S. Sundaram and C. D. Forbes, *Biomaterials*, 1994, **15**, 737.
101. H. O. Ham, S. H. Park, J. W. Kurutz, I. G. Szleifer and P. B. Messersmith, *J. Am. Chem. Soc.*, 2013, **135**, 13015.
102. E. Luong-Van, I. Rodriguez, H. Y. Low, N. Elmouelhi, B. Lowenhaupt, S. Natarajan, C. T. Lim, R. Prajapati, M. Vyakarnam and K. Cooper, *J. Mater. Res.*, 2013, **28**, 165.
103. S. Bauer, P. Schmuki, K. von der Mark and J. Park, *Prog. Mater. Sci.*, 2013, **58**, 261.
104. S. Sheikh, D. Y. Yang, C. Blaszykowski and M. Thompson, *Chem. Commun.*, 2012, **48**, 1305.
105. J. M. Anderson, *Annu. Rev. Mater. Res.*, 2001, **31**, 81.
106. B. D. Ratner and S. J. Bryant, *Annu. Rev. Biomed. Eng.*, 2004, **6**, 41.
107. M. Thompson, S. Sheikh and C. Blaszykowski, *Austin J. Nanomed. Nanotechnol.*, 2014, **2**, 2.
108. J. N. Adkins, S. M. Varnum, K. J. Auberry, R. J. Moore, N. H. Angell, R. D. Smith, D. L. Springer and J. G. Pounds, *Mol. Cell. Proteomics*, 2002, **1**, 947.

109. J. Ladd, Z. Zhang, S. Chen, J. C. Hower and S. Jiang, *Biomacromolecules*, 2008, **9**, 1357.
110. S. P. Nichols, A. Koh, W. L. Storm, J. H. Shin and M. H. Schoenfisch, *Chem. Rev.*, 2013, **113**, 2528.
111. P. H. Kvist and H. E. Jensen, *J. Diabetes Sci. Technol.*, 2007, **1**, 746.
112. J. Wang, *Electroanalysis*, 2001, **13**, 983.
113. A. P. McGuigan and M. V. Sefton, *Biomaterials*, 2007, **28**, 2547.
114. J. M. Johlin, *J. Biol. Chem.*, 1929, **81**, 99.
115. R. G. Macfarlane, *Nature*, 1964, **202**, 498.
116. E. W. Davie and O. D. Ratnoff, *Science*, 1964, **145**, 1310.
117. J. M. Anderson, G. Cook, B. Costerton, S. R. Hanson, A. Hensten-Pettersen, N. Jacobsen, R. J. Johnson, R. N. Mitchell, M. Pasmore, F. J. Schoen, M. Shirtliff and P. Stoodley, Host Reactions to Biomaterials and their Evaluation, in *Biomaterials Science: An Introduction to Materials in Medicine*, ed. B. D. Ratner, A. S. Hoffman, F. J. Schoen and J. E. Lemons, Elsevier Academic Press, USA, 2nd edn, 2004, pp. 293–338.
118. K. N. Stevens, Y. B. Aldenhoff, F. H. van der Veen, J. G. Maessen and L. H. Koole, *J. Biomed. Biotechnol.*, 2007, **2007**, 29464.
119. B. Nilsson, K. N. Ekdahl, T. E. Mollnes and J. D. Lambris, *Mol. Immunol.*, 2007, **44**, 82.
120. B. R. Young, W. G. Pitt and S. L. Cooper, *J. Colloid Interface Sci.*, 1988, **124**, 28.
121. G. J. Murphy and G. D. Angelini, *J. Cardiovasc. Surg.*, 2004, **19**, 481.
122. A. M. Murray, *Adv. Chronic Kidney Dis.*, 2008, **15**, 123.
123. N. Stroobant, G. Van Nooten, Y. Van Belleghem and G. Vingerhoets, *Chest*, 2005, **127**, 1967.
124. D. Pavithra and M. Doble, *Biomed. Mater.*, 2008, **3**, 034003.
125. L. Tang and W. Hu, *Expert Rev. Med. Devices*, 2005, **2**, 493.
126. T. Riedel, Z. Riedelová-Reicheltová, P. Májek, C. Rodriguez-Emmenegger, M. Houska, J. E. Dyr and E. Brynda, *Langmuir*, 2013, **29**, 3388.
127. G. Gunkel and W. T. S. Huck, *J. Am. Chem. Soc.*, 2013, **135**, 7047.
128. Y. Wei, A. A. Thyparambil and R. A. Latour, *Colloids Surf., B*, 2013, **110**, 363.

CHAPTER 3

Interaction of Cells and Tissue with Substrate Surfaces

3.1 Cells and Substrate Physicochemical Properties

A key issue with respect to sensors and medical devices of many types, which are to be exposed to blood, other biological fluids and tissue, is the effect that a substrate may have on the cellular structure and behavior, and *vice versa*. (A detailed evaluation of the *medical* consequences of such interactions is covered in detail in the next chapter.) The number of such substrates with respect to composition is expected to be immense, ranging from inorganic materials such as silicon, silicon nitride, stainless steel, aluminum nitride, and titanium to a host of polymers such as polyvinyl chloride, polycarbonate, and polysulfone. There will also be a wide variety of applications such as biosensors for the clinical biochemistry laboratory and point-of-care devices, the components of medical bypass circuitry (bypass surgery and renal dialysis), and structures to be implanted on a corporeal basis for the purpose of regeneration and repair. It is to be expected that many substrate factors will influence the response of a particular cell to a device. These include surface chemical functionality, free energy, morphology, crystal structure, and interfacial charge characteristics, with a number of these properties being inextricably linked. The existence of such surface parameter inter-dependency strongly implies that the nature of a specific cellular response cannot *necessarily* be ascribed to influence by any one factor. With respect to how the various physicochemical factors influence cells will obviously depend on the structure and biochemical components that define the biological moiety. Essentially, eukaryotic cells are composed of a plasma membrane that surrounds the cytoplasm and organelles which

RSC Detection Science Series No. 8
Biological Fluid–Surface Interactions in Detection and Medical Devices
By Michael Thompson, Christophe Blaszykowski, Sonia Sheikh, Cesar Rodriguez-Emmenegger and Andres de los Santos Pereira

Published by the Royal Society of Chemistry, www.rsc.org

include the cell nucleus, mitochondria, Golgi complex, *etc*. They are embedded in a material termed the extracellular matrix (ECM), which serves as a scaffold for the support of cellular structures. Over recent years it has become abundantly clear that interaction of substrate surfaces with the various cellular and extracellular components can affect a large number of parameters, including not only biochemical behavior, but also viscoelastic and mechanical properties. Such influences, at least in general terms, can be considered to be instigated initially by interaction of the substrate with the surface of the cell. (At a later stage in the interaction, internal cellular machinery may be affected because of signaling processes transduced *via* the plasma membrane components.) Another issue of importance is interaction of inorganic and organic nanoparticles with various cells. Research in this field is often concerned with *theranostics*, which is defined as medical therapy combined with detection. However, the ingestion of such particles into the cytoplasm of cells can lead to serious deleterious effects. Research in this area has spawned a relatively new discipline, that of *nanotoxicology*.

In this section, we review concisely the properties of substrate surfaces, introduced in Chapter 1, that are expected to play a role in the interactive response of cells. The surface of biological cells of whatever type will display a plethora of "exterior" functional structures, composed from protein, lipid, and saccharide moieties, together with domains of polar and hydrophobic character orchestrated in a highly dynamical fashion. Accordingly, with respect to their interaction with substrates, the presence of interfacial functional groups and their spatial distribution, surface morphology, free energy, charge, and rigidity/elasticity are all expected to play pivotal and, importantly, *concerted* roles. As will become abundantly clear, a particularly significant effort in terms of research has been the correlation of cellular behavior with substrate surface properties, as examined by a variety of analytical techniques. Prevalent among the armamentarium employed by surface scientists in this regard are X-ray photoelectron spectroscopy (XPS), Auger electron emission spectroscopy, atomic force microscopy (AFM), scanning tunneling microscopy, secondary ion mass spectrometry (SIMS), electron microscopy (EM), confocal fluorescence microscopy, and contact angle (CA) measurement.[1,2] For example, AFM and EM are important for the assay of surface morphology, whereas XPS and SIMS are more likely to be employed for chemical functional group characterization. With regard to surface chemistry, the two major forces involved in cellular interaction will be intermolecular hydrogen-bonding, which is an electrostatic interaction,[3] and the hydrophobic effect, which is generally thought to originate from entropic considerations associated with the exclusion of water and disruption of water-based hydrogen bonds. CA measurement is interesting in this respect, since it is used as an indicator of the substrate surface free energy governed by the level of hydrophilicity/hydrophobicity, which in turn are connected to the surface functional groups and morphology. Since the surface components of a cell exhibit both polar and non-polar domains, the overall device surface free energy constitutes an important factor in the

Table 3.1 Basic equations for the energy of electrostatic interactions.[a]

Interaction	Potential energy (U)
Charge–charge	$U = q_1\,(q_2/r)$
Charge–dipole	$U = q\,\mu\,\cos\theta/(Dr^2)$
Dipole–dipole	$U = -\mu_1\,\mu_2\,(\cos\theta_{12} - 3\cos\theta_1\cos\theta_2)/4\,\pi\varepsilon_0 r_{12}{}^3$
Charge–induced dipole	$U = -1/2\,\alpha\,q^2/(r^4 D^2)$

[a]Key: q is the formal charge, r is the distance, μ is the dipole moment, D is the dielectric constant, ε_0 is the permittivity, α is the polarizability, θ is the angle between the charge and/or the dipole (center). Subscripts 1 and 2 label each one of the two dipoles.

cell–substrate interaction. In addition to the electrostatic hydrogen bond, there are potentially other charge-based interactions that can occur between the exterior of a biological cell and a substrate surface, especially when the latter is expected to be charged, as may be the case for electrochemistry-based biosensors. A compendium of the energy of these forces and their distance dependence is given in Table 3.1. Charge at the surface of a substrate is expected to interact either adhesively or repulsively with the cell membrane surface dipole electrical potential. The latter arises from the dynamic array of dipoles present on lipid headgroups.[4] This type of interaction has been largely ignored, but could be an important factor in terms of electrochemical devices, which display electrical double-layer phenomena.

The physical morphology of the substrate surface is an extremely important factor, given that a variety of cells clearly recognize the roughness of a surface in terms of their biological response to a solid material. As will become evident later in this chapter, surface roughness is a parameter that is crucial in determining the level of biocompatibility exhibited by, for example, implanted structures such as stents and other devices. This observation has led to the orchestrated design of substrate morphology by photolithographic and other techniques, for example to produce pillars and wells, which influence cell adhesion and response. Another factor that has received relatively scant attention is the rheological character and, especially, the elasticity of the substrate. Experiments in recent years have shown that various cells are capable of "sensing" the rigidity or elastic properties of the substrate.

We now turn to a concise discussion of the make-up of cells specifically intended as an introduction for the materials/surface scientist, not readers from the disciplines of biology, biochemistry, biotechnology, *etc.* The emphasis is on the exterior of the eukaryotic cell, cells present in blood, and cells which have been employed extensively in fundamental studies of their interaction with surfaces of various devices, all of which are topics of high relevance to this text.

3.2 The Eukaryotic Cell and Extracellular Matrix

Animal eukaryotic cells have a dimension of the order of 5–70 μm in diameter and possess shapes that range from spherical to flat. Certain cells

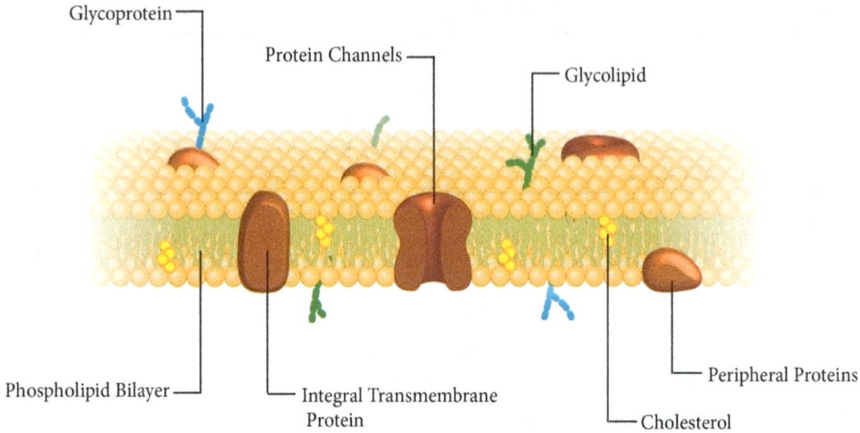

Figure 3.1 The lipd bilayer membrane and membrane components.

such as neurons display long "extensions" that can be up to a meter in length. A plasma membrane of thickness around 8 nm surrounds the cell (Figure 3.1) and is composed of a mixture of amphipathic phospholipids, proteins, and sterols. The wide variety of proteins present in the membrane have several functions, including molecular transport *via* channels and biochemical signaling through receptor binding of ligands.[5] In certain cases, such binding events are transduced to chemical machinery in the interior of the cell. (As mentioned above, this is one of the reasons that surface interactions of cells with solids can result in cellular behavioral changes.) An additional function of membrane proteins is their capability to attach to entities on other cells, which leads to the formation of tissue.

The interior of the eukaryotic cell is divided in two compartments, the *cytoplasm* and *nucleus* (Figure 3.2). The latter is composed of an interior, the *nucleoplasm*, which is surrounded by two concentric membranes. Pores are present in these membranes for molecular transport processes. The nucleus contains the hereditary information molecule, DNA, and chromosomal proteins (histone and non-histone). Each nucleus in the cells of humans has 46 chromosomes, made up of a combination of a linear strand of DNA with associated chromosomal proteins. The cytoplasm contains *ribosomes*, vesicles, and a number of *organelles*, the latter being compartmentalized *via* membranes. The former are responsible for protein synthesis through tRNA and mRNA released from the nucleus. Ribosomes can be attached to membranous sacs, which constitute either the *endoplasmic reticulum* or the *Golgi complex*. The chemical energy *via* production of ATP required by eukaryotic cells is produced by a membrane-bound organelle called the *mitochondrion*.[6]

There are other structures present in the eukaryotic cell such as *microbodies* (organelles), which are responsible for oxidative reactions, and *microtubules* and *microfilaments*, which are associated with cellular movement. In addition to motile properties, microtubules and microfilaments, in concert with *intermediate filaments* composed from protein, form an internal

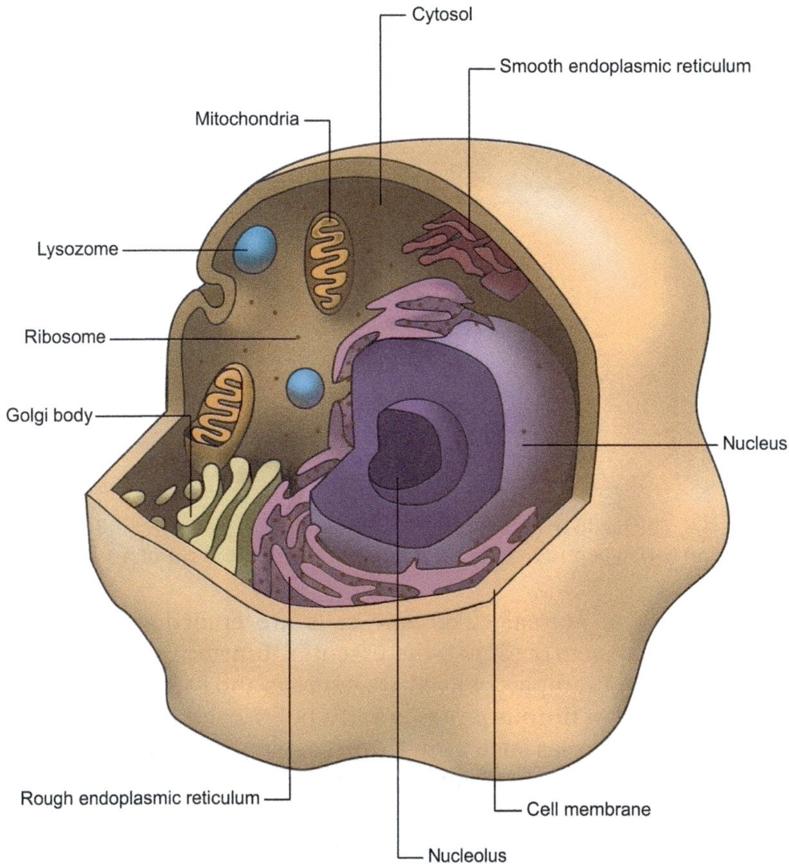

Figure 3.2 A schematic of the eukaryotic cell.

support system for the cytoplasm, called the *cytoskeleton*. These fibers, which are in the region of 10 nm diameter, are also involved in muscle contraction (see Section 3.3.3). It is important to note that when it comes to the external environment of the cell, the cytoskeleton also plays a crucial role with respect to the property of adherence.

We now turn to the "exterior" environment of the eukaryotic cell, which, of course, will be crucial in terms of the interaction of the cell with solids. A number of molecules extend from the plasma membrane into the extracellular space and are generally composed of glycolipids or glycoproteins. These species have several roles including cellular recognition, chemical signaling to the interior of the cell, and adherence to other cells. The signaling process involves moieties termed molecular receptors which selectively bind molecules such as hormones and other entities, which in turn lead to a variety of cellular responses. The mechanism associated with exterior-to-interior of cell signaling is highly complex, an example being the receptors which control protein kinase activity.

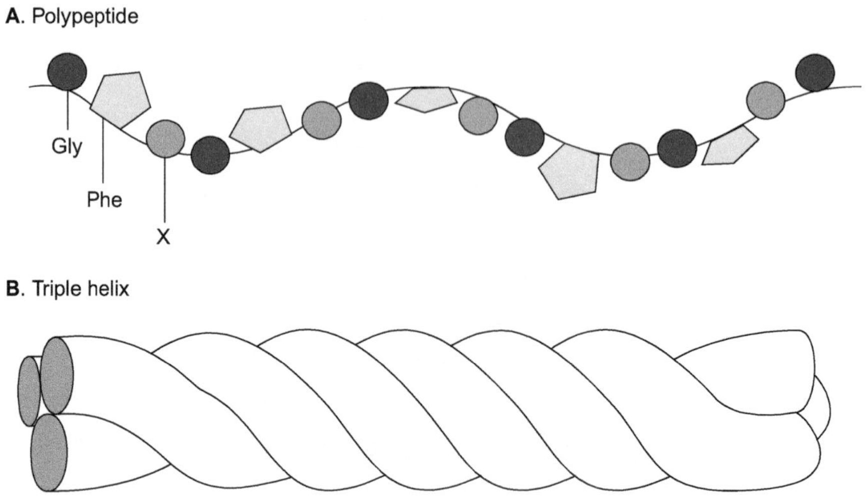

Figure 3.3 (A) The basic structure of the glycoprotein collagen present in the form of an α-polypeptide chain. (B) Collagen takes the form of a triple helix.

A wide variety of animal eukaryotic cells are embedded in materials called *extracellular matrix* (*ECM*), the essential function of this being the support of the cell. Examples of ECM are tendons and cartilage. The matrix, which is composed of fibrous proteins and a polysaccharide gel-like matrix, also possesses other roles such as in the processes of cellular proliferation, adhesion, and motility.[7] It should be noted that the ECM in connective tissue has incorporated cells relatively sparsely spaced, whereas in the epithelium they are in close proximity with each other. A main structure of the ECM is the fiber collagen. This polymeric molecule (Figure 3.3A) is composed primarily of glycoprotein with the protein part being represented by glycine, hydroxylysine, and hydroxyproline, and the carbohydrate section incorporating the sugar molecules glucose and galactose. The molecule is linear in character and exists in the form of a triple helix (Figure 3.3B), which in certain versions incorporates hinge regions which confer flexibility. The fiber has several forms, which are given the designations I through XIV; some of these are outlined with their biological location in Table 3.2.

The collagen species are surrounded by a group of glycoproteins, called *proteoglycans,* where the molecular composition is generally dominated by carbohydrates. The molecules incorporate a linear polypeptide chain around which carbohydrate chains (mostly *glycosaminoglycans*) are arranged. There is a very large distribution of proteoglycans in eukaryotic cells, and the molecular size of these molecules is highly variable ranging from 250 K to enormous sizes of over 3 million Daltons. The highly hydrated proteoglycans are attached to collagen fibers in a network, with the amount of each one determining the physical nature of the complex, for example, there is a very high presence of collagen in tendons.

Table 3.2 Some collagen types and example biological locations.[a]

Type	Dimension (nm)	Location
I	300	Skin, bone, teeth, tendon
II	300	Cartilage, vitreous humor of eye
III	300	Skin, tendons, blood vessel walls
IV	400	Basal laminae, eye lens
V	300	Cornea, interstitial tissue
VI	105	Connective tissue - cartilage
VII	450	Skin, tongue, cornea, sclera

[a]There are also types VIII–XIV not included here.

The attachment of the network outlined above to cells occurs *via* binding to cellular receptors through a recognition process, or by binding directly to the cell plasma membrane itself. Adherence is also assisted by linker molecules such as *fibronectin* and *laminin*, both of which are glycoproteins. Fibronectin is composed of two polypeptide chains linked at the C-distal end by a disulphide group and a relatively small amount of carbohydrate. The molecule is capable of binding several entities such as collagen, heparin and cell surface receptor species. The latter interaction involves the cell binding region of the protein which possesses the RGD (arginine–glycine–aspartic acid) sequence.[8] As will become evident later in the text, this well-known sequence has figured prominently in experiments concerned with attempts to study the behavior of cells on the surface of substrates. The molecule also possesses functions other than adherence – it is involved in blood clotting through its ability to attach fibrin. Laminin is a cross-shaped glycol protein of molecular weight 900 K Daltons, which contains three polypeptide chains with 13% of carbohydrate. As for fibronectin, this molecule is capable of binding heparin, collagen, and cell surface receptors.

Although there is direct interaction between the collagen–proteoglycan complex and the cell surface receptors, as outlined above, the interaction of various cells with the ECM is also mediated by membrane-bound proteinaceous species known as integrins.[9] These proteins effectively connect together the cytoskeleton with proteins in the ECM such as laminin, fibronectin, and collagen. They are present in the lipid membrane in the form a dimer and take a variety of forms capable of attaching to ECM proteins (Figure 3.4). As mentioned for fibronectin, all the ECM proteins bind to integrin *via* their conserved RGD sequence. The interaction between integrin and ECM components is complex and involves conformational changes in the former, often described in terms of *focal adhesions*.

ECM chemistry is highly relevant to protocols for the modification of surfaces with experiments on detection technology and biocompatibility in mind. For example, in order to promote solid-substrate binding of various cells, the interface is often treated with ECM proteins on a so-called inert background. Such efforts will be discussed in more detail later in this chapter.

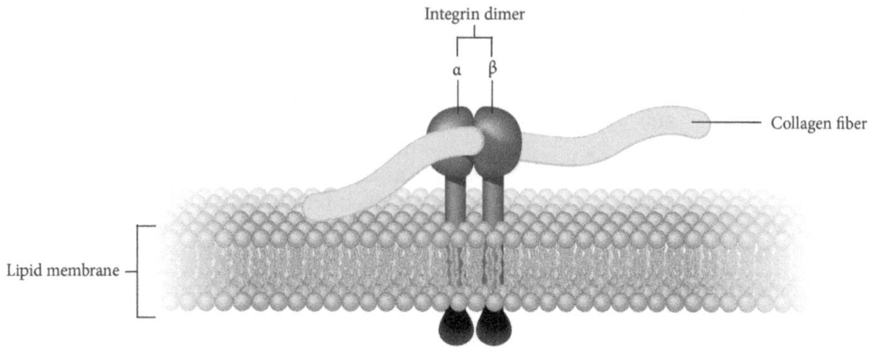

Figure 3.4 Integrin subunits embedded in a lipid membrane with connection to the ECM.

3.3 Cells

Among the myriad of animal cells, certain types have been incorporated more than others into studies of cell response and behavior on contact with solid interfaces. The reason for this lies in the relevance of particular cells in, for example, issues of biocompatibility, development of clinical biosensor, and detection technology, and tissue regeneration. Examination of the cells present in blood is an obvious area of research in view of processes connected to clotting and other phenomena. (For an excellent look at cell types present in blood, the reader is directed to ref. 10.) Other cells such as fibroblasts have figured as "model" species in terms of their response to solid surface morphology. In recent years, there has been very significant interest in combining neurons or neural tissue with a variety of electronic devices such as field-effect transistors. The purpose of this research ranges from efforts to understand the electrical behavior of the cell to attempts at regeneration and repair of damaged neural tissue. Accordingly, the nature of the device – neuron interaction – is a crucial component of these activities. For these reasons we provide a concise look at the nature of the aforementioned cells targeted to the non-biologist reader.

3.3.1 Cells in Blood

3.3.1.1 Leucocytes

Leukocytes, also called white blood cells, are part of the immune system that protects the body from infection. They are present in both the blood circulatory and lymphatic systems and are produced from *hematopoietic* stem cells in the bone marrow. Leukocyte count in the blood ranges from 4 to $11\times10^9\ L^{-1}$, although this figure will depend heavily on the presence or absence of an infectious disease. There are five types, namely *neutrophils, eosinophils, basophils, monocytes*, and *lymphocytes.* The lymphocytes include T- and B-cells and killer cells. The cells, which are all nucleated, are depicted in schematic form in Figure 3.5.

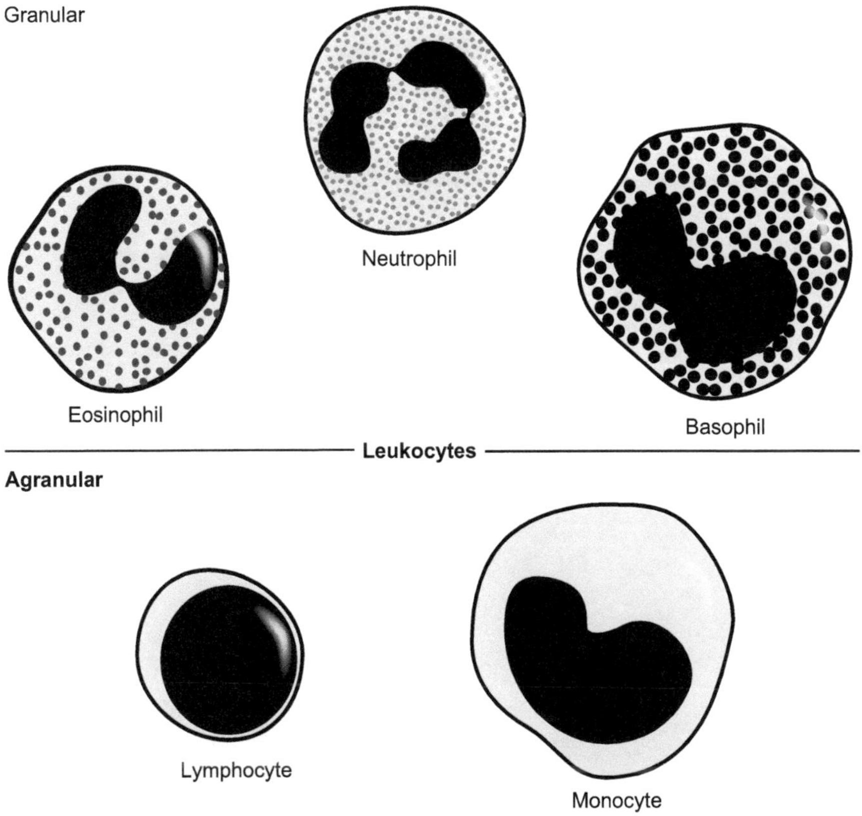

Figure 3.5 Types of leucocyte cells.

Neutrophils are polymorphonuclear cells that constitute some 40–70% of white blood cells in humans. They possess a diameter of 8–9 μm and a life span of 5–6 days. The inactivated cell is close to spherical, with its nucleus having a number of lobes, and is thus termed multilobular. Neutrophils are activated by chemical signals such as interleukin-8 from infection sites associated with foreign invaders. Following interaction with a cytokine receptor, the cells find their way to the site by the process of chemotaxis (motile progression through detection of chemical concentration gradient). They also undergo a change of shape on activation, including the presentation of *pseudopods* or cellular extensions. On infection, neutrophils will also recruit other cells in the immune response *via* secretion of cytokines and other molecules. With regard to defense against bacteria, for example, the cells have three distinct strategies – the processes of *phagocytosis* and *degranulation*, and the formation of *extracellular traps*. The first of these involves the initial process of opsonization of the pathogen (process of attachment of recognition antibodies), followed by ingestion of the foreign particle to produce a *phagosome* (Figure 3.6). The pathogen, such as a bacterium, is then destroyed in due course by released oxygenated entities such HClO.

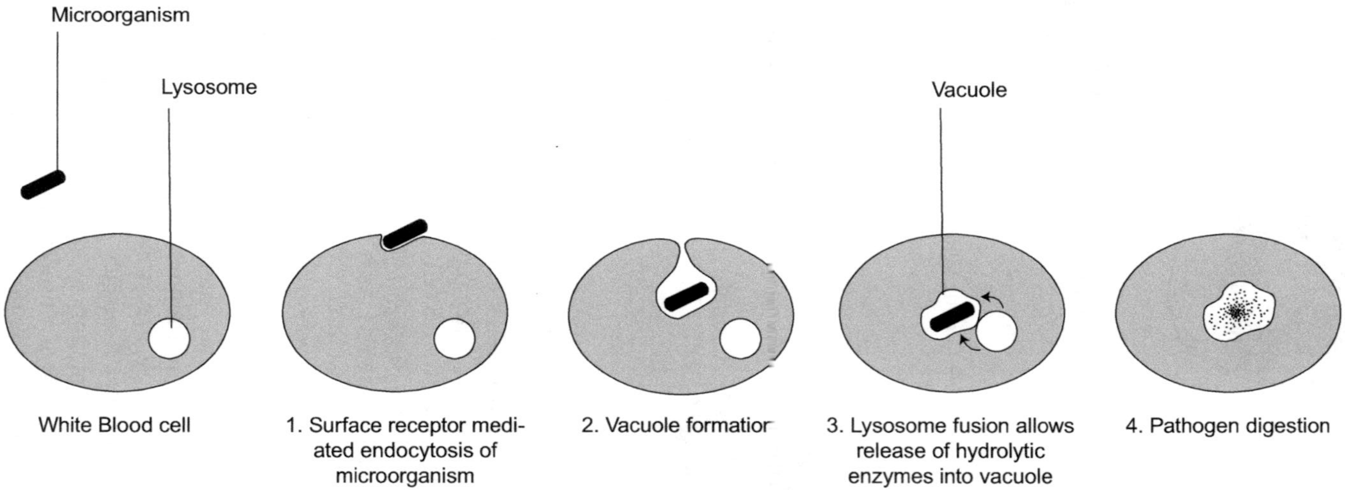

Figure 3.6 Schematic of the process of phagocytosis.

Eosinophils account for 2–5% of white blood cells and are found at sites of foreign invasion. The particles possess surface receptors for a variety of proteins present at sites on invaders, such as immunoglobulins and complement. Eosinophils are located in tissue rather than circulating blood and, as for neutrophils, they release species that are capable of the destruction of invaders. Basophils constitute only 0.5% of the white blood cell count. As is the case for neutrophils, they are capable of the process of chemotaxis with their apparent main mechanism being the release of heparin (an anticoagulant) in the proximity of sites of foreign invasion (minimization of blood clot formation). Basophils also produce histamine, which assists in the transfer of other leucocytes from the circulation into tissue. Monocytes make up about 5% of the white cell population and are immature when first injected into the blood circulation. The particles are transferred to tissue where they become macrophages, which contain increasing counts of *lysozymes.* These entities possess enzymes that are capable of the digestion of foreign invaders. Macrophages are also present in the lymphatic system and are a very important component of the immune response to an invader.

Lymphocytes account for some 25% of the white blood cells and are present in circulating blood in addition to the lymph nodes. The cell volume is almost all occupied by the nucleus (Figure 3.7). There are two types of lymphocytes, the T- and B-cells, which are programmed to activate and recognize antigens (invaders) to produce the immune response. T-cells are released from the hematopoietic bone marrow, as for other leucocytes, but require maturation in the thymus before they are capable of recognizing an invader. The molecular receptors on the surface of the T-cell recognize antigens presented to them by specialized antigen-presenting cells (APCs). This constitutes the *cell-mediated*

Lymphocyte

Figure 3.7 Much of the cell volume of the lymphocyte is occupied by the nucleus.

immune response. In this mechanism, T-cells have several functions – they work with other cells in reacting to an invader, lyse antigenic species, and stimulate B-cells to produce antibodies. B-cells, largely located in the lymph nodes, directly recognize antigens or are stimulated by T-cells. On activation, they differentiate into plasma cells, which are capable of producing antibodies against an invader. This process is termed the *humoral immune response*.

3.3.1.2 Erythrocytes

Erythrocytes, red blood cells, are produced in the bone marrow in a process termed *erythropoiesis* involving hormonal protein initiation (*erythropoietin*). The immature cells become enucleated on injection into the circulating blood. The cell count is approximately $6\times10^6\ \mu L^{-1}$ in humans. The life span of the cell is in the region of 100–120 days in adults and, after this time, they become *senescent* and are removed from the blood stream. The cell has a diameter dimension of 7 μm, and a thickness dimension of around 1 μm. Human erythrocytes have a biconcave-disk shape. They are flat in the middle and possess a dumbbell-shaped cross section. The cell is flexible, which facilitates transport through capillaries, with its shape conferring high surface area for gas transfer and maximization of laminar flow in blood vessels.

The main function of the erythrocyte is the transport of oxygen. Oxygen diffuses through the cell membrane and then is complexed by the metalloprotein *hemoglobin* (Hgb). Following perfusion, Hgb releases O_2 to tissues and the various bodily organs. The Hgb molecule is composed of two α- and two β-chains with an iron-containing heme moiety attached to each one. The spectral character of the cell is associated with the chemistry of the reversible $Fe\text{-}O_2$ reaction. The cell also has other functions, such as the dilation of blood vessels and synthesis of the signaling molecules NO and H_2S. Finally, there are at least 50 or more known erythrocyte membrane proteins, which, from a physical point of view, confer flexibility, but that are also responsible for ionic and molecular transport, adhesion, and recognition of other cell types.

3.3.1.3 Thrombocytes (Platelets) and a Summary of the Coagulation Cascade

Platelets are a crucial component of the biological process of *hemostasis*, that is, the prevention of blood loss, and are consequently heavily involved in the blood coagulation cascade.[11] Since the cell is an integral part of the natural clotting or coagulation system (thrombus formation), which also tends to occur on foreign surfaces exposed to blood, its behavior constitutes a major topic of research for substrate biocompatibility and anti-thrombogenicity. (This is one of the reasons why patients are heparinized prior to certain types of surgery and also in the case of renal dialysis.) Here, we briefly describe the nature of the cell and follow this with a brief summary of the coagulation process. The latter is described in more detail in Chapter 4.

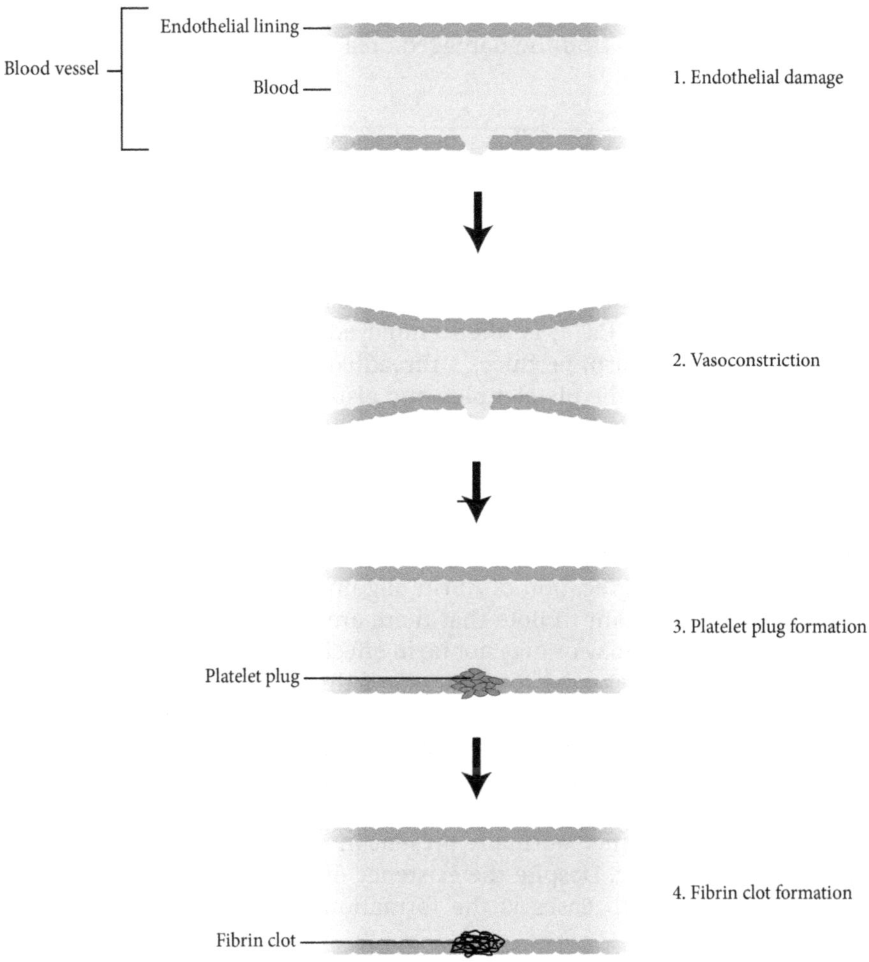

Figure 3.8 Steps involved in the formation of a plug at a damaged vessel.

The inactivated, anucleate cell has a dimension in the range of 2–4 μm, with a count of 150 000–400 000 μL^{-1} in circulating blood. They have an average lifetime of some 9–12 days once introduced into the circulation. Non-functional platelets are removed by the spleen and liver by phagocytosis. The mechanism of hemostasis is composed of three overall overlapping steps – *vascular spasm*, formation of a platelet plug at the blood leakage site, and initiation of the *coagulation cascade.*[11] The first of these involves the constriction of veins (not arteries) in order to reduce blood loss. Formation of a plug at a damaged vessel is considered to involve five steps (Figure 3.8):

1. Platelets contact the site of vessel damage at the endothelial lining.
2. The cells adhere to the damaged area.
3. Platelets change shape and spread out and flatten at the damage site.

4. Following this shape change, they release ADP, which is a signal for other platelets to find the damaged area.
5. Platelets aggregate.

The coagulation cascade follows platelet aggregation at the vessel damaged site (Figure 3.9). Phospholipids produced by activated platelets stimulate clotting factors, which are circulating in the blood on a continuous basis. Many proteinaceous species produced in the liver are activated in a chain-like process, which also involves the calcium cation. This ion is capable of forming a bridge between phospholipids and certain clotting factors. (Citrate, which binds Ca^{2+}, is often employed as an anti-coagulant.) This complex process, in turn, produces a threadlike protein called *fibrin*, which forms a network over the platelet plug and also stabilizes it. The production of fibrin begins with the glycoprotein, *fibrinogen*, which is synthesized in the liver. This molecule is composed of three pairs of polypeptides linked together with 29 disulfide bonds. The soluble glycoprotein, which is also stabilized by Ca^{2+}, contains a binding site for thrombin that cleaves fibrinopeptides from the structure. This process, in turn, involving a number of factors, allows the aggregation of fibrin oligomers.

Finally, it is important to note that there are two pathways in the coagulation cascade which may or may not be in effect at the same time during the clotting mechanism. Crucial to the topic of this text, the *intrinsic pathway* involves activation by not only damaged endothelium, but also by contact of blood with a foreign surface. In this pathway, activated platelets provide the necessary phospholipid moieties, which are associated in several ways with clotting factors. The *extrinsic pathway* involves the release of the phospholipid, *thromboplastin* (not normally present in circulating blood), from the site of damaged tissue. Despite the existence of the two different pathways, the end result in both cases is the formation of a fibrin mesh over the platelet aggregate.

3.3.2 The Neuron

Contemporary neuroscience includes not only research on the human nervous system and brain through a variety of techniques involving neurophysiology, cognitive science, molecular biology, imaging, *etc.*,[12–17] but also the detailed study of single and small collections of nerve cells. As emphasized above, one aspect of the latter is research on the combination of neurons with electrical devices. This research ranges from efforts conducted at a fundamental level to understand medical conditions such Alzheimer's disease, epilepsy, and effects of drugs on neurons, to work on "repair", such as that involved with retinal prosthetic interfaces and other implants.[18]

The neuron is an electrically excitable cell that conveys information through chemical and electrical signaling mechanisms. The cells are incorporated into complex networks which display massive interconnectivity. Various cells are associated with a host of functions ranging from detection

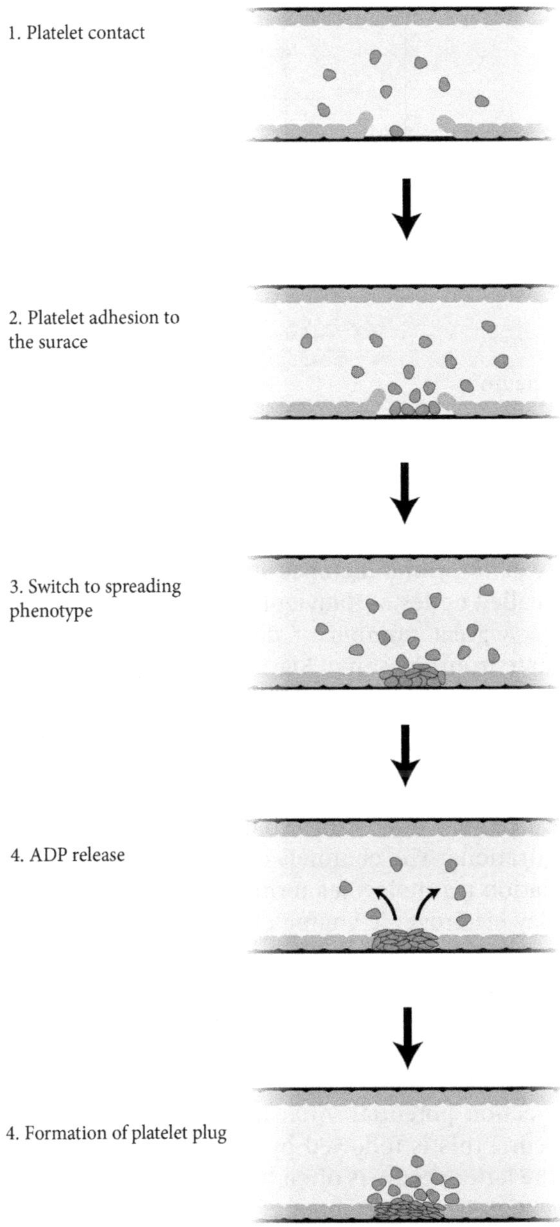

Figure 3.9 Coagulation cascade following platelet aggregation at the site of vessel damage.

of information from the outside world (*e.g.* olfactory neurons and the sense of smell) to motor neurons that signal to muscle tissue.[19] Cells that relay information about their environment for transmission to the central nervous system (CNS) are called afferent neurons, whereas those that orchestrate

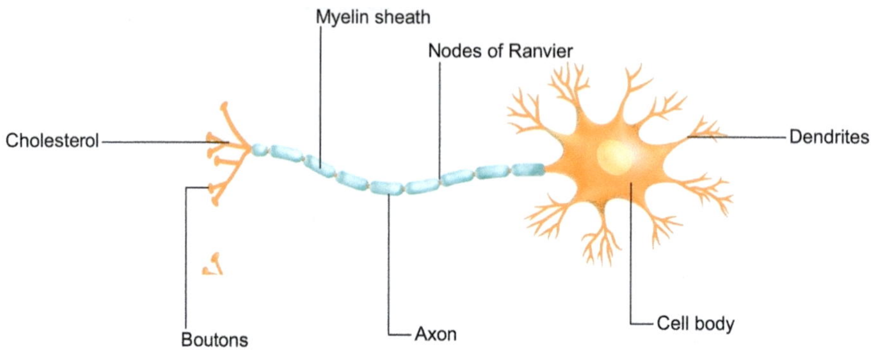

Figure 3.10 The neuron.

signals originating from the CNS are termed efferent neurons. The cell is composed of three parts: soma or cell body (which contains the nucleus), dendrites, and axon (Figure 3.10). In close proximity to the soma is the axon hillock, where there is a high density of Na-ion channels. The axon is a projection that extends some distance in comparison with the soma, contains structures called nodes of Ranvier (constrictions), and is "coated" with a myelin sheath. A great number of dendrites displaying a high level of branching emanate from the soma. Signals from other neurons are received by the soma *via* the dendrites, which are described as being unipolar, bipolar, or multipolar, whereas signals from the axon terminal are transmitted to other cells. The signaling mechanism has components of electrical and chemical processes, which occur *via* the synapse, at which location the membranes have protein assemblies that constitute ion channels and ion pumping configurations. The channels can be gated by voltage effects or by chemical stimulation *via* molecules termed neurotransmitters.

Neurons display electrogenic character with an electrical potential which varies between −10 and −100 mV. The cell membrane is permeable to Na^+, K^+, and Cl^- *via* specific channels, although active transport *via* the sodium–potassium pump is in play. The hydrolysis of ATP yields the energy required for the operation of active transport. Impulse stimulus of the axon of a neuron results in a number of electrical potential changes collectively referred to as the action potential. After a latent period, a depolarization of about 15 mV occurs; this is followed by a further rapid depolarization. The point at which the latter occurs is often termed the firing level. The potential then rises above zero in an overshoot effect, before eventually returning to the resting potential. Obviously, in the various types of nerve tissue, the overall behavior will reflect the compounding of a vast number of action potentials and signals, the most complex of this being the human brain.

3.3.3 Some Cells Used in Model Substrate Studies

Fibroblasts are active cells that maintain tissue metabolism. (The cells outlined in this section are described in detail in ref. 20.) The term "blast" is generally associated with their stem cell origin. They are termed *fibrocytes* when in an inactive condition. Fibroblasts have a branched cytoplasm surrounding an elliptical-shaped nucleus, whereas fibrocytes are smaller and spindle shaped. Their lifetime is in the region of several days. Fibroblasts produce many of the species responsible for cellular structural integrity alluded to with respect to the ECM, that is, molecules such as collagen and the glycoproteins. When tissue becomes damaged, fibrocytes are activated to act as fibroblasts, which are also capable of migration.

Endothelial cells line the internal surface of blood and lymphatic vessels in a monolayer structure. Accordingly, they are in direct contact with blood or lymph. They are present in the entire vascular system including the heart. These cells perform a variety of functions, for example, the transit of a host of moieties in to and out of the circulating blood. With regard to antithrombogenicity, the cells are intimately connected to inactivation of several of the co-factors involved in the coagulation cascade outlined above in reference to hemostasis. Additionally, these cells are responsible for the control of blood pressure, and the repair and formation of new blood vessels (the latter is called *angiogenesis*). Interestingly, certain aspects of endothelial cell dysfunction are connected with both atherosclerosis and coronary heart disease.

Smooth muscle cells, an integral component of smooth muscle tissue, are involuntarily controlled, not striated, not branched, and singly nucleated. The cells are located in many parts of the body including blood vessels, reproductive, gastrointestinal and respiratory tracts, kidneys, skin, and iris of the eye. They are long and thin (Figure 3.11), with much of the cytoplasm being occupied by the protein fibers *myosin* and *actin.* These molecules, which stretch both across and between smooth muscle cells, are largely responsible for the process of contraction. The latter is caused by filaments of myosin and actin sliding over each other. Actin filaments are anchored *via* dense bodies and intermediate filaments. Excitation of a non-spontaneous contraction in certain muscles is induced by neurotransmitters generated by the autonomic nervous system. In other tissues, such as blood vessels, various hormones are associated with contraction.

Osteoblasts are the cells that are intimately connected with the synthesis of bone in vertebrate, although a number of other cells also take part in the

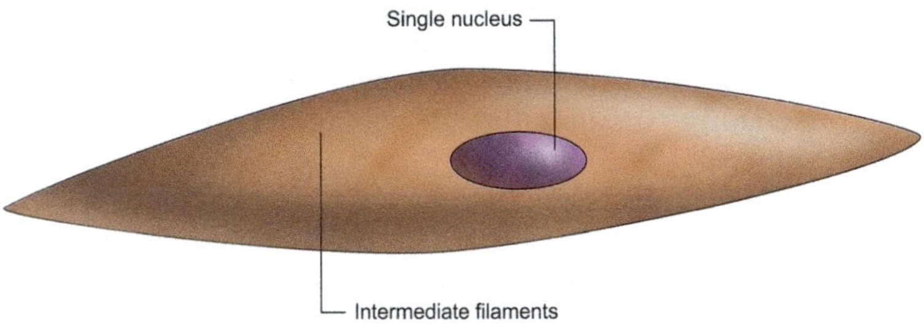

Figure 3.11 The smooth muscle cell.

synthetic process. The origin of the cell is from certain stem cells in the bone marrow (mostly), where eventual conversion to a single nucleus osteoblast involves a number of factors in a highly complex process. Cells that are buried in the bone matrix are generally called osteocytes. In bone, the cells are connected by both tight and gap junctions, which allow both separation from the extracellular matrix and function as a whole unit in concert. Osteoblasts produce a very dense organic matrix, mostly of collagen (see above), which is a major component of bone. The matrix is mineralized by hydroxyapatite, a calcium phosphate-based salt, to provide the compressive strength of bone in contrast with the tensile strength associated with collagen. Production and maintenance of bone is a highly dynamical process, which involves continual generation and resorption. A preponderance of the latter can lead to breaks. The reader will be familiar with this result in terms of the medical condition of *osteoporosis*.

3.4 Interaction of Cells with Substrates: Adhesion, Growth, Guidance, and Proliferation

Aside from the basic question of the role of surfaces in terms of biocompatibility, a relatively recent focus of research on cell–solid substrate interactions has been the production of a "scaffold" that allows the structural support and maintenance of healthy cells with the overall purpose of tissue repair in mind. (This technology is often referred to as "tissue engineering" or "regenerative medicine", although engineering and regeneration hardly seem to be correct descriptions of what is involved.) Two approaches have generally been employed to achieve the goal of tissue repair – modification of substrate surfaces on a chemical basis, and micrometer and nanometer-scale patterning of surface topography. It should be noted that the two strategies are certainly not mutually exclusive. The former method involves both coating of substrates by various entities and surface modification by covalently-bound species, whereas patterning is associated with physical changes achieved by micro- and nano-fabrication techniques. With respect to surface chemistry, the cues provided by the nature of the extracellular matrix have proven to be invaluable in the study of cellular behavior, adhesion, spreading, migration, proliferation, and differentiation. Much of the work on patterning is achieved through the use of engineered modification of the topography by techniques such as photolithography and reactive etching, in an effort to mimic the native topographical environment of a cell.

The main purpose of the rest of this chapter is to concisely review *examples* (only) of the various approaches that have been employed to attempt the retention of cellular integrity, healthy growth, and proliferation. A look at the role of substrates with respect to thrombogenicity and in nanotoxicology is also included in the chapter.

3.4.1 Bare and Treated Polymer and Metal Substrates

A major aspect of research on the interaction of cells with bare surfaces, whether of metals or organic polymers, has been the comparison with coated substrates in order to evaluate the role of such films from the perspective of enhanced biocompatibility. Here, we distinguish coating technologies, to be discussed in subsequent sections, from studies that involve surface treatments, electrochemistry based, for example, including the deliberate attachment of cells in order to reduce occlusion on substrates and overall thrombogenicity. Although not strictly associated with non-coating efforts, these investigations are included in this section. As it pertains to metals, we also outline an interesting reverse aspect of biocompatibility, that of the corrosion caused by cells on surfaces.

3.4.1.1 Poly(dimethylsiloxane)

A polymer that has been very widely employed in microfluidic and "lab-on-a-chip" devices, is poly(dimethylsiloxane) (PDMS).[21] This is also the case for various other polymers belonging to the siloxane chemical family. This material has been considered to possess good properties in terms of biocompatibility,[22] although it must be said that the evidence for this is rather weak. The polymer is also easy to subject to fabrication protocols, exhibits low permeability to water, and minimal electrical conductivity.[23] For these reasons, it has been very much the material of choice for those working with microfluidic channel-based devices, including numerous experiments involving bioanalytical chemistry, cell handling and sorting, *etc.* Given the preponderance of use of PDMS in various aspects of device technology, it is unsurprising that there have been a number of basic studies of cellular interaction with the polymer. One example is the detailed study of Whitesides and co-workers,[24] who examined the behavior of a number of mammalian cells on various forms of the polymer. Four different types of cells were investigated – primary human umbilical artery endothelial cells (HUAECs), transformed 3T3 fibroblasts (3T3s), transformed osteoblast-like MC3T3-E1 cells, and HeLa (transformed epithelial) cells. Several PDMS substrates were prepared for exposure to the cells with an emphasis on the ratio of base-to-curing agent employed. Polymer samples where the surface was treated with extracting solvents and subjected to plasma treatment were also examined. The samples were characterized by surface contact angle measurement and by Young's modulus assessment of stiffness. Following culture and growth of the various cells on polymer slabs, these were visualized after fixing through imaging achieved by fluorescently labeling the cell nuclei. There was some dependence on the cell type studied, but stiffness appeared to have a marginal effect. The conclusion was reached that PDMS appears to have advantages as a surface for culturing cells in comparison with silicon, for example. Interestingly, however, the study avoided any role played by the necessary underlying layer of fibronectin in terms of *its* interaction with the polymer surface.

In a second example, the common procedure of altering the surface properties of PDMS by plasma treatment was examined with emphasis on the effect on the mechanical properties.[25] The latter was achieved by application of spherical indentation testing coupled with atomic force microscopy. The results were interpreted in terms of a two-layer material model composed of a thin, stiff plasma-oxidized layer (296 nm and 3.66 MPa, respectively) superimposed on a thick layer of bulk polymer (elastic modulus of 10.5 kPa). The work then proceeded to examine the effect of such surface alteration on the interaction of the polymer with cells. Interestingly, the value of a cell elastic modulus, neglecting the mechanical effects of plasma treatment, were more than an order of magnitude lower than estimates accounting for the surface layer (9.6 ± 4.2 kPa *vs.* 124 ± 55 kPa, respectively).

As a final comment with regard to the polymer, we note that the literature abounds with the notion (as implied above) that instigation of a hydrophilic surface renders PDMS "bio-inert". In reality, caution is clearly necessary with respect to this concept, since biological macromolecules and cells have been shown to be influenced by the polarity at the surfaces of substrates just as much as with the hydrophobic ones. A case in point is the molecule avidin, often used for surface attachment of biomolecules, which is conformationally compromised by polar interfaces but not by hydrophobic surfaces.[26] Apparently, interaction of the sugar groups on the molecular surface is strongly influenced by the polarity. Interestingly, the structural change does not occur in the aglycosylated version, neutravidin.

3.4.1.2 *Polymer Thrombogenicity*

Extracorporeal circulation of blood is mandatory during medical procedures such as coronary bypass surgery and hemodialysis with regard to patients suffering from renal failure. These protocols require the use, for example, of conduits, membranes, and oxygenators constructed from a variety of polymers such as polycarbonate and polyvinylchloride. Interaction of blood components, including cells, with such artificial surfaces may however stimulate biological processes orchestrated by the immune and coagulation systems, much as outlined above, with deleterious outcomes. For example, there is the possibility for an increase in surface-induced thrombogenicity, that is, the increased risk of clot formation in the vasculature, even though heparin will almost certainly been employed in an attempt to minimize the risk of the occurrence of such. One aspect of thrombus formation, for which there is evidence with regard to extracorporeal circuitry interaction with blood, is the possible future occurrence of cognitive deficit experienced by the patient. It has been shown by Stroobant *et al.*[27] that many patients who undergo cardiopulmonary bypass surgery present post-surgical cognitive defects. The mechanism of this process is thought to involve cerebral microstrokes caused by thrombi release from, for example, the perfusion apparatus due to platelet activation and protein coagulation. In a similar vein, it has also been shown that patients who receive continual hemo-dialysis treatment for renal failure exhibit a progressive decline in cognitive function.[28] Again, at least part of this decline is thought to involve foreign surface instigation of inflammatory effects.

Given the possible deleterious effects caused by foreign surfaces, it is not surprising that a number of commercial organizations have appeared over the years that claim surface treatment cocktails that enhance the anti-thrombogenicity of substrates, although the evidence for such a property is often less than clear. One example of a scientific study conducted in an attempt to produce an anti-thrombogenic polycarbonate surface is that involving a bio-inspired surface treatment with a highly hydrated carbohydrate-rich mesh together with a polysaccharide coating.[29] In our own work on polycarbonate, we have developed an ultrathin (0.5 nm) anti-thrombogenic coating based on monoethylene glycol silane surface chemistry,[30] Anti-thrombogenicity was assessed in a real-time manner using confocal fluorescence microscopy together with a perfusion chamber and whole human blood, labeled with a cell-permeant fluorescent dye dispensed for 5 min at a controlled (high) shear rate of 1000 s^{-1}. Platelet adhesion, aggregation, and thrombus formation on the coated surface was greatly inhibited (>97% decrease in surface coverage) compared to the bare substrate. This remarkable result is, in part, associated with structured interstitial water trapped between the silane chains covalently attached to the polymer surface,[31,32] the so-called "kosmotrophic" effect.[33] In view of its great importance in terms of the blood–surface interaction, this topic will be re-visited later in Chapter 6.

3.4.1.3 *Medical Grade Stainless Steel and Relevance to Cardiovascular Stent Technology*

With regard to metals there has been particular interest in interactions of biological moieties with the surface of medical-grade stainless steel, ranging from individual cells to whole blood. Research in this area has been very much influenced by use of the metal for the construction of stents (see later) and implants used in bone surgery and heart valves. With regard to the biocompatible behavior of stainless steel in terms of its interaction with various types of cells, numerous studies have appeared with respect to the electrochemical pretreatment of the metal. (These are reviewed comprehensively in ref. 34 and 35.) In one example, stainless steel (316LS) was subjected to passivation by an electrochemical process with the goal of examining the responses of fibrinogen, platelets, endothelial and smooth muscle cells.[36] Interestingly the electrochemical treatment reduced the accumulation of platelets on the steel surface (by around 30–50% depending on the exposure time). There was also evidence of lower activation of platelets by the modified surface, compared to the bare substrate.

Another recent study dealt with an effort to prevent the adhesion and proliferation of cells on medical grade steel by coating the metal with polymer films.[37] Again, the culturing of murine fibroblasts and human umbilical vein endothelial cells on the metal was successful and polymer films were capable of reducing the numbers of such cells. The authors pointed out such coatings could have advantages in terms of anti-thrombogenic behavior compared to other methods for biologically passivating steel.

The effect of cells on steel, as distinct from what the metal surface does to cells, has also been subjected to study. An example is the investigation of the corrosion of steel by osteoblasts.[38] The study found that the cells could be cultured successfully on bare surgical grade stainless steel. After several days of exposure to the cells, corrosion-instigated pits were observed on the steel wafers by scanning electron microscopy. Analysis of the culture supernatant liquid revealed the presence of the expected metal ions derived from the metal, *e.g.* Ni^{2+} and Mn^{2+}, which were not present in the control samples. Although the precise mechanism of corrosion remains obscure, the authors specified that it must have an "electrochemical" basis. From the medical point of view it is interesting that the steel–cell system resulted in the generation of cytokines, which are pre-inflammatory species.

As mentioned above, one of the premier medical uses of stainless steel is in the fabrication of stents. This technology for the treatment of coronary artery disease was first introduced by Sigwart *et al.*[39] as an alternative to bypass surgery and balloon angioplasty. Stents are meshed cylindrical scaffolds which are typically inserted *via* balloon- or self-expandable technology into the vascular lumen and expanded into contact with the diseased portion of the arterial wall, restoring adequate blood supply to the heart muscle (Figure 3.12). Fabrication of the device from medical grade stainless steel is very common due to its excellent mechanical properties and resistance to corrosion, although other metals and organic polymers have also been employed in their construction. There are five basic types of stent configuration: coil, helical spiral, woven (braided or knitted), ring (individual or sequential), and cell (closed or opened). The devices differ slightly in strut pattern, width, length, diameter, inter-unit connections, geometry, flexibility, radial-strength, radiopacity, surface area coverage, metal content, and composition. These changes have been rung in order to address, in part, issues associated with blood flow and long-term contact of the metal with the medium. There are two relevant aspects of steel biocompatibility in this case – thrombogenicity, which we have outlined previously, and neointimal hyperplasia (NIH), which is the principal cause of in-stent re-stenosis, that is a re-narrowing of the artery or blood vessel. (NIH is associated with the proliferation of smooth muscle cells within the intima – the layer of endothelial cells in contact with the circulating blood.) Better stent delivery methods and the use of anti-platelet drug therapies after bare stent implantation has effectively reduced the incidence of stented-vessel blockage by thrombus to $\leq$1%, but the serious problem of in-stent restenosis as a result of NIH remains. This issue has been the subject of intense research over recent years, comprehensively reviewed in ref. 35, and has generally

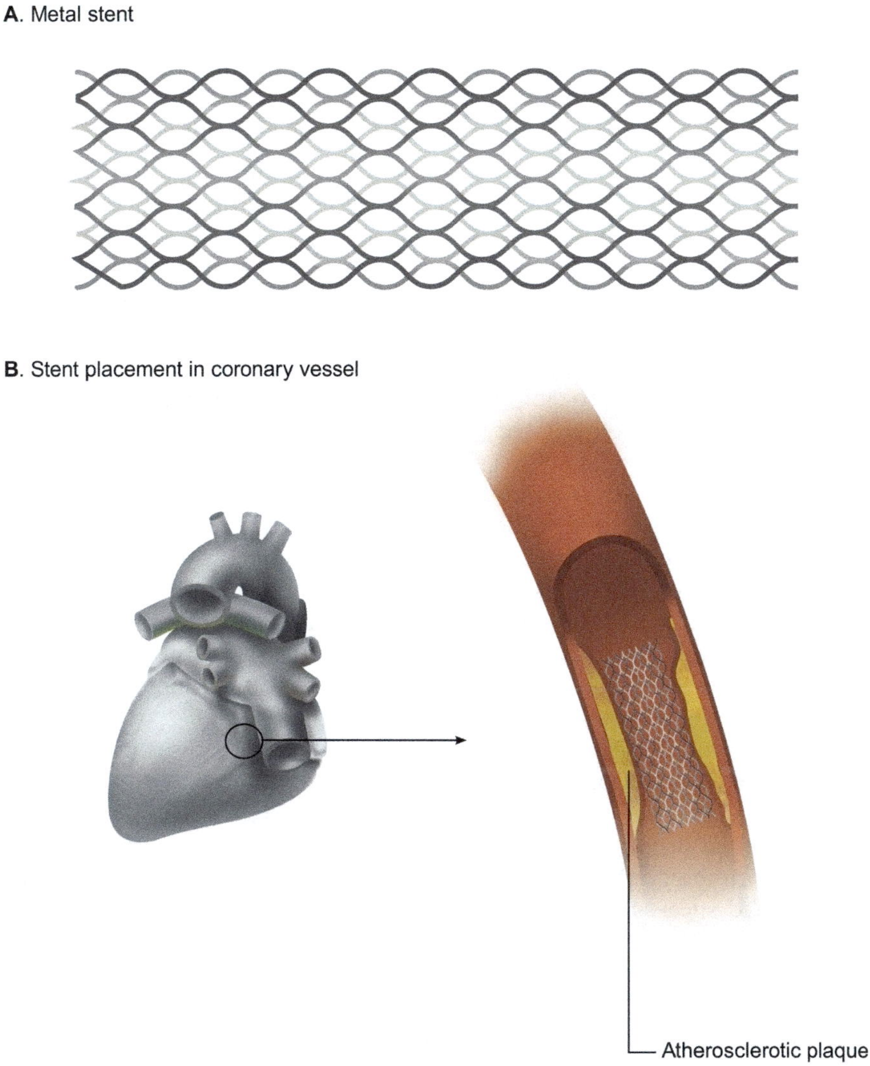

Figure 3.12 Typical stent and insertion into a coronary artery.

been addressed by alteration of the stent material and overall design as alluded to above, and use of a variety of surface coatings. Two coatings among those that have been employed in order to reduce the occurrence of NIH are the "natural" molecules, heparin and phosphorylcholine. In addition to these surface films, attempts have been made to use synthetic polymers such as polyurethane with some success. However, it is the drug-eluting stent (DES) that has become a standard technology in this medical field. In this technology, therapeutic agents are loaded by adsorption into pre-fashioned pores on metallic stent surfaces, or mixed with non-degradable or degradable synthetic polymers and coated onto stent surfaces, where release occurs *via* diffusion. Despite the advantages offered by this approach, there still appears to be a subsequent risk of thrombosis and restenosis since the permanent polymers are constantly in contact with blood, both during drug-elution and *after* the drug has been completely exhausted from the stent.

A particularly interesting strategy to enhance stent biocompatibility is that of re-endothelialization, *via* progenitor cells (EPCs) of the device surface in order to generate a "natural" environment.[40,41] This approach represents an interesting adjunct to the usual philosophy for the cell–surface studies outlined above in the sense that, in this case, cells on a surface are being employed to *enhance* the biocompatibility. (The approach has been recently reviewed extensively by Ravindranath, Romaschin, and Thompson in ref. 42.) Endothelial progenitor cells are circulating moieties in the blood that display specific cell–surface species. These cells are capable of differentiating into endothelial cells. Endothelial cells line the blood vessel and are essential for recovery from injury. EPCs are commonly used as "seeding" cells for endothelialization to occur. Earlier work involved simple seeding with endothelial cells prior to and after implantation, but this approach can lead to cell loss. Unsurprisingly, monoclonal antibodies have represented a strategy for the capture of circulating EPCs or actual endothelial cells. Examples of such species are anti-CD133, anti-CD34, anti-VE-cadherin (anti-CD144) and anti-VEGFR-2, which are specific binders for the surface proteins of cells CD133, CD34, VE-cadherin, and VEGFR-2, respectively. Li *et al.*[43] explored the efficacy of using anti-CD133 in capturing EPCs, by incubating stents immobilized with anti-CD133 antibody in CD133+ cells for 14 days. These authors were able to show that stents coated with anti-CD133 resulted in improved EPC capture and attachment, and the spindle-like cell morphology observed under fluorescence microscopy suggested that the cells had successfully bound to the antibody-modified stent. Similarly, Lee *et al.*[44] found that stainless steel stents coated with VE-cadherin antibody was more selective and specifically captured EPCs and human umbilical vein endothelial cells (HUVECs) *in vitro*, compared to stainless steel stents coated with CD34 antibody. *In vivo*, stents that were coated with anti-VE-cadherin promoted re-endothelialization and reduced neointimal hyperplasia by 25% (42 days after implantation) in comparison to the CD34 antibody-coated stent.

One key issue that has been discussed in recent times is the influence that orientation of a particular surface-bound antibody will have on its capability to bind antigenic species. If whole antibodies are immobilized in a disorientated fashion, their antigen-binding site may become inaccessible, clearly hampering the cell capture by the stent substrate. Furthermore, it has been suggested that random orientation of antibodies can lead to denaturation and loss of antibody function. In our own work, we have studied the possibility that increased biocompatibility of the steel surface can be combined with *oriented* antibodies *via* use of a covalently-bound Fab 1 fragment.[45] The overall strategy employed to achieve this aim is depicted in Figure 3.13.

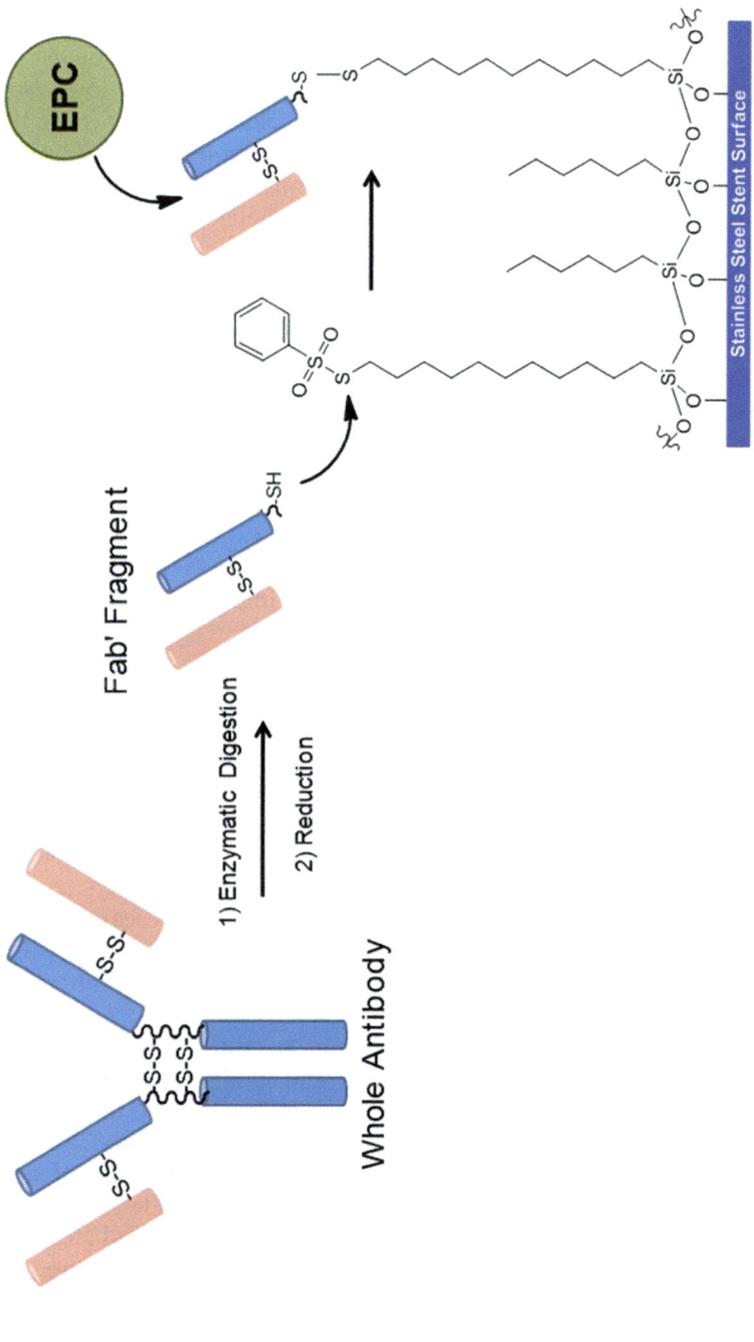

Figure 3.13 Strategy for surface attachment of endothelial progenitor cells *via* oriented Fab 1 fragment of antibody. The fragment is a selective binder for cell surface protein.

Finally, we note that, aside from the immunochemical-based capture of EPCs, a number of other strategies have been suggested, such as the use of aptamer and ECM-based chemistry.

Aptamers are artificially synthesized oligonucleotides (DNA or RNA) which are often regarded to replicate the functions of naturally occurring antibodies. They are around 70–90 nucleotides long, and are able to recognize target molecules and cells with selectivity and specificity approaching those of conventional antibodies. The use of aptamer technology for cell capture by cardiac stents has shown tremendous promise as a method to reduce neointimal hyperplasia and accelerate re-endothelialization of the stent. Although not on steel, Hoffmann *et al.*[46] attempted to use DNA aptamers to selectively bind EPCs (*i.e.* CD31 and CD144 markers) onto polytetrafluoroethylene (PTFE) and polydimethylsiloxane disks, coated with a six-arm star-shaped PEG coating. Once the aptamers were selected using SELEX and immobilized onto the substrate, they were immersed in whole anticoagulated porcine blood with the goal of capturing EPCs, using a modified Chandler loop model. By measuring the fluorescence intensity, they were able to confirm that EPCs were captured by the immobilized aptamers on the disks. In another study, cobalt-chromium stents were coated with aminoparylene, and aptamers specific to EPCs were immobilized on the substrate prior to implantation into a porcine model. After 30 days, it was shown that neointimal thickness was significantly lower in the stent coated with the aptamers.

3.4.1.4 *Gold*

An attractive approach for the study of metal–cell interactions is the use of biosensor detection technologies based on acoustic wave physics and surface plasmon resonance (SPR). The thickness–shear mode (TSM) acoustic wave biosensor is excited by an alternating potential applied to gold electrodes deposited on each face of the device.[47] The device offers sensitive, label-free detection in a flow-through format. In our own research, we have employed the sensor to study the behavior of cells in terms of their interaction with the bare electrode and cellular response to stimulants. A few examples of this research are presented here. Attachment of rat aortic smooth muscle cells (ASMC) to bare gold and electrodes coated with extracellular matrix species such as fibronectin and laminin were studied in a quantitative fashion. Additionally, ions (Ca^{2+} or Mg^{2+}) were used to affect the cell morphology, and a depolarization event instigated by KCl was also detected.[48] The oxidant hydrogen peroxide is well known to cause cell morphological changes, so this molecule was used as a probe to ascertain if the acoustic wave technique was capable of detecting structural alterations. Cells were harvested from a Lewis rat thoracic aorta and sub-cultured prior to introduction *via* an on-line system to the TSM gold surface to result in a count of approximately 4000 cells per electrode. The response of the acoustic sensor showed clearly that AMSC attach to bare gold successfully and are not

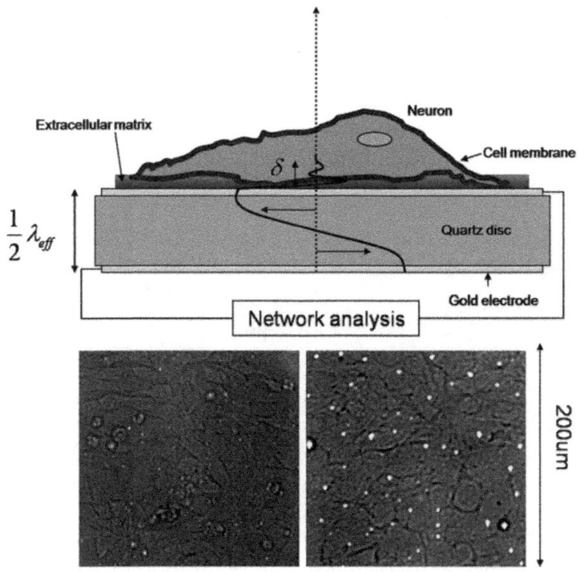

Figure 3.14 Top: A representation of shear wave propagation into a single neuron on the TSM resonator surface. (Diagram not to scale). Bottom: mHypoE-38 neurons (left) and mHypoE-46 neurons (right) on gold electrodes of the TSM quartz crystal.

removed by washing. Rather surprisingly, the bare metal surface yielded the best conditions with respect to both surface capacity and fastest kinetics of cell attachment compared to the laminin- or fibronectin-coated electrodes (on-line). This observation allowed the study of cell detachment by trypsin and *in situ* destruction by various reagents such as peroxide.

Acoustic wave experiments with immortalized mouse hypothalamic neurons have been conducted with regard to interfacial adhesion to the device surface, detection of depolarization events[49] (Figure 3.14), and response to stimulants such as glucagon.[50] Hypothalamii were harvested and dissected from mouse embryo and grown in primary cultures. The cultures were infected with a replication-deficient retrovirus. The minced hypothalamic cultures were then subcloned to yield useable clones for various experiments. Immortalized neurons were generally cultured directly onto the gold electrode surface to a confluence of 80–100% (Figure 3.15). Using this configuration, the behavior of the cells in terms of various parameters and effects were examined. With respect to adhesion and deposition of neurons, changes in the acoustic parameters are clearly correlated to the amount of adhered neurons on the sensor surface, whereas non-adhered neurons do not produce any significant change in the monitored parameters. In the absence of serum proteins, initial cell adhesion is followed by subsequent cell death and removal from the sensor surface. The TSM technique was successful in detecting cytoskeletal changes instigated by the drugs nocodazole, colchicines, cytochalasin B, and cytochalasin D, for which the

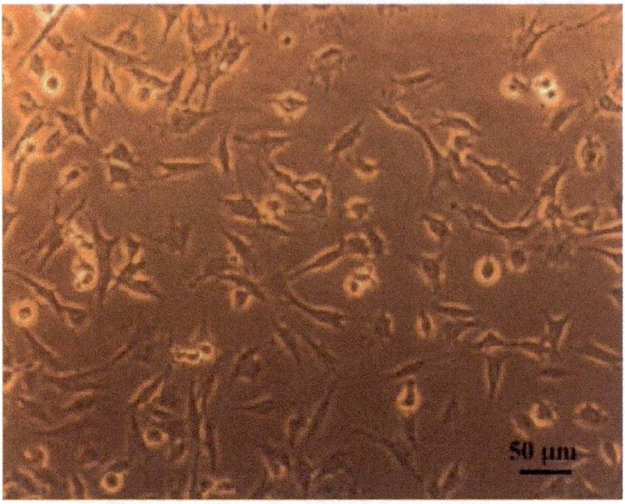

Figure 3.15 Optical micrograph of mouse N-38 hypothalamic neurons cultured on an electrode of TSM illustrating the importance of cell confluence.

neuronal responses were validated by phalloidin actin-filament staining. Stimulation by the addition of KCl to the neurons resulting in cellular depolarization has also been examined with the TSM device. Although the event was successfully detected by the system, the precise relationship between acoustic parameters and cell behavior remains obscure. Research has also been conducted on the synchronization of the circadian rhythm generator of neurons with reference to the effects of glucagon.[50] Interestingly, the process of partial and full synchronization of the cells resulted in different acoustic responses.

In an analogous fashion to the acoustic wave biosensor, the surface plasmon resonance technique also commonly incorporates gold as the primary substrate, which allows detection by plasmon physics. An excellent example of the type of information regarding surface–cell interactions that can be obtained from SPR is that published recently on the role played by laminar shear stress.[51] In this study, a gold nano-slit array-based SPR microfluidic device was employed to measure the dynamic response of a small number of cells upon mechanical stimulation in real time without the need for labeling (Figure 3.16). Detection was performed by directly observing cell adhesion changes prior to the detachment of cells from the substrate. A large-area gold nanostructure consisting of 500 nm-period nano-slits was fabricated on a plastic film using a thermal-annealed template-stripping method. Human lung cancer (CL1-0), human lung fibroblast (MRC-5), and human dermal fibroblast (Hs68) were cultured directly on the gold nano-slits and, following flow removal of suspended cells, their dynamic responses to laminar shear stress were measured. Changes in the SPR signal, that is, alteration in the optical mass density, were correlated with

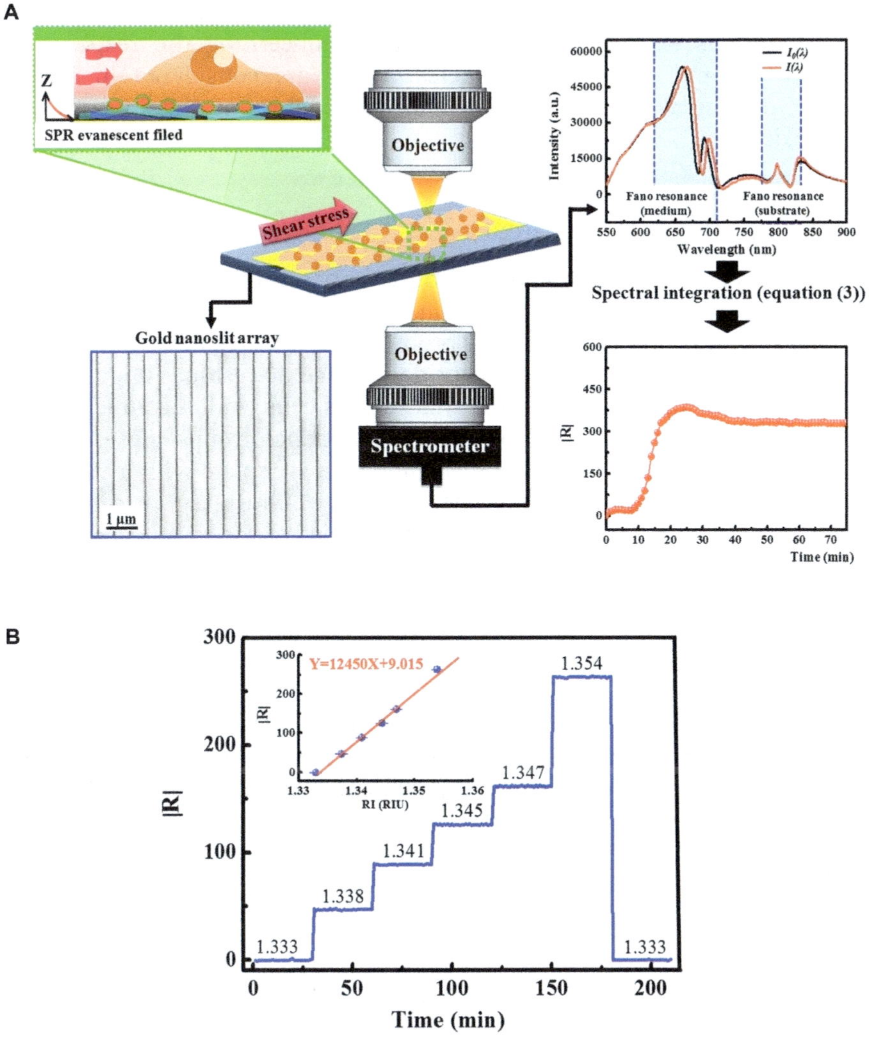

Figure 3.16 A schematic diagram of the instrumental arrangement with gold nano-slits for the study of shear-flow induced cell adhesion changes in real-time with label-free measurements.
(Reprinted from ref. 51 by kind permission of the Public Library of Science.)

cell–surface interactions as governed by focal adhesion. Interestingly, the results of this work indicate that cells can sense different strengths of mechanical force in fluidic shear stress. The cancer cells showed an increase in focal adhesion under fluidic shear stress stimulation. On the contrary, the fibroblast cells lost their adhesion in shear flow. The lung fibroblast is more sensitive to shear stress than the skin fibroblast. The cell responses to shear stress are different depending on the cell type.

Finally, with regard to the gold–biological cell interaction, we note here that numerous experiments have appeared concerning the response of cells to Au (and other) nanoparticles. These are considered in a later section dealing with such particles in the context of theranostics and nanotoxicology.

3.4.2 Polypeptide Coatings

A plethora of different coatings and films imposed on surfaces have been employed in order to attempt the enhanced adhesion of cells to a substrate, with poly-L-lysine (PLL) being highly prominent in this regard. This molecule is a positively charged homo-polypeptide of some 30 residues (Figure 3.17). Over 30 years ago, it was suggested that various cells could be attached to a PLL-coated substrate *via* the polyanionic functional groups present at the cell surface.[52] Today, the commercial literature is replete with a variety of available micro-plates and other surfaces with PLL in place for cellular attachment. Not only has the molecule been applied to the area of cell surface attachment, it also appears to figure prominently in DNA and protein microarray technology for the glass surface binding of protein and nucleic acid probes for multiplexed assay of biochemical interactions.[53,54]

One older example of research involving PLL is the study of human K562 erythroleukemic cells grown in polylysine-coated culture containers.[55] In this case, changes in both membrane conductivity and permittivity were apparently altered by the nature of the cell–PLL interaction. Redistribution of cellular species after exposure to polylysine was investigated using flow

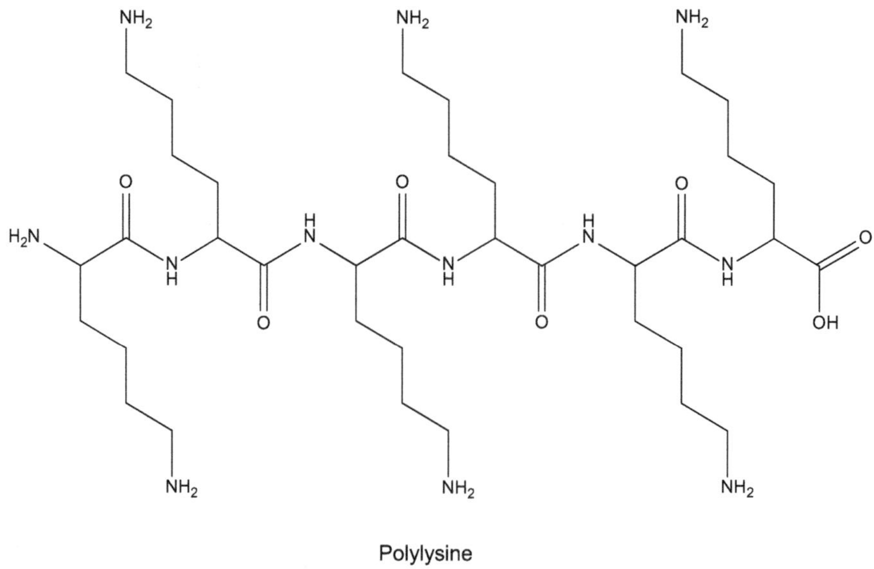

Figure 3.17 The homopolymer poly-L-lysine.

cytometry and immunofluorescence microscopy. The results seem to indicate that such redistribution may, in part, be responsible for cellular adaptation to the new growth environment of K562 cells and for the variations in membrane electrical properties observed.

More recent times have seen attempts to adhere cells to substrates in a 3-dimensional fashion.[56] For example, thermo-formable polymer films were fabricated to produce micro-structured scaffolds. The surface of a polylactic acid membrane was coated with a photo-patterned layer of poly-L-lysine and hyaluronic acid (VAHyal) to attempt spatial control over cell adhesion. Human hepatoma cells (HepG2) and mouse fibroblasts (L929) were used to demonstrate the so-called guided cell adhesion. PLL has also been used recently in conjunction with a polyethylene glycol-based hydrogel to fabricate a superior neuro-electrode interface.[57] Here, a technique was described involving enhancement of the interface between polymeric brain mimetic coatings and neural tissue using PLL. Polymer-modified PEG-based hydrogels were synthesized, characterized, and shown to promote neural adhesion using a PC12 cell line. Interestingly, it was observed that the polymeric materials were adhered to the electrodes for at least 4 weeks.

In a similar vein to that discussed above in terms of polymer thrombogenicity (Section 3.4.1.2), an elastin polypeptide coating has been studied with a view improve the blood compatibility of cardiovascular devices such as vascular conduits and arterial/venous catheters.[58] Platelet adhesion and activation were studied *in vitro* using three synthetic materials: polyethylene terephthalate, a poly(tetrafluoroethylene/ethylene) copolymer, and polyurethane. Coated with the adsorbed polypeptide, all three synthetic materials demonstrated reduced platelet activation and adhesion in platelet rich plasma *in vitro*. *In vivo*, in a rabbit model, evaluation of polyurethane catheters coated with the polypeptide showed a marked increase in catheter patency and a significant decrease in fibrin accretion and embolism when compared to uncoated controls.

3.4.3 Extracellular Matrix Proteins and Derived Peptides

With respect to substrate–cell interactions, a significant volume of research has focused on the surface modification of a wide variety of both inorganic and polymer substrates with ECM proteins and related molecules. (The chemical nature of various moieties included in this category has been described in modest detail above.) Most work has focused on laminin, collagen, and fibronectin, where either the whole protein complex (or cocktails thereof) or peptide sequences are so-called "grafted" onto a substrate – "grafting" being a term often used by the polymer science and biomedical engineering communities. Extremely common in terms of this chemistry has been research on the binding of the aforementioned RGD sequence to hydrogels and other polymer-based entities.[59,60] The obvious rationale behind the employment of such ECM species for surface modification, particularly with respect to cell adhesion and growth, is that cells interacting *via* the presence of integrin will be viable in function, proliferation, morphology, *etc.*, when exposed to such "compatible" layers rather than, say, bare substrates. The attachment of ECM protein or peptides to particular substrates by non-covalent interactions is based conventionally on either hydrophobic, van der Waals, or electrostatic forces (*e.g.* hydrogen bonding). Compared to techniques involving covalent binding of biological macromolecules, such strategies are advantageous because of their ease of application since no chemical modification is required prior to immobilization. The drawbacks are protein denaturation due to uncontrolled interactions between the proteins and the surface (sometimes referred to as "fouling") and undesired protein desorption during the study period. A more stable means of protein immobilization is to link a protein to the surface covalently *via* a chemical bond. The following describes a few examples of the employment of ECM-based species on the behavior of a variety of cells on substrates.

It is fair to state that this research field, for the most part, got underway in the 1980s as a consequence of the actual discovery that cell adhesion is mediated by the RGD sequence present in ECM proteins.[61] One of the earliest works was that of Brandley and Schnaar,[62] who covalently attached a synthetic nonapeptide (Tyr-Ala-Val-Thr-Gly-Arg-Gly-Asp-Ser), to polyacrylamide gel surfaces utilizing *N*-succinimidyl active esters. Immobilization was approximately 80% efficient and the authors performed an approximate calculation which indicated that the gel surface contained around 17.5 nmol peptide per cm^2. Fibroblast cells (see above) adhered readily to the peptide-derivatized surfaces, even in the absence of serum. Surfaces derivatized with a control nonapeptide having *no RGD* sequence were not supportive of cell attachment or growth. Subsequently, the notion that the surface spacing of peptides is important in terms of cell behavior was introduced by Massia and Hubell.[63] In this work, the synthetic peptide Gly-Arg-Gly-Asp-Tyr (GRGDY) was attached to a glass surface with different concentrations of peptide to purportedly "space out" the molecules on the surface. Using fibroblasts, it was shown the spreading of cells occurred at a minimum spacing of 440 nm, whereas focal contact formation occurred at 140 nm. Although these results on a 2-dimensional substrate indicated the importance of peptide spacing, it should be stressed that little or no surface characterization was included in the experiments. This is not an uncommon feature of the early research efforts in this field.

Much later, a far more rigorous experiment was published in which the deliberate, so-called nanopatterning of RGD peptides was employed with the specific goal of examining cell spreading.[64,65] Copolymers in micellar structural form contained gold nanoparticles and were then deposited on glass slides. The polymer was removed for this system *via* plasma treatment, leaving gold nanoparticles of 5–8 nm diameter. The gold particles were organized in hexagonal patterns and separated by 28–110 nm. The space between gold nanoparticles was then covered with polyethylene glycol, supposedly to prevent cell and protein adhesion. The interaction between Au and sulfur was used to attach the thiolated cyclic peptide c-RGDfK to the gold particles. Cellular adhesion experiments were conducted using integrin-transfected fibroblasts. The results of this work indicated strongly that, when exposed to the nanoparticle-modified surfaces, the cells were able to "recognize" spacing effects. Using a phase-contrast microscopy protocol, it was shown that fibroblasts spread well for the case where nanodots were spaced by 50 nm, in contrast to the situation with 110 nm spacing, where limited spreading was observed. In addition to this observation, it was noted by microscopy that quiescent cells are rounded, whereas migrating cells exhibit a polarized shape. The role of integrin $\alpha_v\beta_3$ during adhesion to RGD-nanopatterns at different spacings was also examined *via* this technique, which involved the antibody-based blocking of receptors prior to seeding cells on the surfaces. The number of attached cells was evaluated 1 h after seeding using formaldehyde fixation and toluidine blue staining. Interestingly, the number of cells adhered at both spacings were almost identical after such a blocking procedure. In addition to these experiments the authors also studied the recruitment of integrin-associated molecules in terms of their dependence on the distance between integrin ligands. Example images from this type of research are shown in Figure 3.18.

More recent research has demonstrated that not only the fundamental spacing between RGD sites can affect cellular interfacial behavior, but the actual ordering of nanopatterns on the surfaces can, in addition, influence cell adhesion.[66] Ordered and disordered nanopatterns using the gold particle technology were produced on a substrate that was prepared to be bio-inactive. Various types of ordering were instigated through the use of an "interference reagent" (a co-polymer) during nanopattern formation, and cellular adhesion studies involved osteoblasts. Ordering was characterized through an order parameter in addition to particle diameter and spacing on the surface, as evidenced from atomic force microscopy. An important aspect of this study was that the nature of the nanoparticles meant that only one integrin complex could be attached to an RGD-treated particle. After allowing the cells, which were fluorescently labeled, to interact with the various substrates, the number of attached cells and area occupied by osteoblasts was determined. It was found, on ordered patterns, that the number of cells and area occupied by them decreased with the increasing distance between nanoparticles. This was not the case for disordered patterns, the result being ascribed to the greater variety of ligand (peptide) densities. Moreover, the crucial nature of the interaction or clustering of integrin moieties in the cell membrane is more likely to be exhibited in the case of disordered surface nanodots.

With regard to nanopatterning, Gooding *et al.*[67] have specifically stressed the fact that microfabrication processing of silicon, normally used to produce electronic devices, can be utilized with great success to provide combined topography *and* RGD presentation to cells. (The sole influence of topography is considered in the next section.) One example is the etching of silicon (100 crystal plane) with various concentrations of KOH in order to produce random pyramidal structures at both the nano- and micro-scale. The presence of pyramids was confirmed by atomic force microscopy, which

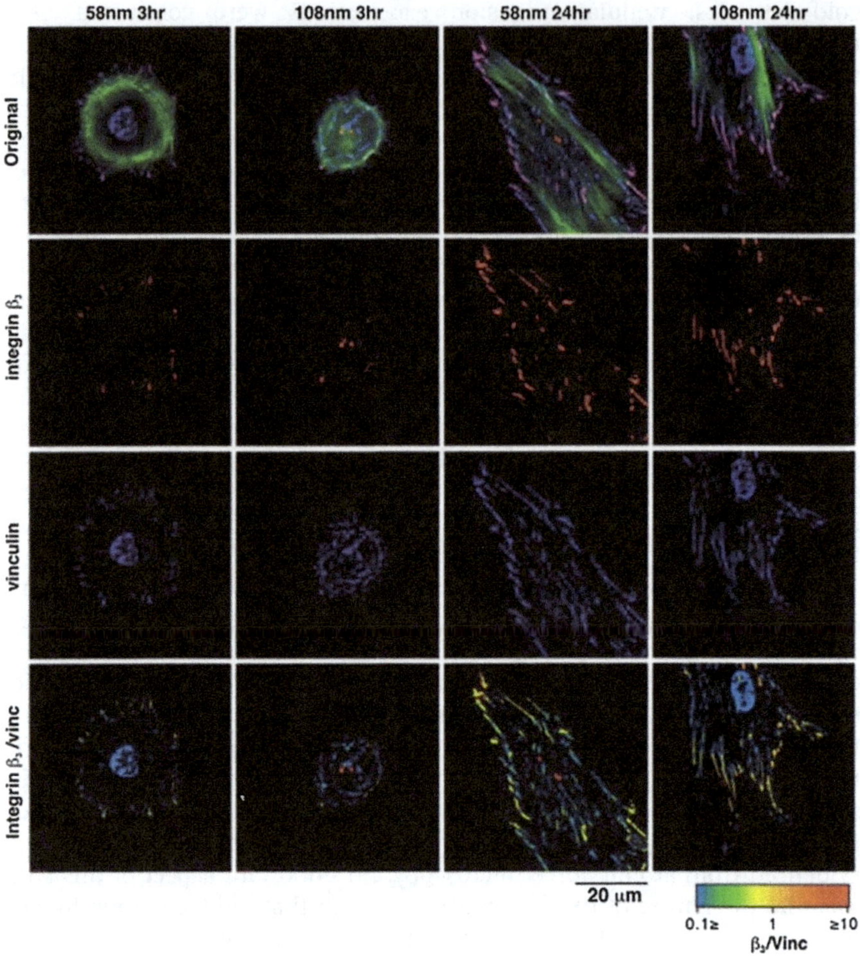

Figure 3.18 Immunostaining and fluorescence ratio images (FRI) of focal adhesion proteins. REF52 cells transfected with GFP-integrin β3 fixed and immunostained with primary antibody against vinculin, followed by Cy5-conjugated secondary antibodies. Actin filaments were visualized with phalloidin-TRITC. Cells on 58 nm and 108 nm RGD-nanopatterns observed at 3 h and 24 h after plating. The rows present the images with integrin β3 in red, vinculin in blue, and actin in green. The last row shows the ratio between integrin β3 and vinculin intensities. FRI are presented in a spectrum scale as indicated in the lookup table. (Reprinted from E. A. Cavalcanti-Adam, T. Volberg, A. Micoulet, H. Kessler, B. Geiger and J. P. Spatz, Cell Spreading and Focal Adhesion Dynamics Are Regulated by Spacing of Integrin Ligands, *Biophys. J.*, **92**, 2964–2974, Copyright 2007 The Biophysical Society with permission from Elsevier.)

showed clearly that the structures could be characterized in terms of both root mean square roughness and pyramid height. Following removal of the oxide layer by HF acid, the silicon wafers were subjected to a hydrosilylation

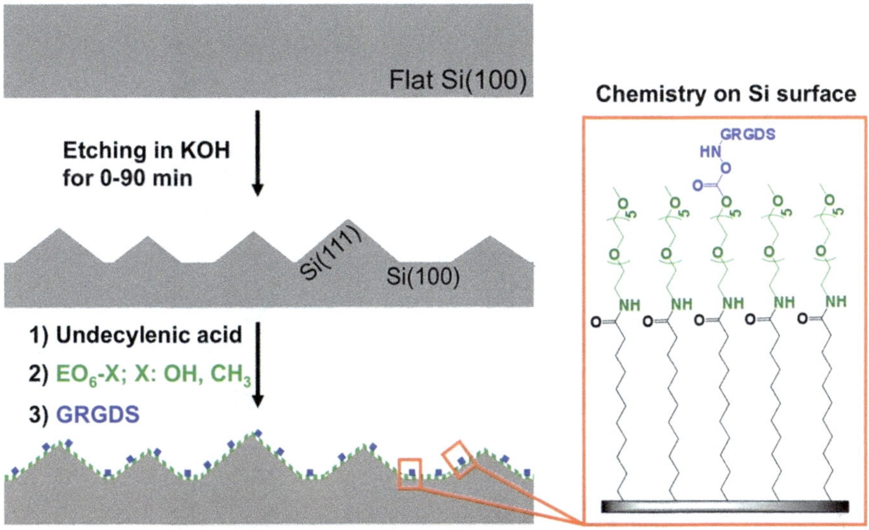

Figure 3.19 Depiction of the adhesive substratum on which surface features and chemistry could be independently controlled. Etching times (0–90 min) in KOH determine the size of pyramidal features and hence surface roughness. Subsequent functionalization with self-assembled monolayers control the RGD densities. Monolayer chemistry consists of a base layer of undecylenic acid that couples 1-amino hexa(ethylene glycol) moieties (EO_6-X) to the surface. GRGDS peptides were grafted to activated EO_6-OH. RGD surface densities therefore were controlled by the ratio of EO_6-OH and EO_6-CH_3.
(Reprinted from ref. 67 by kind permission of the Public Library of Science.)

process involving undecylenic acid to produce monolayers (Figure 3.19). Notably, this process generated highly stable films, with incorporation of distal, reactive –COOH groups because of the Si–C bonds. Using standard surface EDC-NHS (1-ethyl-3-(3-dimethylaminopropyl)-carbodiimide-*N*-hydroxy-succinimide) chemistry, the peptide Gly-Arg-Gly-Asp-Ser was attached to the surface at various densities of surface population (6×10^2–6×10^{11} RGD mm^{-2}). This surface was then allowed to interact with fluorescently labeled bovine endothelial cells. Interestingly, it was found that flat or nano-scaled surfaces adhered more cells than those with rougher surfaces. Cell spreading was more controlled by the population density of the peptide than by the substrate morphology. The very important conclusion was reached that initial contact of cells with a substrate may be controlled by the topography, whereas the engagement of cell surface receptors is dominated by the surface chemistry, *i.e.* the RGD surface density.

One important research activity involving cells on surfaces with incorporation of ECM chemistry is that of binding of neurons and neurites at interfaces – usually considered to be in the field of *neuro-regeneration.* (Attachment of neurons to electronic devices will be discussed briefly in the next section, since there is a morphological aspect to this type of research.)

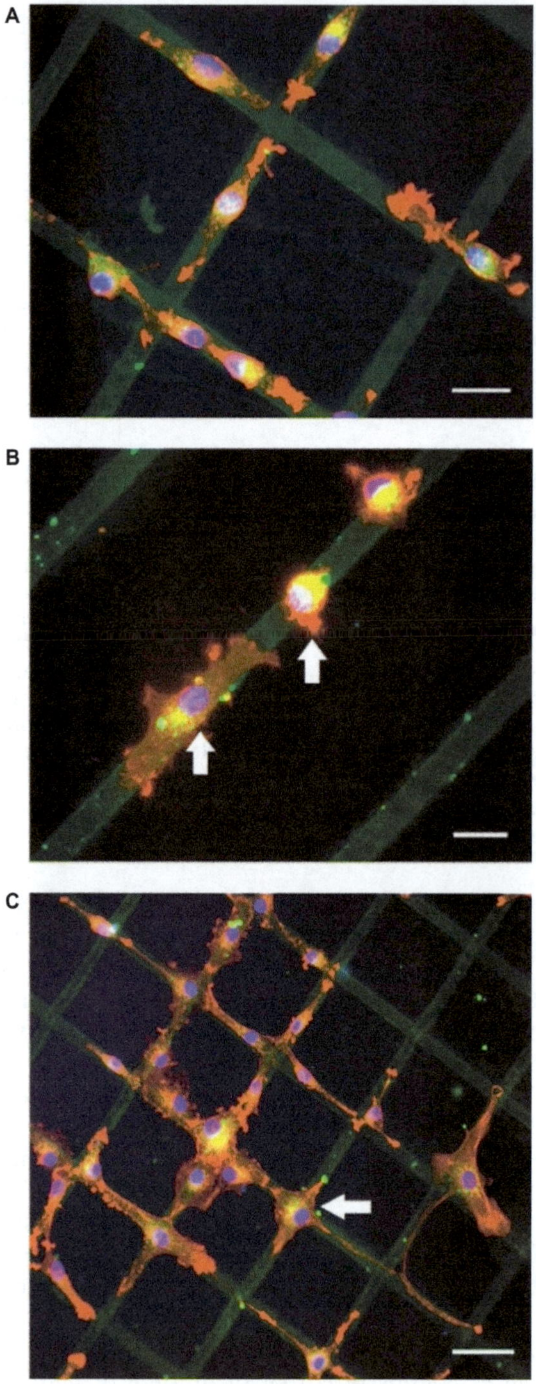
A
B
C

In a similar vein to the hydrogel experiments outlined above, ECM proteins have been employed in conjunction with "3-dimensional" polymers involving dextran and methacrylate.[68] As usual, PEG has also been used, presumably in order to avoid fouling effects. The peptides studied were CRGDS or a mixture of CDPGYIGSR and CQAASIKVAV, with which a significantly enhanced growth of primary embryonic chick dorsal root ganglia on peptide-treated polymers was observed. In a similar fashion, guidance of neurite growth using ECM-based peptides incorporated again into a hydrogel was attempted.[69] Cylindrical channels in the hydrogel were produced, where GRGDS oligopetides could be attached and concentrate within the channels. The interaction of the hydrogel with neural cells derived from rat dorsal root ganglia was studied to examine the possibility for peptide-guided neurite growth following cell migration within the channel volumes. The remarkable result obtained was that the GRGDS gradient across the x–y plane of the biochemical channels affected neurite outgrowth such that neurites extended *preferentially* up the concentration gradient, towards the central core of the peptide channel. The authors pointed out that the result constituted the first time that guided neurite growth was achieved in such a fashion.

The well-known streptavidin–biotin interaction has been employed as a method for binding ECM proteins and peptides to hydrogels.[70] Acrylamide-based hydrogels were photo-polymerized onto silicon wafers in the presence of an acroyl-streptavidin monomer to create planar, functionalized surfaces capable of binding biotin-labeled proteins. Soft protein lithography (microcontact printing) of proteins was used to transfer the biotinylated extracellular matrix proteins, fibronectin and laminin, and the laminin peptide biotin-IKVAV, onto the modified surfaces. In terms of cell culturing, it was found that both LRM55 astroglioma and primary rat hippocampal neurons adhered to the areas patterned with biotin-conjugated proteins, whereas LRM55 cells were observed to attach only to protein-stamped regions of the hydrogel (Figure 3.20).

Figure 3.20 Cell attachment to protein-functionalized hydrogel surfaces. (A) At 24 h, rat astroglioma LRM55 cells were seen to attach and extend on regions stamped with either the laminin (left to right) or the fibronectin (top to bottom) proteins. Cells were stained using Alexa 568-conjugated phalloidin (cytoskeleton, red) and Hoescht 33342 (nuclei, blue). Scale bar = 25 μm. (B) Patterned LRM55 cells were shown to incorporate BrdU, denoting that cells were actively synthesizing DNA prior to cell division. BrdU-positive cells are indicated by arrows. Scale bar = 25 μm. (C) At 4 weeks, *in vitro* cells were found to maintain compliance with the stamped patterning of the hydrogel. Cells were observed to extend along both the laminin and the fibronectin stampings, or they spread where the two proteins intersected (arrow), indicating that μCP proteins retain long-term functionality. Scale bar = 50 μm.
(Reprinted from M. R. Hynd, J. P. Frampton, N. Dowell-Mesfin, J. N. Turner and W. Shain, Directed cell growth on protein-functionalized hydrogel surfaces, *J. Neurosci. Methods*, **162**, 255–263, Copyright 2007 with permission from Elsevier.)

3.4.4 Substrate Morphology

In this section, we concentrate in a concise fashion on the influence of physical micro- and nano-surface structuring of materials on the response of cells. This property is obviously of high relevance for materials employed in medical implants and extra-corporeal device technology, where there is an obvious interaction of cells with blood. We also include a look at the specialized area of interaction of neurons with various devices, which often involves fabricated surface structures and their interaction with this category of cells.

A variety of techniques has been employed to create 2D and 3D nanostructured features such as photolithography, focused-ion-beam lithography, e-beam lithography, nano-imprint lithography, interference lithography, reactive ion-etching, glancing angle deposition, physical vapor deposition, electro-spinning, self-assembly patterning, colloidal lithography, polymer demixing, co-block phase separation, two-photon polymerization, and chemical etching or oxidation.[71,72] The features instigated on a number of different substrates include gratings, posts, pits, and island geometries in both the micro- and nanometer range, as shown in Figure 3.21. One feature that appears to result from the voluminous number of studies with the "model" blood-based cells mentioned above is that considerable variability is observed in terms of cellular surface-motional dynamics, adhesion, proliferation, and morphology.

With respect to inorganic substrates, such as NiTi alloys, TiO_2, and Ti, there are indications that endothelial cells (ECs) display both increased migration speeds and enhanced adhesion and proliferation on various

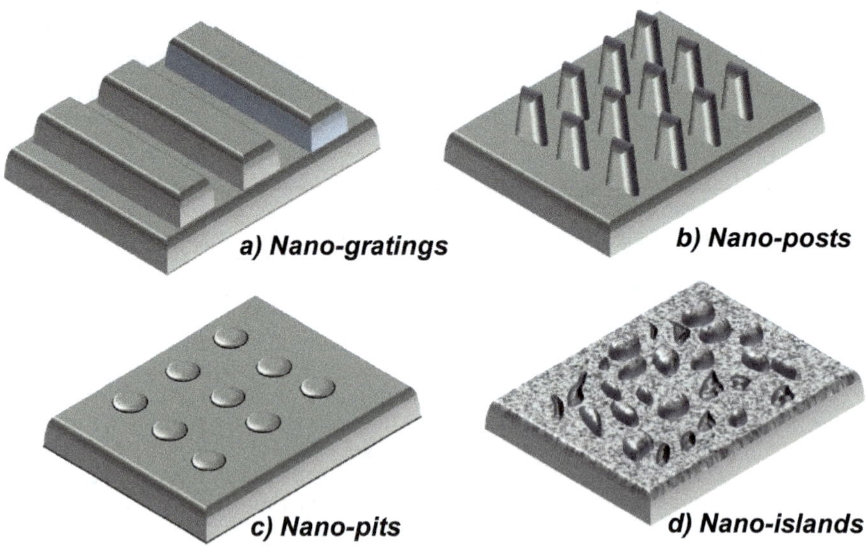

Figure 3.21 Schematic of typical surface topographies produced for study of cell adherence and behavior: (a) nano-gratings, (b) nan-posts, (c) nano-pits and (d) nano-islands.

nanopatterned surfaces.[73,74] Additionally, this sort of property is often connected with the fact that line gratings on substrates are somewhat analogous to natural endothelium.[75–77] On the other hand, there have also been a number of reports which indicate that such patterns result in reduced cellular proliferation. The reason for this observation is very unclear at the present time and clearly requires further investigation. Such discrepancies have been similarly been observed on nano-island geometries imposed on polymer surfaces.[78,79] A limited number of studies on the response of ECs to nano-posts have been reported. For example, Kim *et al.*[80] demonstrated that nano-post PEG surfaces fabricated using capillary lithography actually enhanced the focal adhesion of ECs. This result was attributed simply to an increase in the material surface area and adhesion sites for cells. Zawislak *et al.*[81] demonstrated that ECs on 3D nano-post silicon surfaces impale themselves on the pillar to the extent that even some of the cell nuclei were penetrated by the pillar tips.

Micro- and nano-grated topographies have also been shown to influence the response of smooth muscle cells (SMCs) on surfaces, particularly those composed of polymers such as PDMS and PMMA. The general trend in these findings is that micro-grated polymers yield increased alignment, elongation, and orientation, but with reduced proliferation.[82,83] With regard to nano-pits, the consensus appears to be that there is no substantial influence on the behavior of SMCs.[84,85] Finally, numerous studies have examined the response of SMCs on random nano-topographies. Yet again, some authors indicate increased adhesion and proliferation on such surfaces,[86] whereas others observe the opposite response.

Fibroblast cells (FCs) have been widely studied as a model to investigate the influence of micro- and nano-grating islands, pits, and post geometries on cellular function. As for the studies on ECs and SMCs outlined above, the emphasis has been placed on cellular adhesion, alignment, elongation, and contact guidance on a variety of substrate materials with micro-grating topographies.[87,88] On both inorganic and polymer surfaces, the trend is that micro-gratings result in enhanced proliferation and cell viability.[89,90] Many studies have been conducted on the FC response to nano-island geometries. As specified already, quite marginal differences in geometry can render significant differences in proliferation, spreading, and biological function.[91,92] Several studies have looked at the response of FCs to nano-pits. One such investigation utilized arrays of nano-pits produced by e-beam lithography The results were increased cell spreading and filapodia interactions on 120 nm pits, as compared to 75 nm pits, while cells on 35 nm pits had a similar number of filopodia to those grown on control samples.[93] The effect of nano-post geometries on FC function has been the subject of much research and these structures generally seem to enhance adhesion and proliferation. Green *et al.*[94] found that posts of 2 and 5 μm-high resulted in increased cell proliferation compared to 10 μm-high posts and smooth surfaces. Milner[95] observed FC adhesion and proliferation on PLA surfaces patterned with 400 nm and 700 nm posts *via* replication molding. Their results demonstrated increased FC adhesion and decreased cell proliferation on surfaces with 400 nm textures, as compared to 700 nm textures and smooth surfaces. The effect of FC adhesion on polycarbonate and poly (etherimide) surfaces with micro-post features of varying dimensions, that were generated by laser treatment demonstrated no cellular orientation with respect to posts. FCs spread and elongate whether in contact with posts or micro-smooth materials.[96] Similarly, Dalby *et al.*[97] demonstrated reduced FC adhesion on PMMA substrates with nano-posts (prepared by colloidal lithography) that were 100 nm in diameter, 160 nm in height, and having a pitch of 230 nm. An increase in endocytosis was also noted on nano-pits using clathrin staining, indicating that these nanoscale features are in the same size range as those features with which the *in vivo* cells interact. FCs have also been studied on silicon surfaces possessing nano-posts with a height ranging 50–100 nm and with a pitch of 230 nm that were fabricated by interference lithography and deep reactive-ion etching.[98]

Finally, we turn for a brief look at the case of neurons on electronic device surfaces. Measurement of electrical activity in a brain slice or cultured neuronal network with micro electrode arrays (MEAs) and specialized metal-oxide semiconductor (CMOS) technology in an *in vitro* fashion is enabling further understanding of the brain and neuronal network function.[99] In addition to passive measurement of neuronal potentials, active applied experiments through excitation of neurons and observation of changes can be used for toxicology studies, drug screening, and applications in cellular biosensors. Unsurprisingly, conventional photo-lithographic protocols applied to various electronic devices, much as described in the two previous sections, have been employed in order to examine the electrical behavior of neuron networks. For the most part, these techniques try to mimic or amplify the natural topographical effects of the cell's base membranes and artificially create a designed substrate with nanostructures that stimulate directed growth of neurons. One example of such an experiment is shown in Figure 3.22, which illustrates the growth of a neuron over a transistor.[100] A key issue in the fabrication of such a combined neuron-device system is the perfect dimensional matching between electrodes and neurons, the main purpose being that the ionic conduction in the cellular matrix is transferred to electronic conduction in the sensing device. With respect to the use of microelectrodes in conjunction with electronic devices, careful selection of the electrode positions at statistically significant locations and avoidance of overlap in the sensing area of the electrode constitute the most experimentally sound methods to achieve selectivity and resolution.

In addition to the study of neuron populations on various CMOS devices,[100–102] there has been significant interest in experiments involving MEAs and the combination thereof with transistor-based devices. Multi-electrode arrays allow the study of distributed patterns of electrical activity in cultured neuron-based networks *via* non-invasive extracellular electrodes built into the substrate. These electrodes can also be used to stimulate neurons extracellularly and non-destructively, thus enabling a long-term

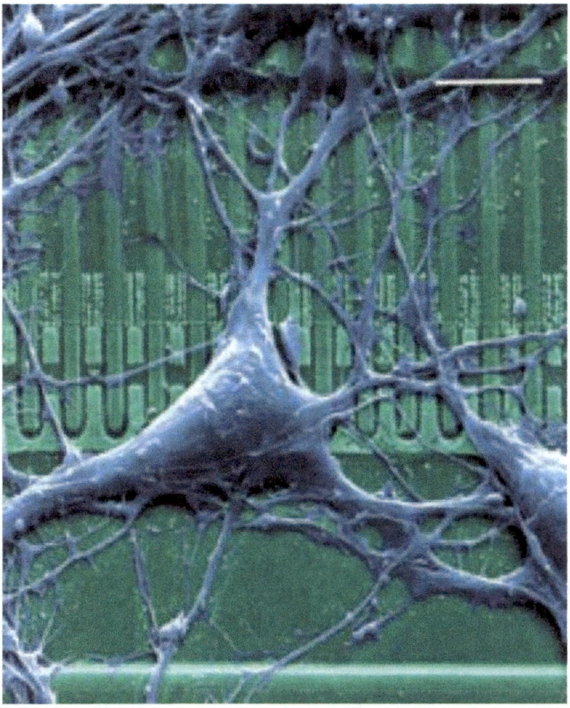

Figure 3.22 Image of a neuron grown on an integrated field effect transistor, which illustrates the dimensional match between neurons and microelectronics. (Reprinted from A. Cohen, J. Shappir, S. Yitzchaik and M. E. Spira, Experimental and theoretical analysis of neuron–transistor hybrid electrical coupling: The relationships between the electro-anatomy of cultured Aplysia neurons and the recorded field potentials, *Biosens. Bioelectron.*, **22**, 656–663, Copyright 2006 with permission from Elsevier.)

two-way connection between a cultured neuronal network and a computer. Quite recently, commercial configurations have become available where computer and data storage technology had rendered it feasible to record continuously 60 electrodes at sampling rates over 20 kHz per channel. Despite these advances, significant challenges remain that must be overcome with respect to the detection strategies presented thus far for MEAs. For instance, the low signal-to-noise ratio due to the resistance between the neurons and the field-effect transistor or microelectrode requires improvement in terms of acquisition and processing of multiple noisy signals, which can be achieved *via* the use of low-noise transistors and CMOS technology.[103,104]

3.4.5 Substrate Rigidity and Elasticity

In sharp contrast with the very significant research effort on the role played by surface morphology on cellular response, the rigidity and elasticity of substrates has received relatively little attention. By far the majority of the work performed in this area has been concerned with the employment of polymer substrates for the purpose of examining rigidity in particular. Often, the thrust of this type of research has been directed at cell locomotion, which is considered to be very important in terms of overall cell movement in immune response and wound healing, *etc.* We outline some examples in this section.

In an earlier seminal study, polyacrylamide layers were employed in a rationale investigation of cell response to substrate flexibility.[105] In this work, normal rat kidney epithelial and 3T3 fibroblastic cells were cultured on a collagen-coated polyacrylamide substrate that allowed the substrate flexibility to be varied, while ostensibly maintaining a constant chemical environment. The latter presumably involved an attempt at the retention of the same surface functional groups. Cells were fixed and studied by standard fluorescence microscopy and immunoblotting protocols, with polymer flexibility being characterized by Young's modulus determination. Compared to cells on rigid substrates, those on flexible substrates showed reduced spreading and increased rates of motility or lamellipodial activity. Focal adhesions on flexible substrates were irregularly shaped and highly dynamic, whereas those on firm substrates had a normal morphology and were much more stable. Cells on flexible substrates also contained a reduced amount of phosphotyrosine at the adhesion sites. Treatment of these cells with phenylarsine oxide, a tyrosine phosphatase inhibitor, induced the formation of normal, stable focal adhesions similar to those on firm substrates. Conversely, treatment of cells on firm substrates with myosin inhibitors, 2,3-butanedione monoxime or KT5926, caused the reduction of both vinculin and phosphotyrosine at the adhesion sites.

In a similar study to that outlined above, fibroblasts were cultured on flexible polyacrylamide sheets again coated with type I collagen.[106] However, in this experiment a transition in rigidity was actually introduced into the central region of the polymer layer by a deliberately introduced discontinuity in the concentration of the bis-acrylamide cross-linker. Cells approaching the transition region from the "soft" zone were able to migrate across the boundary, with a concurrent increase in spreading area and traction forces. In contrast, cells migrating from the "stiff" side, interestingly, reversed direction or retracted as they reached the boundary. In connection with this phenomenon, the authors termed the effect "durotaxis." Additionally, cell movement could also be guided by manipulating the flexible substrate to produce mechanical strains in the front or rear of a polarized cell.

Variation of polymer properties has been employed in order to examine the effect of substrate mechanical behavior on neurite extension of dorsal route ganglion in 3D culture experiments.[107] Gels with different concentrations of agarose were prepared and characterized in terms of their mechanical properties by standard rheometry. The rate of neurite extension for the various gel substrates was perfumed using time-lapse microscopy, and it was shown that the higher the gel concentration the less is the rate of neurite extension. A mathematical model was applied to the results in an attempt to derive a relationship between polymer mechanical properties and ganglion behavior.

Collagen-coated polyacrylamide gels were used yet again in a study of the spreading and other properties of aortic smooth muscle cells in a very similar approach to those mentioned above.[108] In this case, the elasticity of the polymer substrates was determined using an atomic force microscopy method. Gel stiffness was also examined *via* a macroscopic tension method. Cell spreading, cell shape, cytoskeletal or focal adhesion assembly, and fluorescent intensity were studied by fluorescence microscopy following the usual cell fixing and staining protocols. The SMC spread significantly higher on "rigid" collagen-coated glass compared to the gels in agreement with the work outlined above on epithelial and fibroblast cells. (As seems to be the norm in this field, any role played by the collagen layer in terms of stiffness was ignored.) The cell spreading properties found in this work are reproduced in Figure 3.23. Interestingly, the authors ascribed the peaks observed in the plots to SMC movement capability associated with cell ligand density. The biphasic phenomena exhibited in the plots is attributed to cell crawling being initially limited at low ligand densities whereby a cell cannot form adequate attachments to pull itself forward or spread. However, at high ligand densities, a cell cannot detach from enough ligand to bring its rear forward.

Finally, the now standard protocol of using a polymer gel to produce substrates exhibiting variable rigidity has been employed to examine the surface behavior of neurons and astrocytes.[109] A fibrin-based scaffold was also included in the study in an attempt to compare cell growth, *etc.* with that in natural tissue. The dynamic shear moduli of both gels were measured *via* spectrometric rheometry and the cell interactions with the matrices were examined by standard immunocytochemistry. It was found that, on soft gels, astrocytes do not spread and have disorganized F-actin compared to the cytoskeletons of astrocytes on hard surfaces. Neurons, however, extend long neurites and polymerize actin filaments on both soft and hard gels. Laminin-coated soft gels induce attachment and growth of neurons, whereas they suppress astrocyte growth. The number of astrocytes on soft gels is lower than on hard ones, even in the absence of the mitotic inhibitors normally used to temper the astrocyte population. The stiffness of the material required for optimal neuronal growth, characterized by an elastic modulus of several hundred Pa, is in the range measured for intact rat brain.

3.5 Nanoparticles, Theranostics, and Nanotoxicology

Nanoparticles (NPs) have been studied for over 50 years, although in past times they may have been included in the field of colloid chemistry. In other words, there has occurred a certain level of rebranding over recent years. Some of these, such as drug-delivery vesicles, have been known for a considerable time, whereas others such as quantum dots and carbon nanotubes have appeared more recently. NP technology has become intimately linked

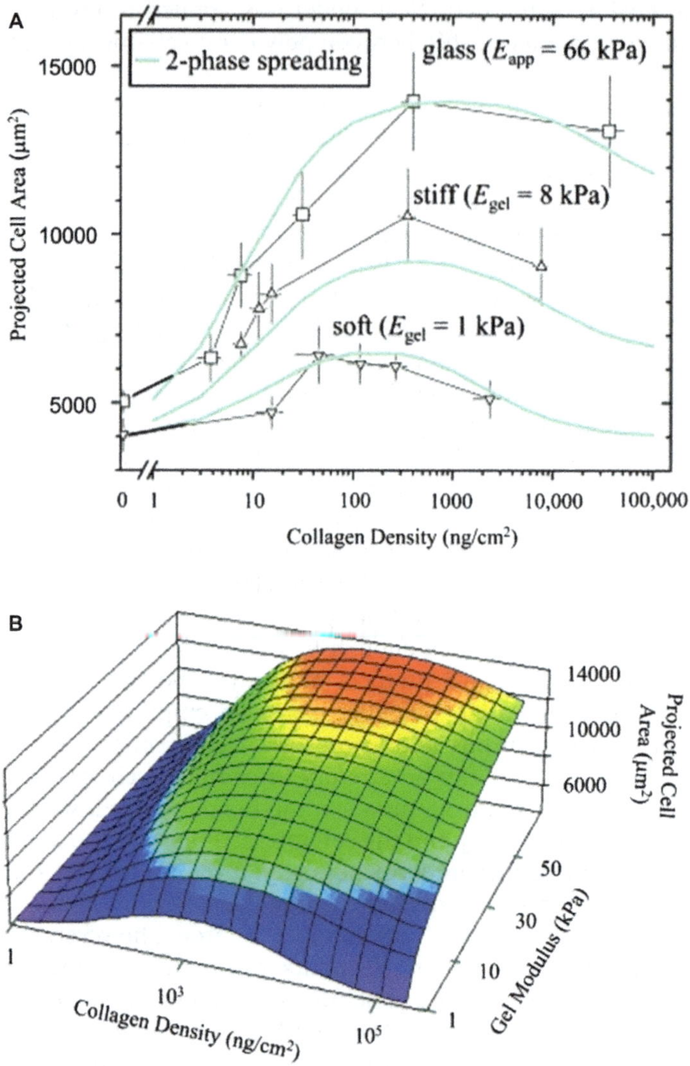

Figure 3.23 Spread cell area as a function of ligand density on soft, stiff, and rigid substrates. (A) The projected cell area was determined 4 h after plating ($n > 10$ per datapoint), giving the indicated average (mean ± SE). The smooth curves are calculated from a model for two-phase spreading (see Appendix) expressed in terms of both E (or E_{app}) and collagen density. Note that cells respond strongly to increasing collagen density on glass and hardly at all on soft gels. (B) Curved surface in three dimensions that fits SMC spreading.
(Reprinted from A. Engler, L. Bacakova, C. Newman, A. Hategan, M. Griffin and D. Discher, Substrate Compliance versus Ligand Density in Cell on Gel Responses, *Biophys. J.*, **86**, 617–628, Copyright 2004 The Biophysical Society with permission from Elsevier.)

to the relatively new medical protocol termed *theranostics*, which is defined as the combination of diagnostic and therapeutic agents on a single platform, such that both modalities are applied in order to provide a simultaneous diagnosis and treatment. This concept first introduced by Funkhauser[110] has been reviewed comprehensively by Thompson *et al.*[111] in the context of personalized medicine. One critical aspect of this technology is the interaction of tailored nanoparticles with cells and tissue of diseased organs. Theranostic nanoparticles, which can be composed of many materials such as metals, polymers, gas bubbles, and biological species, must be multifunctional in terms of carrying a therapeutic agent, as well as a contrast agent for imaging purposes.[112] Given their high surface-to-volume ratio, NPs can be loaded with both drugs and contrast agents in an especially orchestrated ratio.[113] Ideally, the contrast agent would allow monitoring of the treatment process, even in blood circulation, on passage to the target site. Additionally, the NP surface can be functionalized with targeting moieties, such as a substrate for a receptor that is known to be over-expressed in a certain type of cancer.[114] Targeting ligands may be combined so that one ligand may allow the nanoparticle passage into the cell by endocytosis, and the other then targets an internal receptor before drug release.[112] Some examples of the various approaches that have seen use of NPs in theranostics are depicted in Figure 3.24.[113] Finally, we note that a number of different imaging modalities have been employed in recent years in order to detect NP-based agents – these are summarized in Table 3.3 together with some of their relevant properties.[111] We now turn to a few examples of the practical use of the theranostic nanoparticle platform.

We begin with vesicle technology, which has taken two forms: liposomes and polymersomes.[115–118] The former of these species is generally composed

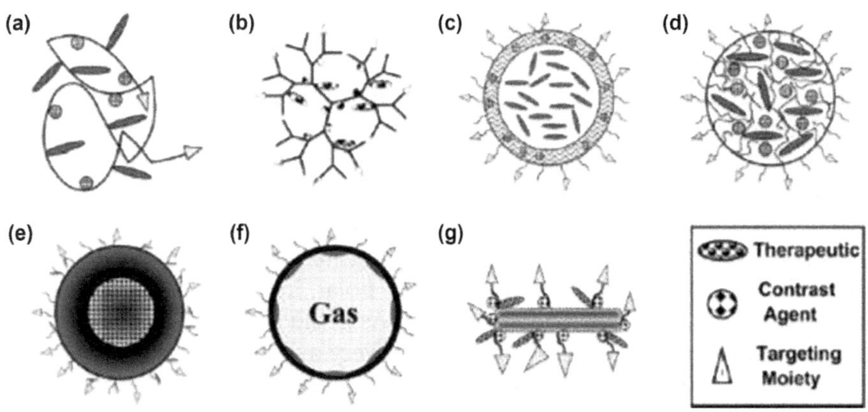

Figure 3.24 Structural representations of general theranostic platforms based on nanoparticles.
(Reprinted from S. M. Janib, A. S. Moses and J. A. MacKay, Imaging and drug delivery using theranostic nanoparticles, *Adv. Drug Delivery Rev.*, **62**, 1052–1063, Copyright 2010 with permission from Elsevier.)

Table 3.3 Imaging modalities for theranostics.

Modality	Characteristics
Magnetic resonance imaging (MRI)	Spatial resolution of 25–100 μm Radiation free Low sensitivity: 10^{-3}–10^{-5} M Contrast agents can be toxic
Computed tomography (CT)	Spatial resolution of 50–200 μm Requires X-radiation Good differentiation of tissue type Sensitivity not studied extensively
Positron emission tomography (PET)	Spatial resolution of 1–2 mm Produces 3-dimensional images Requires γ-radiation High sensitivity: 10^{-11}–10^{-12} M
Single photon emission computed tomography (SPECT)	Spatial resolution of 1–2 mm Produces 2-dimensional images Requires γ-radiation High sensitivity: 10^{-10}–10^{-11} M
Ultrasound (US)	Spatial resolution of 50–500 μm Radiation free 3-Dimensional images possible Real time imaging possible Considered effective, inexpensive
Optical imaging (OI)	Spatial resolution of 2–5 mm Radiation free Sensitivity 10^{-9}–10^{-12} M fluorescence Sensitivity 10^{-16} M bio-luminescence Inexpensive and facile application

of "sacks" fabricated from phospholipids, where drugs and/or contrast agents can be either encapsulated within the interior of the liposome or attached to the surface of the vesicle. One example of a clinically successful liposome is that incorporating the drug, doxorubicin, employed to treat certain forms of cancer.[119] The liposome is shielded by polyethylene glycol and has been shown to accumulate at cells in tumors more readily than freely circulating drug in blood. In terms of relevance to imaging, liposomes have also been functionalized with a variety of specialized agents such as iodine, manganese, and radionuclides. Polymersomes are produced from the self-association of diblock and triblock polymers.[120,121] In an analogy with vesicles generated from natural chemicals, the interior and outside surface of the polymersome can be arranged to be polar in character, whereas the membrane structure itself is hydrophobic. As is the case for liposomes, these moieties have been both used for drug delivery and for imaging purposes.

A similar approach to those described above is that of micelle-based technology.[122] Polymeric micelles are made of block copolymers like vesicles, except micelles are formed by one layer where the hydrophobic blocks associate inside the micelle to form a hydrophobic core while the hydrophilic blocks associate to form the outer surface known as the corona. One advantage of this strategy for encapsulating drugs is that the hydrophobic interior of the particle can be employed successfully to deliver water-insoluble drugs. Notably, this type of micelle is in clinical trials again for the treatment of cancer. As for vesicles, micelles have also been used extensively for imaging purposes, especially In-111 and Tc-99 for gamma scintigraphy and Gd for MRI.[123] Yet another example of the use of polymers to generate nanoparticles is the dendrimer approach.[124,125] Dendrimers are formed by standard polymerization reactions and begin as a planar molecule. As the polymerization reaction carries on through multiple generations, the planar molecules branch out and eventually become spherical and large. The polymer backbone and many side chains form a web-like structure within the sphere and on its surface. This type of hyper-branched structure yields many potential binding sites for targeting moieties, contrast agents, drug molecules, and chelators. The major problem with dendrimers is that the polymers used for their construction tend to be too toxic for use in humans.

Carbon nanotubes (CNTs) are small sheets of carbon in graphite-like arrangements that are rolled into hollow tubes consisting of a single layer; other layers of carbon can be added to produce multi-walled tubes. The properties of CNTs can be tuned by altering their size and number of walls. For example, they can be synthesized to exhibit a high absorption coefficient and a broad excitation profile that covers a wide range of wavelengths.[126] CNTs have not been studied to any great degree as a contrast nanoparticle. However, the tube itself has shown potential as a contrast agent in photoacoustic imaging and, especially, in near-infrared and Raman spectroscopy.[127] Most of the research surrounding the use of CNTs in the medical area has been associated with drug delivery. The reason for this lies in the observation that the particle is readily taken up by cells *via* the process of endocytosis.[128] An interesting possibility with respect to CNT chemistry is the alteration of the particle surface for various purposes. Such a mechanism, for example, can aid in endocytosis and also render a level of aqueous solubility. However, as for a number of other NPs, the significant drawback for their use in medicine is their toxicity and the fact that they do not breakdown in the body.

Turning to core–shell particles, there has been a very high level of research activity with regard to the likes of iron oxide-, quantum dot-, gold-, and silica-based NPs in recent years. Iron oxide particles (IONPs) appear to possess considerable potential for use in MRI in the light of their magnetic properties. One method for their preparation is synthesis by high temperature decomposition in an organic solvent.[129] The problem with this new method is that the IONPs are encased in a thick organic top-coat that is hydrophobic, although there have been many attempts to impart water solubility to these particles through surface modification.[130] Since the particles are magnetic or superparamagnetic, they can be employed with success to produce an image on a T2-weighted map of the body.[131] One such NP has been commercially exploited for imaging the spleen and liver.[132] Additionally, an external magnetic field can be used to interact with the magnetic properties of IONPs to direct the particles to an area of interest in the body.

Quantum dots (QDs) are nanocrystals of semiconducting materials such as CdSe/CdTe, CdTe/CdSe, Cd_3P_2, InAs/ZnSe, and InAs/InP/ZnSe, that are usually coated in ZnS.[133,134] The particles are noted for their bright fluorescence with a narrow emission spectrum, and for their photo- and chemical-stability. QDs are synthesized by heating a solution of the necessary organometallic precursors in a high boiling point organic solvent with surfactants like trioctylphosphine and trioctylphosphine oxide to control particle growth. Species like mercaptoacetic acid, glutathione, and cysteine, which can form disulfide bonds with the QDs, are often used to enhance the water solubility of the particles. One example of an attempted use of QDs to transport drug molecules is the work presented by Bagalkot *et al.*,[135] in which a QD–aptamer–doxorubicin conjugate was used for simultaneous imaging, therapy, and monitoring. An RNA aptamer was used as a biovector targeting prostate-specific membrane antigens. The aptamer was covalently bound to the surface of CdSe/ZnS QDs, and then, doxorubicin was allowed to intercalate into the aptamer sequence forming the complete conjugate. Interestingly, doxorubicin and the QDs are independently fluorescent but, in the aptamer conjugate, both fluorescent emissions were quenched until doxorubicin was released gradually once the QD conjugate reached the tumor tissue. As doxorubicin was released, it became fluorescent again and so did the QD

Gold nanoparticles (Au NPs) are well established with a variety of shapes such as spheres, cubes, rods, wires, and cages being available.[136] In a rather similar vein to the development of the acoustic wave and SPR-base biosensors mentioned above, the particle surface can be modified directly with thiolated biomolecules, or *via* linkers through the strong Au–S bonds. If the linker is bifunctional, biomolecules such as antibodies can be attached with facility. This allows the functionalization of Au NPs with both drugs and targeting agents.[137,138] With regard to theranostic possibilities, gold has certain advantages in view of the fact that it displays strong plasmon absorption. This property facilitates its use in photoacoustic and surface enhanced Raman spectroscopies. One interesting example of a theranostic Au NP platform in clinical development is a product which is used for photothermal ablation of tumors.[139] In this approach, a gold nano-shell is arranged around a colloidal silica core together with thiolated polyethylene glycol on the surface so as to provide supposed immune system "cloaking". The particles are administered by injection for accumulation in the offending tumor, where they are imaged by MRI to ensure they have collected in the correct area. Then, the particles are irradiated with near-infrared light, and the surface plasmon resonance effect produces heat that destroys the tumor. The problems with the use of gold are its inherent cost and, secondly, the inevitable Au–S chemistry is vulnerable to reducing chemicals present in the body.

Finally, in terms of core–shell entities, we will mention silica NPs, which are usually synthesized by the hydrolysis and condensation of tetraethyl orthosilicate (TEOS).[140] The synthesis process often involves co-precursors like aminopropyltrimethoxysilane or mercaptopropylmethoxysilane in order to introduce amine and thiol groups to the particle surface, respectively.

Functional groups on the particle surface can then be used to bind fluorescent dyes for optical imaging and chelating metals like Gd for MRI imaging. One interesting characteristic of silica NPs is the possibility that they can be fabricated to have mesoporous properties, which means that many channels in the particle matrix can be instigated *via* special synthetic protocols.[141] Other NPs have been incorporated into the silica version, one example being the work of Vivero-Escoto *et al.*,[142] who employed Au NPs in the context of drug delivery. In this case, the silica NP was loaded with the drug paclitaxel in its channels, followed by pore closure with Au NPs. The interaction between the Au NPs and the silica was designed to be photolabile so that, when the Au NPs were irradiated with light, they would be removed, thus opening the channels to release the drug. In an analogous experiment with respect to drug delivery, a luminescent mesoporous silica nanoparticle was used to present doxorubicin in an *in vitro* model.

A common feature of a significant portion of research on NPs to be potentially applied in the medical field is that the particles are often considered to be "bio-inert" or "biocompatible". However, there is a tacit recognition that this may not be the case since a number of studies invoke "cloaking" by species such as polyethylene glycol, which is generally thought to enhance the level of biocompatibility. The reality that biological moieties may be adversely influenced by NPs has resulted in increased interest into this aspect of the particle–cell/tissue interaction, a research area now incorporated in the field of nanotoxicology.[143,144] Research in this category has developed to the level that new journals and special issues, which include this topic, have appeared in recent times (Table 3.4). In summary, the biological introduction of nanoparticles can lead to the generation of reactive oxygen species such as, for example, free radicals. These species can lead to

Table 3.4 Example journals and special issues on "Nanotoxicology".

Publication	Publisher	Type
Nanotoxicology	Taylor and Francis	Journal
Journal of Nanotoxicology and Nanomedicine	*IGI Global*	Journal
Nanotoxicology	Informa Health Care	Journal
Nanomaterials	MDPI	Special issue 4, 2014
International Journal of Molecular Sciences	MDPI	Special issue 13, 2012
Archives of Toxicology	Springer	Special issue 85, 2011
Small	Wiley	Special issue 9, 2013
Angewandte Chemie	Wiley-VCH	Special issue 53, 2014
Nature Nanotechnology	Nature Publ. Group	Special issue Online 2011
ACS Nano	ACS	Virtual issue 4, 2010

inflammation and consequent damage to both cellular proteins and DNA alike. Research in relatively recent times has also demonstrated that NPs can damage cells in terms of viability and morphology, possibly leading in turn to apoptosis and necrosis. Given the widely suggested potential for NPs with respect to detection and therapy *via* drug release, *etc.*, it is mandatory that much more research is carried out on the negative consequences of the corporeal introduction of such particles.

References

1. *Surface Analysis Methods in Materials Science*, ed. D. J. O'Connor, B. A. Sexton and R.St.C. Smart, Springer, Berlin and New York, 2nd edn, 2003.
2. *Surface Analysis: The Principal Techniques*, ed. J. C. Vickerman and I. S. Gilmore, Wiley, Chichester, 2nd edn, 2009.
3. Y. Maréchal, *The Hydrogen Bond and the Water Molecule: The Physics and Chemistry of Water, Aqueous and Bio Media*, Elsevier, Amsterdam and Boston, 2007.
4. R. J. Clarke, *Adv. Colloid. Interface Sci.*, 2001, **89–90**, 263.
5. B. Alberts, D. Bray, J. Lewis, M. Raff, K. Roberts and J. D. Watson, *Molecular Biology of the Cell*, Garland Publishing, New York, 3rd edn, 1994.
6. A. Schapira, *Biochemist*, 2005, **27**, 24.
7. *Extracellular Matrix Biology*, ed. R. O. Hynes and K. M. Yamada, Cold Spring Harbor Laboratory Press, New York, 2012.
8. E. Ruoslahti, *Annu. Rev. Cell Dev. Biol.*, 1996, **12**, 697.
9. M. J. Humphries, *Biochem. Soc. Trans.*, 2000, **28**, 311.
10. B. J. Bain, *Blood Cells: A Practical Guide*, Wiley-Blackwell, Chichester and Hoboken, 5th edn, 2015.
11. *Essential Guide to Blood Coagulation*, ed. J. P. Antovic and M. Blombäck, 2nd edn, Wiley-Blackwell, Chichester, UK, 2013.
12. J. F. Stein and C. Stoodley, *Neuroscience: An Introduction*, Wiley, 2006.
13. G. L. Fain, *Molecular and Cellular Physiology of Neurons*, Harvard University Press, Cambridge, 1999.
14. J. G. Nicholls, A. R. Martin, B. G. Wallace and P. A. Fuchs, *From Neuron to Brain*, Sinauer Associates, Sunderland, 2001.
15. D. A. Lovejoy, *Neuroendocrinolgy: An Integrated Approach*, Wiley, 2005.
16. *Molecular and Supramolecular Information Processing*, ed. E. Katz, Wiley-VCH, Weinheim, 2012.
17. *Biomedical Engineering and Cognitive Neuroscience for Healthcare: Interdisciplinary Applications*, ed. J. Wu, IGI Global, Hershey, 2013.
18. M. Thompson, L.-E. Cheran and S. Sadeghi, *Sensor Technology in Neuroscience*, Royal Society of Chemistry, Cambridge, UK, 2013, p. 142.
19. R. Carter, *The Human Brain Book*, DK Publishers, New York, 2009.
20. *Cells*, ed. B. Lewin, L. Cassimeris, V. R. Lingappa and G. Plopper, Jones and Bartlett, Boston, 2007.

21. J. Friend and L. Yeo, *Biomicrofluidics*, 2010, **4**, 026502.
22. J. H. Park, K. D. Park and Y. H. Bae, *Biomaterials*, 1999, **20**, 943.
23. J. C. McDonald and G. M. Whitesides, *Acc. Chem. Res.*, 2002, **35**, 491.
24. J. N. Lee, X. Jiang, D. Ryan and G. M. Whitesides, *Langmuir*, 2004, **20**, 11684.
25. G. Bartalena, Y. Loosli, T. Zambelli and J. G. Snedeker, *Soft Matter*, 2012, **8**, 673.
26. L. M. Furtado, *Interfacial oligonucleotide chemistry studied by an on-line biosensor, radiochemical labeling and nucleic acid microarrays*, PhD thesis, University of Toronto, 2001.
27. N. Stroobant, G. Van Nooten, Y. Van Belleghem and G. Vingerhoets, *Chest*, 2005, **127**, 1967.
28. A. M. Murray, *Adv. Chronic Kidney Dis.*, 2008, **15**, 123.
29. A. S. Gupta, S. Wang, E. Link, E. H. Anderson, C. Hofmann, J. Lewandowski, K. Kottke-Marchant and R. E. Marchant, *Biomaterials*, 2006, **27**, 3084.
30. K. Fedorov, C. Blaszykowski, S. Sheikh, A. Reheman, A. Romaschin, H. Ni and M. Thompson, *Langmuir*, 2014, **30**, 3217.
31. N. M. Pawlowska, H. Fritzsche, C. Blaszykowski, S. Sheikh, M. Vezvaie and M. Thompson, *Langmuir*, 2014, **30**, 1199.
32. S. Sheikh, C. Blaszykowski, R. Nolan, D. Thompson and M. Thompson, *J. Colloid Interface. Sci.*, 2015, **437**, 197.
33. R. M. Davidson, A. Lauritzen and S. Seneff, *Entropy*, 2013, **15**, 3822.
34. F. Nazneen, G. Herzog, D. W. M. Arrigan, N. Caplice, P. Benvenuto, P. Galvin and M. Thompson, *J. Biomed. Mater. Res., Part B*, 2012, **100B**, 1989.
35. P. Benvenuto, *Self-assembled monolayer-based modification of stainless steel for oriented antibody immobilization. Towards improved stent biocompatibility*, PhD thesis, University of Toronto, 2015.
36. A. Shahryari, F. Azari, H. Vali and S. Omanovic, *Acta Biomater.*, 2010, **6**, 695.
37. P. P. Vicario, Z. Lu, I. Grigorian, Z. Wang and T. Schottman, *J. Biomed. Mater. Res., Part B*, 2009, **89B**, 114.
38. D. Cadosch, E. Chan, O. P. Gautschi, H. P. Simmen and L. Filgueira, *J. Orthop. Res.*, 2009, **27**, 841.
39. U. Sigwart, J. Puel, V. Mirkovitch, F. Joffre and L. Kappenberger, *N. Engl. J. Med.*, 1987, **316**, 701.
40. T. Shirota, H. Yasui, H. Shimokawa and T. Matsuda, *Biomaterials*, 2003, **24**, 2295.
41. J. Aoki, P. W. Serruys, H. van Beusekom, A. T. L. Ong, E. P. McFadden, G. Sianos, W. J. van der Giessen, E. Regar, P. J. de Feyter, H. R. Davis, S. Rowland and M. J. B. Kutryk, *J. Am. Coll. Cardiol.*, 2005, **45**, 1574.
42. R. R. Ravindranath, A. Romaschin and M. Thompson, *Clin. Biochem.*, 2016, **49**, 186.
43. J. Li, D. Li, F. Gong, S. Jiang, H. Yu and Y. An, *Biomed. Res. Int.*, 2014, **2014**, 1.

44. J. M. Lee, W. S. Choe, B.-K. Kim, W.-W. Seo, W.-H. Lim, C.-K. Kang, S. Kyeong, K. D. Eom, H.-J. Cho, Y.-C. Kim, J. Hur, H.-M. Yang, H.-J. Cho, Y.-S. Lee and H.-S. Kim, *Biomaterials*, 2012, **33**, 8917.
45. P. Benvenuto, M. A. D. Neves, C. Blaszykowski, A. Romaschin, T. Chung, S. R. Kim and M. Thompson, *Langmuir*, 2015, **31**, 5423.
46. J. Hoffmann, A. Paul, M. Harwardt, J. Groll, T. Reeswinkel, D. Klee, M. Moeller, H. Fischer, T. Walker, T. Greiner, G. Ziemer and H. P. Wendel, *J. Biomed. Mater. Res., Part A*, 2008, **84**, 614.
47. M. Thompson and D. C. Stone, *Surface-launched Acoustic Wave Sensors: Chemical Sensing and Thin Film Characterization*, Wiley-Interscience, New York, 1997.
48. X. Wang, J. S. Ellis, C.-D. Kan, R.-K. Li and M. Thompson, *Analyst*, 2008, **133**, 85.
49. S. Cheung, L. J. Fick, D. D. Belsham, D. A. Lovejoy and M. Thompson, *Analyst*, 2011, **136**, 4412.
50. S. Cheung, L. J. Fick, D. D. Belsham and M. Thompson, *Analyst*, 2011, **136**, 2786.
51. S.-H. Wu, K.-L. Lee, R.-H. Weng, Z.-X. Zheng, A. Chiou and P.-K. Wei, *PLoS One*, 2014, **9**, e89522.
52. D. Mazier, G. Schatten and W. Sale, *J. Cell Biol.*, 1975, **66**, 198.
53. *DNA Microarrays: A Practical Approach*, ed. M. Schena, Oxford University Press, Oxford, 1999.
54. M. Schena, *Protein Microarrays*, Jones and Bartlett Learning, Boston, 2005.
55. G. Rainaldi, A. Calcabrini and M. T. Santini, *J. Mater. Sci.: Mater. Med.*, 1998, **9**, 755.
56. C. Richter, M. Reinhardt, S. Giselbrecht, D. Leisen, V. Trouillet, R. Truckenmüller, A. Blau, C. Ziegler and A. Welle, *Biomed. Microdevices*, 2010, **12**, 787.
57. S. S. Rao, N. Han and J. O. Winter, *J. Biomater. Sci. Polym. Ed.*, 2011, **22**, 611.
58. K. A. Woodhouse, P. Klement, V. Chen, M. B. Gorbet, F. W. Keeley, R. Stahl, J. D. Fromstein and C. M. Bellingham, *Biomaterials*, 2004, **25**, 4543.
59. L. M. Y. Yu, N. D. Leipzig and M. S. Shoichet, *Mater. Today*, 2008, **11**, 36.
60. G. M. L. Le Saux, *Modified silicon surfaces for controlled cell interactions*, PhD thesis, University of New South Wales, 2010.
61. M. D. Pierschbacher and E. Ruoslahti, *Nature*, 1984, **309**, 30.
62. B. K. Brandley and R. L. Schnaar, *Anal. Biochem.*, 1988, **172**, 270.
63. S. P. Massia and J. A. Hubbell, *J. Cell Biol.*, 1991, **15**, 1089.
64. E. A. Cavalcanti-Adam, A. Micoulet, J. Blümmel, J. Auernheimer, H. Kessler and J. P. Spatz, *Eur. J. Cell Biol.*, 2006, **85**, 219.
65. E. A. Cavalcanti-Adam, T. Volberg, A. Micoulet, H. Kessler, B. Geiger and J. P. Spatz, *Biophys. J.*, 2007, **92**, 2964.
66. J. Huang, S. V. Gräter, F. Corbellini, S. Rinck, E. Bock, R. Kemkemer, H. Kessler, J. Ding, and J. P. Spatz, *Nano Lett.*, 2009, 9, 1111.
67. G. Le Saux, A. Magenau, T. Böcking, K. Gaus and J. J. Gooding, *PLoS One*, 2011, **6**, e21869.
68. S. G. Lévesque and M. S. Shoichet, *Biomaterials*, 2006, **27**, 6277.

69. P. Musoke-Zawedde and M. S. Shoichet, *Biomed. Mater.*, 2006, **1**, 162.
70. M. R. Hynd, J. P. Frampton, N. Dowell-Mesfin, J. N. Turner and W. Shain, *J. Neurosci. Methods*, 2007, **162**, 255.
71. J. J. Norman and T. A. Desai, *Ann. Biomed. Eng.*, 2006, **34**, 89.
72. A. A. Tseng and A. Notargiacomo, *J. Nanosci. Nanotechnol.*, 2005, **5**, 683.
73. J. C. Palmaz, A. Benson and E. A. Sprague, *J. Vasc. Interv. Radiol.*, 1999, **10**, 439.
74. J. Lu, M. P. Rao, N. MacDonald, D. Khang and T. J. Webster, *Acta Biomater.*, 2008, **4**, 192.
75. A. Ranjan and T. J. Webster, *Nanotechnology*, 2009, **20**, 305102.
76. S. Y. Hwang, K. W. Kwon, K.-J. Jang, M. C. Park, J. S. Lee and K. Y. Suh, *Anal. Chem.*, 2010, **82**, 3016.
77. C. J. Bettinger, Z. Zhang, S. Gerecht, J. T. Borenstein and R. Langer, *Adv. Mater.*, 2008, **20**, 99.
78. S. H. Hsu, C. M. Tang and C. C. Lin, *Biomaterials*, 2004, **25**, 5593.
79. M. J. Dalby, M. O. Riehle, H. Johnstone, S. Affrossman and A. S. G. Curtis, *Biomaterials*, 2002, **23**, 2945.
80. P. Kim, D. H. Kim, B. Kim, S. K. Choi, S. H. Lee, A. Khademhosseini, R. Langer and K. Y. Suh, *Nanotechnology*, 2005, **16**, 2420.
81. J. D. Zawislak, K. W. Kolasinski and B. P. Helmke, *Phys. Status Solidi A*, 2009, **206**, 1356.
82. G. R. Houtchens, M. D. Foster, T. A. Desai, E. F. Morgan and J. Y. Wong, *J. Biomech.*, 2008, **41**, 762.
83. E. K. F. Yim, R. M. Reano, S. W. Pang, A. F. Yee, C. S. Chen and K. W. Leong, *Biomaterials*, 2005, **26**, 5405.
84. L. Richert, F. Vetrone, J.-H. Yi, S. F. Zalzal, J. D. Wuest, F. Rosei and A. Nanci, *Adv. Mater.*, 2008, **20**, 1488.
85. K. T. Nguyen, K. P. Shukla, M. Moctezuma and L. Tang, *J. Nanosci. Nanotechnol.*, 2007, **7**, 2823.
86. J. Gao, L. Niklason and R. Langer, *J. Biomed. Mater. Res.*, 1998, **42**, 417.
87. A. Ranjan and T. J. Webster, *Nanotechnology*, 2009, **20**, 305102.
88. S. Y. Hwang, K. W. Kwon, K. J. Jang, M. C. Park, J. S. Lee and K. Y. Suh, *Anal. Chem.*, 2010, **82**, 3016.
89. S. W. Lee, S. Y. Kim, I. C. Rhyu, W. Y. Chung, R. Leesungbok and K. W. Lee, *Clin. Oral Implants Res.*, 2009, **20**, 56.
90. S. A. Biela, Y. Su, J. P. Spatz and R. Kemkemer, *Acta Biomater.*, 2009, **5**, 2460.
91. C. C. Berry, M. J. Dalby, D. McCloy and S. Affrossman, *Biomaterials*, 2005, **26**, 4985.
92. M. J. Dalby, D. Giannaras, M. O. Riehle, N. Gadegaard, S. Affrossman and A. S. G. Curtis, *Biomaterials*, 2004, **25**, 77.
93. M. J. Dalby, N. Gadegaard, M. O. Riehle, C. D. W. Wilkinson and A. S. G. Curtis, *Int. J. Biochem. Cell Biol.*, 2004, **36**, 2005.
94. A. M. Green, J. A. Jansen, J. P. C. M. van der Waerden and A. F. von Recum, *J. Biomed. Mater. Res.*, 1994, **28**, 647.
95. K. R. Milner and C. A. Siedlecki, *J. Biomed. Mater. Res., Part A*, 2007, **82A**, 80.

96. J. A. Hunt, R. L. Williams, S. M. Tavakoli and S. T. Riches, *J. Mater. Sci.: Mater. Med.*, 1995, **6**, 813.
97. M. J. Dalby, C. C. Berry, M. O. Riehle, D. S. Sutherland, H. Agheli and A. S. G. Curtis, *Exp. Cell Res.*, 2004, **295**, 387.
98. C.-H. Choi, S. H. Hagvall, B. M. Wu, J. C. Y. Dunn, R. E. Beygui and C.-J. Kim, *Biomaterials*, 2007, **28**, 1672.
99. M. Thompson, L.-E. Cheran and S. Sadeghi, *Sensor Technology in Neuroscience*, Royal Society of Chemistry, Cambridge, UK, 2013, ch. 3.
100. A. Cohen, J. Shappir, S. Yitzchaik and M. E. Spira, *Biosens. Bioelectron.*, 2006, **22**, 656.
101. P. Fromherz, *Chem. Phys. Chem.*, 2002, **3**, 276.
102. F. Heer, S. Hafizovic, T. Ugniwenko, U. Frey, W. Franks, E. Perriard, J.-C. Perriard, A. Blau, C. Ziegler and A. Hierlemann, *Biosens. Bioelectron.*, 2006, **22**, 2546.
103. P. Fromherz and M. Voelker, *Science*, 2009, **323**, 1429.
104. M. Voelker and P. Fromherz, *Small*, 2005, **1**, 206.
105. R. J. Pelham, Jr. and Y. Wang, *Proc. Natl. Acad. Sci. U. S. A.*, 1997, **94**, 13661.
106. C.-M. Lo, H.-B. Wang, M. Dembo and Y.-L. Wang, *Biophys. J.*, 2000, **79**, 144.
107. A. P. Balgude, X. Yu, A. Szymanski and R. V. Bellamkonda, *Biomaterials*, 2001, **22**, 1077.
108. A. Engler, L. Bacakova, C. Newman, A. Hategan, M. Griffin and D. Discher, *Biophys. J.*, 2004, **86**, 617.
109. P. C. Georges, W. J. Miller, D. F. Meaney, E. S. Sawyer and P. A. Janmey, *Biophys. J.*, 2006, **90**, 3012.
110. J. Funkhouser, *Curr. Drug Discovery*, 2002, **2**, 17.
111. N. Crawley and M. Thompson, *Anal. Chem.*, 2014, **86**, 130.
112. E. Haglund, M.-M. Seale-Goldsmith and J. F. Leary, *Ann. Biomed. Eng.*, 2009, **37**, 2048.
113. S. M. Janib, A. S. Moses and J. A. MacKay, *Adv. Drug Delivery Rev.*, 2010, **62**, 1052.
114. W. Cai and X. Chen, *J. Nucl. Med.*, 2008, **49**(Suppl2), 113S.
115. V. P. Torchilin, *Nat. Rev. Drug Discovery*, 2005, **4**, 145.
116. D. H. Levine, P. P. Ghoroghchian, J. Freudenberg, G. Zhang, M. J. Therien, M. I. Greene, D. A. Hammer and R. Murali, *Methods*, 2008, **46**, 25.
117. G. J. Strijkers, E. Kluza, G. A. F. Van Tilborg, D. W. J. van der Schaft, A. W. Griffioen, W. J. Mulder and K. Nicolay, *Angiogenesis*, 2010, **13**, 161.
118. A. M. Ponce, Z. Vujaskovic, F. Yuan, D. Needham and M. W. Dewhirst, *Int. J. Hyperthermia*, 2006, **22**, 205.
119. A. Gaitanis and S. Staal, *Methods Mol. Biol.*, 2010, **624**, 385.
120. B. M. Discher, Y.-Y. Won, D. S. Ege, J. C.-M. Lee, F. S. Bates, D. E. Discher and D. A. Hammer, *Science*, 1999, **284**, 1143.
121. K. K. Upadhyay, J.-F. Le Meins, A. Misra, P. Voisin, V. Bouchaud, E. Ibarboure, C. Schatz and S. Lecommandoux, *Biomacromolecules*, 2009, **10**, 2802.

122. J. Liu, H. Lee and C. Allen, *Curr. Pharm. Des.*, 2006, **12**, 4685.
123. R. A. Schwendener, R. Wüthrich, S. Duewell, G. Westera and G. K. von Schulthess, *Int. J. Pharm.*, 1989, **49**, 249.
124. A. Surendiran, S. Sandhiya, S. C. Pradhan and C. Adithan, *Indian J. Med. Res.*, 2009, **130**, 689.
125. C. C. Lee, J. A. MacKay, J. M. J. Fréchet and F. C. Szoka, *Nat. Biotechnol.*, 2005, **23**, 1517.
126. L. Dennany, P. Sherrell, J. Chen, P. C. Innis, G. G. Wallace and A. I. Minett, *Phys. Chem. Chem. Phys.*, 2010, **12**, 4135.
127. N. W. S. Kam, M. O'Connell, J. A. Wisdom and H. Dai, *Proc. Natl. Acad. Sci. U. S. A.*, 2005, **102**, 11600.
128. N. W. S. Kam, T. C. Jessop, P. A. Wender and H. Dai, *J. Am. Chem. Soc.*, 2004, **126**, 6850.
129. J. Park, E. Lee, N.-M. Hwang, M. Kang, S. C. Kim, Y. Hwang, J.-G. Park, H.-J. Noh, J. Y. Kim, J.-H. Park and T. Hyeon, *Angew. Chem., Int. Ed.*, 2005, **44**, 2872.
130. J. Xie, J. Huang, X. Li, S. Sun and X. Chen, *Curr. Med. Chem.*, 2009, **16**, 1278.
131. Y.-W. Jun, J.-W. Seo and J. Cheon, *Acc. Chem. Res.*, 2008, **41**, 179.
132. G. M. Lanza, P. M. Winter, S. D. Caruthers, A. M. Morawski, A. H. Schmieder, K. C. Crowder and S. A. Wickline, *J. Nucl. Cardiol.*, 2004, **11**, 733.
133. S. Kim, Y. T. Lim, E. G. Soltesz, A. M. De Grand, J. Lee, A. Nakayama, J. A. Parker, T. Mihaljevic, R. G. Laurence, D. M. Dor, L. H. Cohn, M. G. Bawendi and J. V. Frangioni, *Nat. Biotechnol.*, 2004, **22**, 93.
134. R. E. Bailey, A. M. Smith and S. Nie, *Phys. E*, 2004, **25**, 1.
135. V. Bagalkot, L. Zhang, E. Levy-Nissenbaum, S. Jon, P. W. Kantoff, R. Langer and O. C. Farokhzad, *Nano Lett.*, 2007, **7**, 3065.
136. M.-C. Daniel and D. Astruc, *Chem. Rev.*, 2004, **104**, 293.
137. V. Dixit, J. Van den Bossche, D. M. Sherman, D. H. Thompson and R. P. Andres, *Bioconjugate Chem.*, 2006, **17**, 603.
138. N. L. Rosi, D. A. Giljohann, C. S. Thaxton, A. K. R. Lytton-Jean, M. S. Han and C. A. Mirkin, *Science*, 2006, **312**, 1027.
139. J. A. Schwartz, A. M. Shetty, R. E. Price, R. J. Stafford, J. C. Wang, R. K. Uthamanthil, K. Pham, R. J. McNichols, C. L. Coleman and J. D. Payne, *Cancer Res.*, 2009, **69**, 1659.
140. H. Ow, D. R. Larson, M. Srivastava, B. A. Baird, W. W. Webb and U. Wiesner, *Nano Lett.*, 2005, **5**, 113.
141. I. I. Slowing, J. L. Vivero-Escoto, C.-W. Wu and V. S.-Y. Lin, *Adv. Drug Delivery Rev.*, 2008, **60**, 1278.
142. J. L. Vivero-Escoto, I. I. Slowing, C.-W. Wu and V. S.-Y. Lin, *J. Am. Chem. Soc.*, 2009, **131**, 3462.
143. A. Nel, T. Xia, L. Mädler and N. Li, *Science*, 2006, **311**, 622.
144. G. Oberdörster, E. Oberdörster and J. Oberdörster, *Environ. Health Perspect.*, 2005, **113**, 823.

CHAPTER 4

Biological Consequences of the Blood–Surface Interaction

4.1 Biological Response to Biomaterials – General Concepts

A biomaterial is commonly defined as "a non-viable material used in a medical device, intended to interact with biological systems".[1] While the first uses of biomaterials date back to about 32 000 years ago, they were destined to fail since important concepts related to infection and the biological reaction to materials were ignored. The modern era of the biomaterials started in the 1940's when they were implemented for ophthalmological uses. Since then, biomaterials have spanned a plethora of important applications. The design and engineering of biomaterials resides at the interface of material science, chemistry, mechanics, surface science, bioengineering, biology, and medicine. Indeed, advances in materials science and in the understanding of biology and medicine have brought an explosion in the use of biomaterials, as well as dramatic improvements in their performance. Biomaterials are commonly used in tissue-contacting medical devices over a wide range of applications, among them cardiovascular, orthopedic, dental, and ophthalmological prostheses, stents and vascular grafts, as well as extracorporeal circulation devices (*i.e.* hemodialysis, cardiopulmonary bypass, *etc.*). The majority of these biomaterials are based on advanced technologies and designed to serve their user for prolonged durations. Nevertheless, over a period of time, these devices develop problems arising from biomaterial failure, leading to complications such as health issues to the user or even death. Hence, an

RSC Detection Science Series No. 8
Biological Fluid–Surface Interactions in Detection and Medical Devices
By Michael Thompson, Christophe Blaszykowski, Sonia Sheikh, Cesar Rodriguez-Emmenegger and Andres de los Santos Pereira

Published by the Royal Society of Chemistry, www.rsc.org

understanding of the host responses will enable a more rational design of biomaterials possessing improved performance.

The implantation of any biomaterial invariably results in injury, damaging tissues and the vasculature. This, in turn, initiates a series of mechanisms that ultimately control the fate of the biomaterial. These mechanisms aim at maintaining *homeostasis* (the stabilization and restoration of the internal environment). The injury site is first rapidly flooded by blood. Next, as early as seconds to minutes from the contact with the biomaterial, protein fouling occurs, coating the surface of the just-implanted material with a monolayer of proteins in a range of conformations, from native to completely denatured. The conformation of the adsorbed proteins has profound implications on the access to protein binding sites that support subsequent reactions. Protein fouling commonly occurs well before cells arrive to the implanted material. Importantly, it is this protein–biomaterial interfacial layer which plays a central role in the subsequent reactions of the body to the implant. Cell processes at the surface may lead to desirable reactions (cell adhesion, colonization, spreading, differentiation, formation of tissue), as well as undesirable effects such as inflammation, initiation of the coagulation cascade, complement activation, foreign body reaction, and the formation of a fibrous capsule. These processes and their interplay will be discussed in the subsequent sections of the present chapter.

Moreover, cells and tissues respond to mechanical forces such as shear. The shape, roughness, and stress of the material can lead to vastly different healing processes at the site of implantation.

The same processes involved in maintaining the homeostasis usually lead to adverse effects on biomaterials after a given time. *Rejection* is only observed on xeno- and allografts (*i.e.* implants of biological tissues from a different species or a different individual from the same species). Here, the host immune response plays a central role. On the other hand, on artificial biomaterials, a different set of complications is observed (Table 4.1). The effects on the biomaterials and the host tissue are important in mediating complications and device failure. Typical examples of these complications include: chronic inflammation, thrombotic occlusion of medium-sized

Table 4.1 Modes of device–host interaction.

Material on the implant itself	Local on the implant site	Systemic on the host
Fatigue	Inflammation	Material hypersensitivity
Abrasion	Leaching of toxic substances	Thrombosis and embolism (ischemia/stroke)
Chemical/enzymatic degradation	Particle release	Toxicity
Corrosion	Thrombogenicity	Platelet and leukocyte depletion
Calcification	Bacterial infection	
	Tumorigenesis	

grafts, obstruction and restenosis in stents, embolic complications with catheters, prosthetic valves, and artificial hearts, thrombotic complications with cardiopulmonary bypass or angioplasty, tumor formation, and even infection. Furthermore, the environment and function of the biomaterial can also lead to degradation such as the wear effect on orthopedic prostheses. Even if the risk of complications may appear low (*e.g.* 2–10% risk of thrombotic problems in grafts), the occurrence of the complication may have fatal consequences on the host, leading to high costs. While several strategies have been introduced; to date, a perfectly biocompatible biomaterial does not exist. Accordingly, several pharmacological approaches are used in conjunction with the biomaterials to reduce the risks of complications. Unfortunately, most pharmacological strategies only address one of the players in the host responses, leading to only partial clinical benefits.

4.1.1 Protein Fouling: the Initial Step

The contact of any surface (other than healthy vascular endothelium) with blood leads to the rapid adsorption of proteins. This highly complex and not well understood process is ubiquitous to any (bio)material.[2–7] Proteins that are present at higher concentration adsorb first and are subsequently replaced by others of lower concentrations but higher affinity for the surface in a process termed "Vroman effect".[8–10] Additionally, the adsorption of some proteins from blood can be mediated by the adsorption of others, which are biological partners.[11,12] Some examples include the co-adsorption of high molecular weight kininogen or histidine-rich glycoprotein with fibrinogen (Fbg).[11,12]

The adsorbed proteins are responsible for triggering the thrombogenic, immune, and foreign body reactions, while also taking part in the mechanisms of inflammation. Furthermore, they serve as a *conditioning film* for bacterial adhesion and can support the formation of continuous biofilms.

A generally accepted idea is that the complete prevention of protein fouling will prevent the adverse events that follow it, thus endowing the material with hemocompatibility. This has resulted in enormous efforts aimed at developing perfectly non-fouling surface coatings (refer to Chapter 5). Systems based on polymer brushes have resulted in a dramatic reduction of fouling to values lower than 1.0 pg mm^{-2}.[13] However, this strategy requires the permanent prevention of fouling, which so far has been elusive. An alternative strategy is focused in the prevention of the irreversible adsorption and conformational change of only those proteins capable of triggering the detrimental host reactions.[6,14–18] An example of such a strategy is illustrated with Fbg, which has been identified as one of the most important types of adsorbed proteins that induce platelets adhesion.[19] Fbg undergoes adsorption-induced conformational changes by loss of the α-helix structure. This leads to the exposure of binding sites for platelet receptors (*e.g.* GPIIb/IIIa) that are otherwise not accessible to platelet receptors in the native/solution state.[6,16,19] This implies that the condition of absolute resistance to fouling might be too restrictive and, hence, a focus in the development of surfaces which maximally retain the native conformation of proteins might be sufficient to prevent the deleterious effects. Such surfaces have been termed antimetamorphic.[17]

Other examples of proteins for which adsorption leads to conformational changes and their concomitant effects include: fibronectin, von Willebrand factor (vWF), and vitronectin, which mediate platelet adhesion[15,20,21] and the activation of the blood zymogen factor XII (Hageman factor), responsible for intrinsic blood coagulation.[22]

Albumin, the most abundant protein in plasma, is usually utilized in biology to "passivate" surfaces. It rapidly adsorbs on the surface and saturates it, forming a monolayer. The rationale of that passivation is that all sites on the surface for other proteins to adsorb are blocked. However, recent findings have shown that adsorption-induced unfolding of albumin can lead to conformational changes that support specific platelet adhesion and activation *via* GPIIb/IIIa receptors. It is hypothesized that the mechanism for this response is the creation of RGD-like motifs caused by the adsorption-induced separation of salt-bridged amino acid residues. Hence, caution must be taken when utilizing this passivation mechanism to reduce thrombosis.[6,16]

4.1.2 Blood Coagulation and Blood–Material Interactions

The outcome of a device in a tissue-contacting application is firstly mediated by its interactions with blood. The following sections describe the main mechanisms at play in the different steps of the biological response. The coagulatory aspect of the response is mediated mainly by platelets and the coagulation cascade, while the inflammatory features include the response of leukocytes and the complement system. The cellular components of blood are explained in detail in the previous chapter. The following discussion will put them in the context of their interactions with the rest of the systems. It is important to note that the responses of the different systems are highly interconnected. Thus, they are impossible to separate *in vivo*.

4.1.2.1 Introduction to Hemostasis

Blood is a vital part of the human physiology, which transports nutrients and oxygen to sustain living cells, while simultaneously facilitating the removal of carbon dioxide and other waste products from the body. Blood cells and enzymatic systems are also involved in the body's defense to keep foreign materials and organisms from causing harm. In order to maintain these functions, blood must be kept flowing inside the vascular system. The acute response mechanisms that are initiated to stop bleeding when blood vessels are injured form part of the concept of hemostasis (derived from Greek meaning "blood-stop"). Since the blood is contained in a pressurized system, the temporary vessel repair mechanisms of hemostasis have to act fast and efficiently in order to prevent excessive bleeding. On the other hand, such a potent system also requires regulatory mechanisms to prevent highly undesirable and potentially deadly responses. The mechanism for stopping bleeding needs to be activated under the right set of circumstances. If the control mechanisms fail or if the hemostatic mechanisms respond to triggers that are not specific to a vessel injury, the potent hemostatic system can also lead to unwanted thrombosis with subsequent vessel occlusion or embolism, ultimately causing substantial physiological damage. In summary, hemostasis is a complex system tasked with keeping the delicate and critical balance between bleeding and thrombosis (Figure 4.1). In healthy individuals, the system is ready to form a hemostatic plug to prevent severe bleeding in the event of injury of a vessel. After damage of the vasculature has occurred, a set of reactions and events is set in motion to prevent unrestricted bleeding. Platelets (see below) rapidly adhere to the damaged tissue and are activated, which facilitates the recruitment of more platelets to the damaged vessel wall, forming a plug. The subsequent mechanism, coagulation, is partially triggered by the release of compounds by platelets. The coagulation cascade is a complex enzymatic pathway that involves the conversion of several zymogens into enzymes and various amplification events. The result of this cascade is the conversion of fibrinogen into an insoluble fibrin network that will cross-link the platelet plug and become the first scaffold for tissue repair (see inflammation section). Unrestricted formation of the fibrin network could either occlude vessels or lead to thrombi being released or formed downstream with the concomitant risk to life.

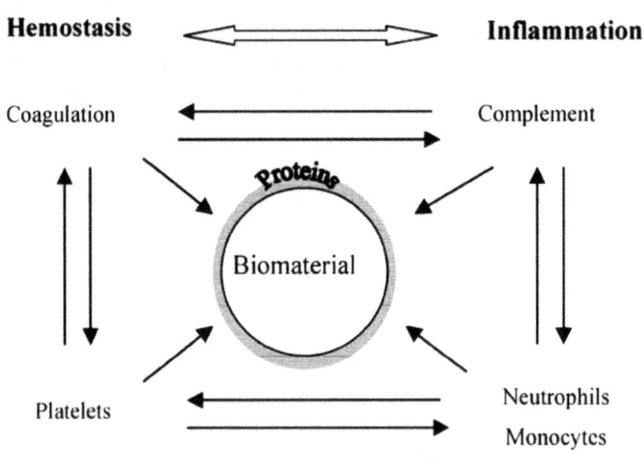

Figure 4.1 Interactions between blood and materials, highlighting the interplay between the main factors affecting the response to biomaterials. Reprinted from M. B. Gorbet and M. V. Sefton, Biomaterial-associated thrombosis: roles of coagulation factors, complement, platelets and leukocytes, *Biomaterials*, **24**, 5681–5703, Copyright 2004 with permission from Elsevier.

Thus, upon recovery of the injured site, the fibrinolytic system operates a series of mechanisms which down-regulate the enzymatic cascade and degrade the fibrin network. Through these series of reactions, the fibrin network becomes digested into soluble fragments that are transported away by the blood flow.

4.1.2.2 Platelet–Material Interactions

The implantation of a biomaterial ultimately leads to damage in the vasculature and the endothelium. In such cases, the damaged vessels need to be rapidly repaired to prevent excessive bleeding. The first mechanism is the formation of a plug which blocks the flow out of the vessel. This plug is formed by the aggregation of platelets, the second most abundant cellular component of blood. Platelets are disk-shaped non-nucleated cells having a diameter of 3–4 μm. They do not replicate but are derived from megakaryocytes in the bone marrow. In addition to arresting bleeding, platelets also play other key roles in hemostasis. They release bioactive compounds which catalyze coagulation reactions leading to the formation of fibrin. The plug formed also releases attractants for leukocytes (see inflammation) and provides a matrix for the healing process to begin. The formation of the plug occurs in the following steps: (i) activation of platelets, (ii) adhesion of platelets, and (iii) recruitment and aggregation of more platelets.

Platelets respond to minimal stimulation and become activated by extracellular stimuli acting on their surface. These include the coupling of agonists to specific receptors on the membrane. Plasma proteins, such as thrombin, fibrinogen, vascular wall components, molecules secreted by inflammatory cells, as well as other platelets, are well known to be strong platelet activators. Furthermore, contact with a thrombogenic surface such as injured endothelium, subendothelium, or most artificial surfaces initiates activation. Even shear forces induced by blood circulation may activate platelets.[23] These stimuli launch a set of reactions leading to profound changes in the platelets. Upon activation, the cytoskeleton, which maintains the circumferential shape in resting state, becomes a strong actuator which changes the shape and supports the secretion of bioactive compounds stored in granules. Platelets contain three types of granules that fuse with the cellular membrane and release their content into the surrounding media. The granules are categorized into: (i) α-granules, (ii) dense granules, and (iii) lysosomes. The α-granules contain a large variety of compounds actively involved in hemostasis (*e.g.* platelet activation and adhesion molecules, plasma coagulation factors, and fibrinolysis proteins). They also contain P-selectin which supports the adhesion of neutrophils and monocytes, which are central in the inflammation process. The dense granules mainly contain Ca^{2+} ions and smaller signal molecules, such as adenosine 5′-diphosphate (ADP), adenosine 5′-triphosphate (ATP), and serotonin. Lysosomes contain acid hydrolase, cathepsin, and lysosomal membrane proteins. A rapid and effective response from the platelets is vital.

The membrane of activated platelets also provides a phospholipid surface which accelerates some important reactions in the coagulation cascade. Furthermore, platelet activation leads to the formation of a spongy, canal-like network which provides the reactive surface to which plasma factors can bind.

After activation, primary adhesion follows. Primary adhesion occurs when an initial layer of activated platelets adheres to the injured endothelium. This is mediated by the synergistic function of several receptors on the platelet surface. The adhesion receptors GPIb on the surface of platelets bind collagen (exposed in the injured endothelium) using vWF as a cofactor. Platelets also adhere to biomaterials using the GPIIb/IIIa membrane receptors which are capable of binding plasma proteins such as fibrinogen, vWF, fibrinonectin, and vitronectin. Resting platelets (non-activated) do not bind to these proteins. The binding is mediated by activation of the platelets and by conformational changes in the proteins after adsorption to the biomaterial surfaces.[6,16,19,24] Once a primary layer of adhesive platelets has covered the exposed subendothelial matrix, subsequent adhesion will continue but in the form of aggregation, in which platelets recruited from flowing blood bind to adhered platelets at the wound site. The mechanism of platelet aggregation is mainly attributed to integrin GPIIb/IIIa and involves two receptors on different platelets that bind to the same fibrinogen molecule. In order for the recruitment of platelets to continue during the aggregation phase, it is necessary that the activated state of the adhered and aggregated platelets is propagated to newly arriving platelets. The release of ADP, calcium, prothrombin, thromboxane B2, and other bioactive compounds and coagulation factors contained in the granules activates other platelets in the vicinity of the forming plug.

When platelets aggregate, their coagulation activity is initiated, including the expression of negatively charged phospholipids, promoting the activation of coagulation factors. Additionally, activated platelets release platelet microparticles. These particles are formed by exocytotic budding and have procoagulant activity. Platelet activation also promotes the conversion of prothrombin into thrombin, a major step in the coagulation cascade (discussed below). Thrombin, in turn, activates more platelets and promotes the polymerization of fibrinogen into fibrin, which adheres to the surface of the platelet thrombus.

4.1.2.3 The Coagulation Cascade and the Fibrinolytic System

After platelet aggregation occurs (primary hemostasis), blood coagulation follows (secondary hemostasis). Blood coagulation involves a series of proteolytic reactions which culminate with the formation of a fibrin clot. When blood coagulation is initiated, the highly specific proteolytic activation of circulating zymogens leads to their transformation into active enzymes. Each enzyme subsequently activates another enzyme starting a reaction cascade. This cascade of reactions is conceptually described as having two pathways (the *intrinsic*/contact activation and the *extrinsic* activation), which

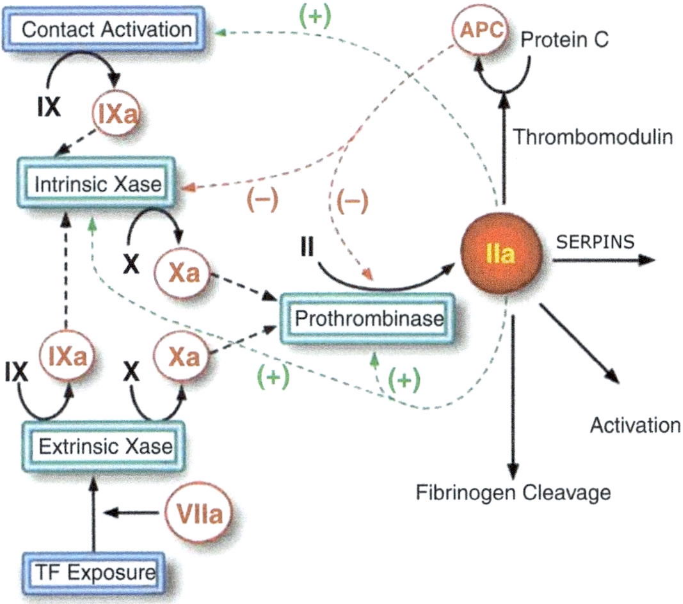

Figure 4.2 Conceptual overview of the biochemical pathways involved in the coagulation cascade, showing the central role of thrombin (factor IIa). Reproduced from ref. 26 with permission from Wolters Kluwer Health.

merge into a common pathway leading to the generation of active thrombin (Figure 4.2). This step consists in the conversion of prothrombin (Factor II) to thrombin (Factor IIa). The thrombin formed as a result of these reactions is the key effector product of coagulation, as it is the proteinase responsible of the cleavage of fibrinogen (Factor I) to form fibrin monomer, which self-polymerizes to form the fibrin network. Moreover, it activates platelets, other blood cells, and the vascular endothelium by cleavage of the protease-activated receptors. It is worth noting that thrombin is a self-limiting product. In the initial stages of the coagulation response, it functions as a positive feedback mechanism by activating factors (XI, XIII) and cofactors (VIII and V). This positive feedback has profound effects in the amplification of coagulation by both pathways. At later stages of the coagulation response, the burst in thrombin generation is counterbalanced by a negative feedback mechanism. Thrombin binds thrombomodulin on endothelial cells forming a complex which activates protein C. Activated protein C proteolytically inactivates cofactors V and VIII, decreasing the rate of production of thrombin. Subsequently, thrombin is eliminated by circulating serine protease inhibitors (discussed below). In summary, the local concentration of thrombin exhibits an initial fast increase, leading to a maximum, and a subsequent decrease (Figure 4.3). This concentration profile is in line with the fibrin concentration which increases until it levels off when the concentration of thrombin is negligible.

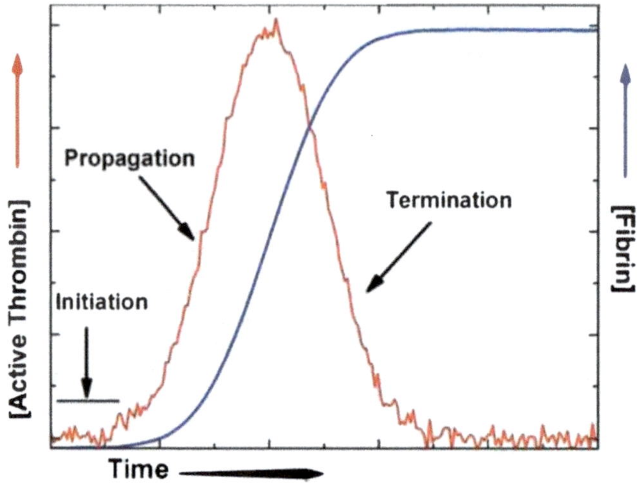

Figure 4.3 Concentration profiles of thrombin and fibrin as they vary in time, showing amplification during the initial low concentration phase followed by inactivation at a later stage.
Reproduced from ref. 26 with permission from Wolters Kluwer Health.

Thrombin also affects the anticoagulant properties of the blood vessel walls. Normally, endothelial cells display a thromboresistant surface, unreceptive to coagulation factors, weakly associating to leukocytes and platelets. These transient interactions allow the biological species to "slide over" the surface of the blood vessel walls, and help to localize leukocytes and platelets to the areas where they are needed. However, thrombin activates endothelial cells *via* protease-activated receptors to effect biological responses, among them contraction, relaxation of interendothelial junctions leading to increased vascular permeability, reduction of thrombomodulin expression (a thrombin inhibitor), and secretion of vWF. The conjunction of all these changes results in the loss of thromboresistance and increase the potential of thrombus formation. Besides coagulation, thrombin plays an important role in inflammation increasing the leukocyte extravasation and transmigration of cytokines.

Most coagulation reactions have four components: the enzyme, the cofactor, the surface, and the Ca^{2+} ions (Figure 4.4). Generally, the enzyme is bound to the cofactor that is also bound to the surface *via* Ca^{2+} ions. The main partners (factors) are listed in Table 4.2. The assembly of these four components into the quaternary complex results in a dramatic increase in enzymatic activity. Before coagulation begins, enzymes are present as zymogens (*i.e.*, inactive precursors) which, after proteolytic cleavage, generate serine proteinases. These enzymes are highly conserved genetically and consist of a serine protease domain linked together with other protein domains to endow different capabilities and functions. Each domain is genetically autonomous.[25,26] Important factors such as II, VII, IX, as well as

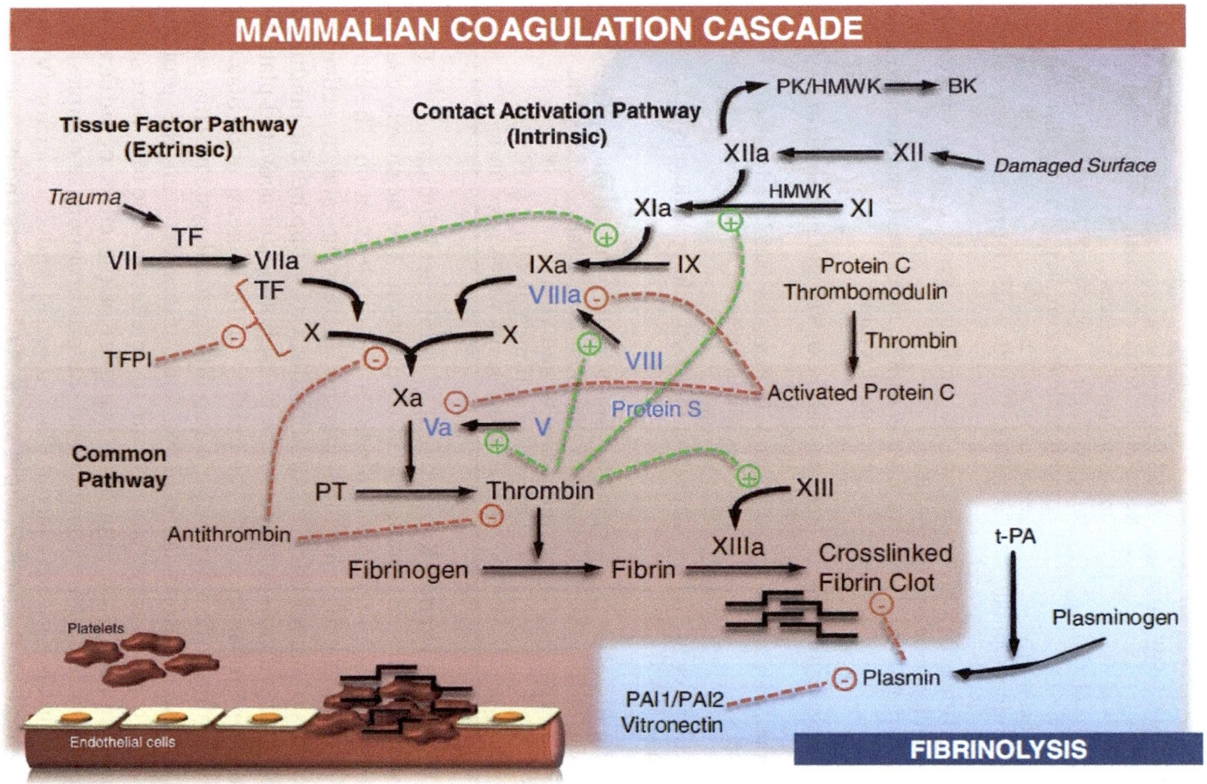

Figure 4.4 Step-by-step scheme of the coagulation cascade showing the individual positive (green) and negative (red) regulation of the reactions.
Reproduced from ref. 26 with permission from Wolters Kluwer Health.

Table 4.2 Main factors involved in the coagulation cascade.

Number/name	Function
Factor I (fibrinogen)	Polymerizes to form fibrin clot
Factor II (prothrombin, PT)	Its active form IIa (thrombin) activates I, V, VII, VIII, XI, XIII, protein C, platelets
Factor III (tissue factor, TF)	Cofactor of VIIa
Factor IV (Ca^{2+})	Required for some coagulation factors to adsorb to phospholipid surfaces
Factor V (proaccelerin)	Its active form is cofactor of Xa to form the prothrombinase complex (generate thrombin)
Factor VII (proconvertin)	Activates IX and X
Factor VIII (antihemophilic factor A)	Cofactor of IX with which forms the tenase complex
Factor IX (Christmas factor or antihemophilic factor B)	Activates X and forms tenase complex with factor VIII
Factor X (Stuart–Prower factor)	Xa together with Va forms prothrombinase complex to generate thrombin.
Factor XI	Activates IX
Factor XII (Hageman factor)	Activates factor XI, VII, and prekallikrein
Factor XIII	Crosslinks fibrin
von Willebrand factor (vWF)	Binds to VIII, mediates platelet adhesion
Prekallikrein	Activates XII and cleaves high molecular weight kininogen
High molecular weight kininogen (HMWK) (Fitzgerald factor)	Activation of XII, XI, and prekallikrein
Antithrombin III	Inhibits thrombin (IIa), Xa, and other proteases
Heparin cofactor II	Inhibits thrombin (IIa), cofactor for heparin
Protein C	Inactivates Va and VIIIa
Protein S	Cofactor of activated protein C (APC)
Plasminogen	Converts to plasmin which lyses fibrin

protein C possess vitamin-K dependent glutamic acid domains.[27,28] These domains play a very important role in coagulation. The negatively charged carboxylic group of glutamic acid binds Ca^{2+}, a divalent cation. The binding of Ca^{2+} results in conformational changes, increasing catalytic activity by several orders of magnitude.[27,29] Additionally, Ca^{2+} supports the binding to negatively charged surfaces such as the phospholipids of activated platelets and platelet microparticles.[26,30] The binding of the protease and cofactor to a surface confines the reaction to a two-dimensional system. This greatly accelerates the reactions, as collision encounters between the protease complex and the non-activated factor are more effective due to a more stable activated complex.[28] Two important examples of the effect of surfaces in coagulation are the tenase complex and the prothrombinase complex. The former comprises activated factor IXa (protease) and its cofactor (VIIIa), which assemble on the phospholipid surfaces of platelets and activate factor X, while the latter consists of factor Xa (protease) and its cofactor (Va) bound to phospholipids. In both cases the binding is mediated by Ca^{2+}. The conversion of prothrombin into thrombin proceeds 1000 times faster when on surfaces.

Cofactors also play an important role in the coagulation cascade. Their binding introduces changes in the catalytic sites, reducing the free energy of the Michaelis complex. This allosteric modulation is an object of current study.

The interplay of different factors leading to the self-limited generation of thrombin and concomitant fibrin network is tightly interconnected and its conceptualization has evolved in the last 50 years. The mechanisms behind the coagulation cascade were initially conceived as being composed of a sequential series of proteolytic reactions including two fully independent pathways that merge into the common pathway.[26,30,31] It was suggested that clotting was initiated by either contact activation *via* the intrinsic pathway or by tissue factor (Factor III) *via* the extrinsic pathway. The names of the pathways have a historical perspective which derives from the way it was studied *in vitro*. In the extrinsic pathway, tissue factor (TF) was added to blood. As TF was not a component of blood, the pathway was termed extrinsic. In the intrinsic pathway, clotting was studied upon contact with a surface. In the latter pathway, all the components were contained in blood. While these simplifications helped in the development and understanding of the biology involved, they failed to describe *in vivo* processes. To date, the TF and factor VIIa are considered the major initiators of *in vivo* coagulation. On the other hand, the contact activation is thought to be linked to the inflammatory response.

In vivo blood coagulation is initiated when blood comes into contact with the subendothelium following vascular injury. The injury makes the TF accessible to factor VII, which flows in blood, and form a potent protease complex. The TF is a transmembrane glycoprotein with two fibronectin type III domains. It is expressed in most cells except for endothelial ones. The binding of TF to factor VII stops immediately without active factor VIIa to cleave factor IX. However, very low concentrations of VIIa are normally present in blood. Factor VIIa binds TF to form TF-VIIa which activates the surrounding TF-VII. Additionally, TF-VIIa activates factor IX and X. In the absence of its activated cofactor, Xa can only produce trace amounts of thrombin by the proteolytic process of prothrombin. This thrombin cleaves factor XI to generate XIa, which, in turn, cleaves more IX to form IXa. Although the amount of thrombin generated in the "initiation" step is not enough for fibrin polymerization, it is sufficient to proteolytically activate profactors V and VIII to generate Va and VIIIa, which are cofactors of Xa and IXa, respectively. These complexes amplify the generation of thrombin in the "propagation" stage, leading to a burst of thrombin activity. These reactions occur in the surface of platelets and are Ca^{2+} dependent. As discussed above, the factors form the tenase complex (IXa-VIIIa) and the prothrombinase complex (Xa-Va), which are then bound to the membrane of platelets anchored to the subendothelial matrix by collagen and vWF.

This increasing amount of thrombin makes it now possible to cleave enough amount of fibrinogen to generate small fibrinopeptides, which polymerize to form fibrin. The formed fibrin is crosslinked by thrombin-activated factor XIII. Factor XIII is a transglutaminase which circulates in blood plasma and catalyzes the formation of amide crosslinks between adjacent fibrin molecules to form a stable clot.

The intrinsic pathway is composed of three proteins: factor XII, prekallikrein and high molecular weight kininogen (HMWK). Remarkably, all steps involved take place in the absence of Ca^{2+}. This pathway is initiated when factor XII is exposed to a surface that activates it to XIIa. This activation has been associated with negatively charged, hydrophilic surfaces.[26,30] Factor XIIa converts XI to XIa, which, in turn, converts IX into IXa. Factor XII also activates prekallikrein to generate kallikrein, which then activates more factor XII and liberates kinins from the HMWK. Kinins such as bradykinin bind to various cells leading to the inflammatory response.

Controversy exists on the relevance of the intrinsic pathway in the coagulation *in vivo*. Under normal physiological conditions, the role of factor XII is questionable, as the rate of activation of factor X *via* the extrinsic pathway is more rapid. However, the presence of a biomaterial may activate enough factor XII upon adsorption.[32] The need for a negatively charged, hydrophilic surface for the activation of factor XII has also been recently challenged.[22,33] Recent findings suggest that a hydrophilic surface can activate the aforementioned factor regardless of the charge.

In particular, the coagulation cascade exists to rapidly stop bleeding. Nonetheless, such an effective system with amplification steps can also dangerously lead to uncontrolled thrombus formation. Hence, the effectiveness of the coagulation mechanism has to be matched by antagonistic processes to prevent uncontrolled thrombus formation and its potential deadly consequences. As discussed previously, thrombin formation is also self-limiting by proteolytical activation of the anticoagulation pathway. Thrombin forms a complex with the endothelial thrombomodulin receptor (TM) which allosterically alters its substrate specificity. The procoagulant substrates (factors V, VIII, and fibrinogen, among others) are no longer proteolyzed to their active form. The thrombin-TM complex, however, is able to activate protein C (forming activated protein C, or APC), which associates with protein S cofactor. This resulting complex degrades factors Va and VIIIa. The activity of thrombin-TM-APC is increased by 20 fold if protein C is bound to the endothelial protein C receptor. Another example of a naturally occurring inhibitor is the tissue factor pathway inhibitor (TFPI), which, upon association with factor Xa, inhibits the complex TF-VIIa. The TFPI is secreted by platelets upon activation by thrombin. Another source of TFPI is the vascular lumen. Antithrombin III is a very effective inhibitor of thrombin cofactors IXa and Xa. Antithrombin III neutralizes these enzymes by blocking their active sites. In the presence of the naturally occurring polysaccharide heparin or heparin sulfate, the activity of antithrombin III is boosted 2000 times.[26] In addition to natural inhibition, dilution and flow play an important role. The blood flow reduces the concentration of precursors and removes activated materials by dilution into a large volume, which is subsequently removed from circulation by the liver. Some reactions only occur at an appreciable rate when catalyzed by a surface, such as the activation of factor X, and reactions that are accelerated by the phospholipid surface of platelets or platelet microparticles.

While formation of the fibrin clot is essential to hemostasis, regulated fibrinolysis is necessary to limit unwanted thrombus formation, to allow blood flow for the healing of the wound to proceed, and to facilitate the reorganization of the tissue during healing. The enzyme responsible for the degradation of fibrin is plasmin, which is generated by enzymatic cleavage of plasminogen (PLG) present in blood plasma. The activation of PLG is catalyzed mainly by tissue plasminogen activator (tPA) and urokinase. The proteolytic cleavage of PLG by tPA is greatly accelerated in the presence of fibrin as a cofactor (but unaffected by fibrinogen), while urokinase does not require fibrin. Several cell types, including monocytes, are known to present a urokinase receptor on their membrane surface. They bind urokinase and help to localize it where it is needed, enhancing its effect. Both tPA and urokinase can be modified into more active forms by plasmin itself. Coagulation factor XIIa has also been found to be a plasminogen activator. Plasminogen activator inhibitors 1 and 2 are able to inactivate tPA and urokinase, playing a role in the regulation of fibrinolysis.

During formation of the fibrin network, PLG is bound to the matrix itself, as it has affinity for fibrin, although it has no catalytic activity in the zymogen form. Moreover, fibrin itself serves as a cofactor in the activation of PLG to plasmin. The action of plasmin cleaves the fibrin polymer into soluble fibrin digestion products, which are then removed by blood.

4.1.3 Inflammatory Reactions and the Foreign Body Response

Implantation of a biomaterial, medical device, or prosthesis results in tissue injury that disrupts the internal environment of the tissue. This initiates the host's defense systems, resulting in a series of responses aimed at restoring homeostasis, which starts with inflammation, followed by wound healing, and finally the foreign body response. The first step, inflammation, is defined as the reaction of vascularized living tissue to injury.[3]

Host-material reactions following implantation include injury, blood–material interactions, provisional matrix formation, acute inflammation, chronic inflammation, granulation tissue development, foreign body reaction, and fibrous capsule formation.[3,14,18,34,35] All these processes are tightly interconnected forming a continuous and sequential series of events.[34] Moreover, the process of inflammation is also influenced by blood coagulation, complement activation, the kinin–kallikrein system, as well as the reaction of various types of leukocytes. The full extent of their interplay is still the object of active research to date, as much remains to be revealed.

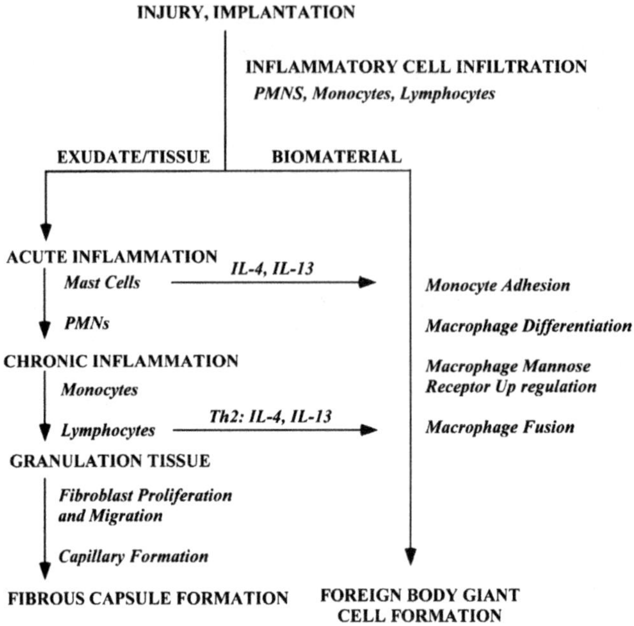

Figure 4.5 Inflammation, wound healing, and encapsulation of bioimplants, showing interleukin mediators released by mast cells and lymphocytes to guide monocyte and macrophage differentiation and fusion into foreign body giant cells.
Reprinted from J. M. Anderson, A. Rodriguez, D. T. Chang, Foreign body reaction to biomaterials, *Semin. Immunol.*, **20**, 86–100, Copyright 2008 with permission from Elsevier.

Injury and implantation of the biomaterial lead to the rapid onset of the *acute* inflammation and formation of the provisional matrix (Figure 4.5). Acute inflammation can last anywhere from minutes up to several days, depending on the extent of the injury and characteristics of the implant. The coagulation and fibrinolytic systems create (and reshape) the network of the fibrin polymer, which serves as a matrix onto which signaling molecules released by the various activated systems and cells are embedded. The provisional matrix serves as a temporary biodegradable scaffold, providing cues for cell adhesion and proliferation, including recruitment of cytokines and growth factors. Vasodilation and increased permeability of the vascular endothelium lead to the flooding of the tissue with plasma (generating an exudate and resulting in edema) and allow the extravasation of various types of leukocytes to the injured tissue. Their migration across the endothelium lining is mediated by chemotaxis – motion against a chemical gradient – as well as by binding to the endothelium *via* selectins, integrins, and immunoglobulins.[36,37] Initially, leukocytes bind to the endothelium loosely, roll along the surface, and then firmly adhere becoming activated, changing their shape from spherical to flat. Adherent leukocytes migrate through the junctions of endothelial cells and move to the injury site. The first to arrive are the neutrophils or polymorphonuclear leukocytes (PMNs). They are the most abundant white cells, representing 40–70% of the leukocyte population. Even though they are short-lived in the inactivated state, they can increase their life time in blood after an inflammation stimulus. Molecules present in the site of injury, such as interleukin-8, interferon gamma, complement-derived fragments C3a and C5a, and leukotriene, support the migration by chemotaxis. Interleukin also supports the phagocytosis of bacteria – the primary tasks of the PMNs – and the release of histamine, which is a vasodilator. As the biomaterial is typically much larger than the leukocyte, only some of the events of phagocytosis can occur. *Opsonins* (mainly immunoglobulin G and C3b) are present among the proteins coating the biomaterial. Opsonins are recognized by receptors on the cell membrane of the leukocytes and serve to facilitate phagocytosis and induce the extracellular release of enzymes in an attempt to degrade the biomaterial.[35]

The resolution of the acute phase of inflammation is followed by chronic inflammation. In contrast to acute inflammation, the predominant leukocytes participating in this phase are monocytes, macrophages, and lymphocytes. Monocytes are the largest type of leukocytes and constitute about 2–10% of them. They possess a short half-life in the circulation, which is extended to several days once activated. They also migrate to the site of inflammation in a similar fashion as PMNs, mediated by chemotaxis, and display specific adhesion and migration through the endothelium.[5,36,37] Once in the tissue, monocytes differentiate into macrophages.[38,39] Macrophages are activated by lymphokines from T-lymphocytes, cleaved complement components (see below), platelets, and the kallikrein system. They produce a large number of biologically active products including proteases, chemotactic factors, reactive oxygen metabolites, complement components, and coagulation factors and cytokines. Macrophages are also phagocytes and interact with the biomaterial surfaces directly. The release of lysosomal molecules, generation of a lower pH, and production of oxidative molecules, such as peroxides and superoxides, can lead to modifications of the surface of the material. This chemical degradation of the material may alter the ability of macrophages to adhere to the materials, thereby affecting the inflammatory response.

Mast cells also play an important role. Their degranulation and release of histamine has been shown to play an important role in macrophage recruitment. Lymphocytes are also present at the site of injury. They are involved principally in adaptive immune reactions and serve as the key mediators of antibody production and delayed hypersensitivity.

Following the action of PMNs and macrophages, fibroblasts and vascular endothelial cells in the implant site proliferate to form granulation tissue. This is a new connective tissue filled with tiny blood vessels that forms on the surface of the injury and fills in the space as part of the wound healing process. It is characterized by an intense red color, high moisture, and granular appearance.

The formation of new blood vessels in the granulation tissue proceeds from existing ones, in a process termed angiogenesis, involving the proliferation of endothelial cells and their arrangement to extend the tissue vasculature. The new extracellular matrix is formed mainly by fibroblasts, which synthesize the proteins it is made of, especially collagen. Another feature of the granulation tissue is its ability to contract for wound closure, which is endowed by the differentiation of some fibroblasts into myofibroblasts, bearing features similar to smooth muscule cells. The extension and proximity of the injury will lead to different healing processes. When the wound is the result of a clean surgical incision and the space is small, then there is almost no scar. However, when granulation tissue is required to fill larger volumes and restitute the parenchyma, large areas of fibrosis and scars may result.

Macrophages are also present during the formation of granulation tissue. The presence of adherent macrophages on the surface of the biomaterial, which attempt frustrated phagocytosis, results in the secretion of hydrolytic products and reduced pH, reaching values as low as 4, promoting their adhesion. Subsequently, the macrophages can fuse to form a multinuclear cell called the foreign-body giant cell (FBGC), which is related to the cross-talk interaction with lymphocytes mediated by lymphokines (in particular interleukins 4 and 13, and integrins).[3] While extensive research has been carried out to shed light on this process, to date, it remains not fully understood. It is believed to share similar pathways with phagocytosis.

The foreign body reaction is composed of the FBGC and the components of granulation tissue. The FBCG can last the whole life of the implants (tens of years). However, it is not known whether the FBGC remains active (releasing lyosomal constituents) or become quiescent. Those macrophages, which are not adhered to the implant, can, in principle, harm the surrounding tissue. A mechanism to prevent this is based on a variation of apoptosis – programmed cell death – which is termed anoikis. This is an apoptosis mediated by cell detachment.

The type of tissue repair at the implant site can follow two distinct processes: (i) the regeneration of functional cells of the original type, or (ii) the formation of connective tissue. The selection between these processes is controlled mainly by the ability of the cells in the tissue to proliferate and the extent of the injury (remains of parenchymal tissue). The adequacy of blood supply (nutrients) and other local and systemic factors may also play a role. The repair on tissues based on cells with no or poor proliferative ability is carried out by connective tissue. Some cells adapt for the overproduction of collagen and extracellular matrix proteins. The end-stage in the healing response to biomaterials is the formation of a fibrous capsule which isolates the biomaterial.

4.1.4 Complement System

The complement system is an important part of the innate immune system. It is a biochemical cascade, comprising several proteins present in blood plasma, many of which are proteases. In a similar fashion to the coagulation cascade, the activation of the system triggers a series of (mostly) sequential proteolytic events, resulting in the amplification of the response. The main effect of the complement system is the killing of invading pathogens. Hence, it is named "complement" as it works together with the recognition of the invading organisms by antibodies. The final event in the cascade is the formation of the *membrane attack complex*, which is a complex of proteins able to directly cause the lysis of microorganisms by disrupting the cellular membrane. However, the system also involves the release of various inflammatory mediators, called *anaphylotoxins,* which cause smooth muscle to contract, enhance vascular permeability, and can trigger the chemotaxis of leukocytes and activate them to release various substances and speed up phagocytosis. Additionally, some components of complement after activation and binding to a surface (C3b and C4b) serve as opsonins, facilitating the phagocytosis of the cell or organism to which they are bound. Moreover, the complement system interacts with the coagulation cascade and the kinin–kallikrein system.[5] Its activation can lead to damage to the tissue surrounding the site where the triggering occurred. Thus, several mechanisms of regulation and inhibition are in place. Besides pathogenic organisms, at least one pathway of the complement system can be set off by any artificial surface, including implanted biomaterials. Keeping that in mind, an understanding of the conditions in which the complement system is activated and the consequences of the process are of major importance for the application of blood- and tissue-contacting devices with minimal adverse response. As will be discussed in Section 4.2.4, the complement system plays a role not only in implanted biomaterials, but also in extracorporeal blood-contacting devices, where systemic inflammatory effects can lead to organ dysfunction.[40,41] The present section provides an overview of the main mechanisms involved in the activation and regulation of the complement system. The reader should note how, besides the generation of the pathogen-killing membrane attack complex, the system is also tightly interconnected with the leukocyte and inflammatory response as well as blood coagulation.[42]

Three pathways can be distinguished for the activation of the complement system: the classical pathway, the lectin pathway, and the alternative pathway. All three pathways require some type of surface to be effective. In all cases, the central mediator of the system is complement component 3 (C3).

The classical pathway is initiated by the binding of C1 to the antibodies IgG or IgM adsorbed on a surface, which, in turn, leads to the activation of the catalytic activity of the C1s subcomponents. They can then cleave C4 into C4a (an anaphylotoxin) and C4b, which itself binds to the surface and C2. C2 is also cleaved by C1 to generate C2b (released to the solution) and C2a, which assembles with the surface-bound C4b to generate the complex C4bC2a, the C3 convertase of the classical pathway.

The lectin pathway generates the C3 convertase in an analogous fashion as the classical pathway, with the exception that no antibodies are involved in the initial step. Mannose-binding lectin is a protein that can function as opsonin when it assembles on the surface of a pathogen, displaying mannose residues in the terminal position of glycoproteins or carbohydrate chains. It performs the same functions as antibody-bound C1, leading to the formation of the C3 convertase (C4bC2a complex).

On the other hand, the alternative pathway can initiate cleavage of C3 without the presence of any other protein previously on the surface. C3 can hydrolyze spontaneously and, in the presence of factors D and B, form a solution-phase C3 convertase, which provides constant production of small amounts of C3b. This can bind to a surface presenting nucleophilic groups (OH or NH_2) and factor B to form the surface-bound alternative pathway C3 convertase (C3bBb). In solution, factors H and I inactivate C3b and the alternative-pathway C3 convertase. Specific inhibitors are also present in the membrane of host cells (decay-accelerating factor, complement receptor type 1, and membrane cofactor protein). However, if the C3bBb complex is bound to any other type of surface (invading microorganism or artificial), it is protected from inhibition and can amplify the activation by cleaving more C3 and forming more C3b, which will allow the generation of more alternative-pathway C3 convertase after binding of factor B.

When C3 is cleaved by a C3 convertase generated during a previous step of one of the pathways, C3b and C3a are released. C3a is the smaller fragment and it is an anaphylotoxin and a chemoattractant to leukocytes. The larger fragment, C3b, serves several functions: (i) it binds factor B to form C3Bb, which is able to cleave more C3 and amplify the response, (ii) it is an opsonin (induces phagocytic cells to attack the surface/particle), and (iii) it can cluster with surface-bound C3 convertase of either pathway (C4bC2a or C3bBb) to form the C5 convertase. The latter complex cleaves C5 to form C5a and C5b. C5a is also anaphylotoxic and is responsible for attracting and activating leukocytes. C5b complexes successively with C6, C7, C8, and several units of C9 to form the membrane attack complex, which can then insert into a membrane (lipid bilayer) and open a ring-shaped pore. By allowing the passage of water driven by the osmotic pressure inside of the penetrated cell, lysis is induced.

In biomaterials, the alternative pathway of activation predominates. However, as immunoglobulins can also adsorb on artificial surfaces, classical pathway activation is also possible. Moreover, proteases of the coagulation cascade, including thrombin, have also been suggested to be active in cleaving C3 and C5, which has been proposed to play a role in hemorrhagic shock and the systemic inflammatory response.[43] The crosstalk between coagulation and the complement system is clearly deeper, as platelets are activated and their aggregation is induced by C3a among other mechanisms.[44] A critical consequence of this crosstalk between the different cascades and recruitment of cells is that it is only possible to evaluate the interactions at a material surface using whole blood.[41] While the separation of the blood response into different cascades and systems is useful for conceptualization, *in vivo* responses cannot be considered independently.

4.1.5 Infection and Biofilm Formation

The advances in biomaterials and their engineering have resulted in the concomitant increase of the number of implants per year. Several tens of millions of implants are used in the USA alone (*e.g.* 720 000 knee replacements, 332 000 hip replacements, 395 000 coronary artery bypass grafts, source: http://www.cdc.gov/nchs/fastats/ inpatient-surgery.htm). However, a significant proportion is colonized by bacteria and becomes the focus of device-related infections. Topical devices (*e.g.* contact lenses) are colonized immediately upon contact, transcutaneous devices (*e.g.* vascular catheters) are more progressively colonized by skin bacteria, and even orthopedic implants can be subjected to infection. Urinary catheters are among the ones with highest incidence of bacterial infection accounting for as much as 30% of all nosocomial infections.[45] Prolonged use of these catheters for over 4 days leads to infection. On the other hand, the chances of developing an infection in orthopedic implants are considerably lower, but its effects are devastating. When the antibiotic treatment fails, surgery to remove the infection (scraping), exchange of the implant, or even amputation may become necessary.

Device-related infections are linked with the ability of the bacteria to colonize the surface of the biomaterial forming a *biofilm*. The biofilm concept was introduced in 1978 and based on the ability of bacteria to adhere, an important factor in their virulence (Figure 4.6).[46-48] Once bacteria are attached to a surface, a series of processes is set in motion, resulting in an adhered microbial community termed biofilm. The biofilm is an accumulation of bacterial cells on a solid surface, embedded in a matrix of excreted extracellular polymeric substances (EPS).[49] Such bacterial communities are largely protected from the immune response and much less susceptible to elimination by antibiotics, therefore increasing the detrimental consequences of the infection. Furthermore, sessile bacteria in the biofilm release antigens stimulating the production of antibodies. The latter are not effective in targeting bacteria in the biofilm and may result in the immune system damaging surrounding tissues.[50,51] Moreover, the pathogens embedded in the biofilm are largely protected even from phagocytes penetrating it, suggesting the presence of other mechanisms disrupting leukocyte function. The biofilm usually is slow growing and after varying periods of time releases planktonic bacteria which can further spread the infection. While these planktonic bacteria can be controlled with antibiotics, the biofilm is generally not disrupted and can continue the release. These give rise to recurrent or chronic infections. Consequently, the inhibition of bacterial colonization of surfaces and the prevention of biofilm formation are a primary research focus.

The increased resistance of biofilms to antibiotics has been explained by (i) mass transfer limitations, (ii) reduced susceptibility of the bacteria, and (iii) phenotypic differences. The antibiotic molecule needs to diffuse into the EPS matrix to access the bacterial cells, and this can lead to huge barriers

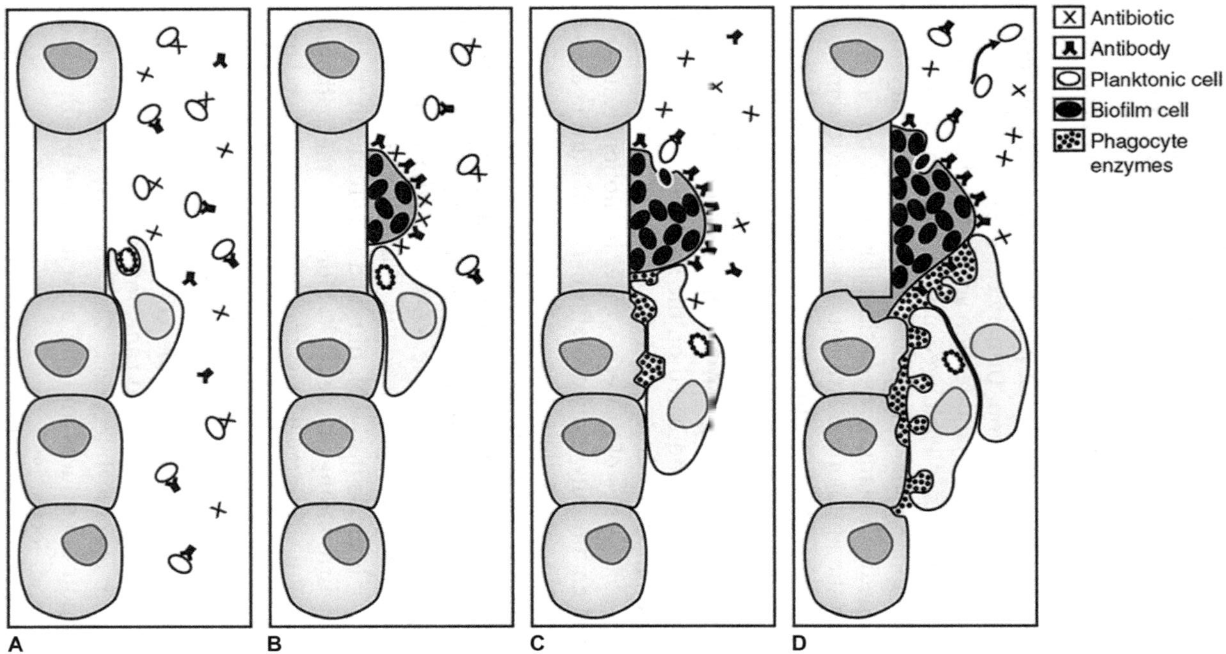

Figure 4.6 Formation of a biofilm on a medical implant. (A) Planktonic bacteria are readily opsonized and phagocytized. (B) Sessile bacteria in the biofilm are protected from antibodies, antibiotics, and phagocytosis. (C) Phagocytes release enzymes into the surrounding medium. (D) Tissue damage and propagation of the infection follow.
Reprinted from ref. 54 with permission from Wolters Kluwer Health, © 2015 by American Academy of Orthopaedic Surgeons.

to diffusion.[52] As a consequence of limited diffusion, the cells in the biofilm experience nutrient limitation and are in a slow-growing state. Cells in this state have been shown to be less susceptible to antimicrobial agents. Additionally, biofilm bacteria usually display a different phenotype as a consequence of a programmed biological response to growth on the surface.[47]

Another challenge posed by biofilms is their early detection. Biofilm-associated infections are often difficult to detect using culturing methods. Moreover, such methods usually lead to large underestimates. Microscopic examination of the device and surrounding tissues is still the only way to definitively demonstrate the presence of biofilms. This can be done using electron microscopy and also confocal microscopy. Another method of detection is based on the detection of bacterial DNA using the polymerase chain reaction (PCR) as an amplification tool. Having said that, the early detection of biofilms remains an important challenge in the effort to minimize the deleterious effects on the host.

The underlying processes leading to the formation of biofilms are common across the variety of different environments in which they occur, with the concomitant differences in the resulting structure. Upon contact of a surface with a medium in which bacteria are present, the first step in the formation of a biofilm is the adsorption of biomacromolecules from the medium. This results in the generation of a layer that acts as a *conditioning film*. The transport of the bacteria to the substrate comes next, leading to adhesion upon contact of the bacterium with the surface.[46,53] The synthesis of the biofilm matrix follows, as the attached bacteria activate the cellular mechanisms for the secretion of EPS. This creates an environment in which the bacteria benefit from improved mechanical stability and protection from adverse chemical conditions.[46,48,54] Therefore, the eradication of the bacteria becomes significantly more challenging once this state has been achieved. The process of bacterial colonization and biofilm formation is further complicated by the complexity and the interplay of both the material and bacteria surfaces.[55]

The adhesion of bacteria onto hydrophobic surfaces is promoted by the conditioning film through a decrease in the interfacial energy when such surfaces come into contact with media containing proteins.[13,56,57] On the other hand, the factors determining the adhesion of bacteria onto hydrophilic surfaces can be separated into specific interactions by receptors on the bacterium surface (mainly adhesins) and general physicochemical interactions.[58] Long-range interactions are responsible for initial adhesion and have been modeled by the extended Derjaguin–Landau–Verwey–Overbeek (DLVO) theory, considering the bacterial cell as a simple colloid. Moreover, the conformation of polymers grafted to the surface has been incorporated in the models.[59–61] The interaction is caused by the Liftshitz–Van der Waals forces and forces originating from the overlap of the electrical double layers.[61] The surface chemistry also plays an important role, as it determines the interactions operating in the short range (few nm), including electrostatic interactions as well as the formation of hydrogen bonds, and in a longer range (typically 5–10 nm) through hydrophobic attractions.

Two general strategies have been employed for the prevention of biofilm formation based on (a) inactivation of any bacterial cells which come into contact with the surface (bactericidal coating), and (b) avoidance of bacterial adhesion (antifouling). Interestingly, coatings combining both approaches have also been introduced.[62–65] Bactericidal coatings are based on molecules, polymers, or nanoparticles which are toxic for the bacteria.[66–73] Additionally, antibiotic-impregnated cement spacers have been used for hip or knee arthroplasty.[74] Nevertheless, intrinsic cytotoxicity is usually an impediment for the use of this approach in biomedical applications.[75,76] Furthermore, the adsorption of bacteria after inactivation leads to reduced effectiveness of the coating, as it can serve as an anchoring point for more bacteria to attach.[55] On the other hand, an ideal non-fouling coating should prevent the formation of the conditioning film as well as the bacterial attachment. The inhibition of the conditioning film formation requires coatings capable of preventing the adsorption of proteins (see Chapter 5), while the bacterial adhesion can be prevented by minimizing the forces driving the bacteria into contact with the surface.

According to the extended DLVO theory, attractive long-range interactions are minimized by introduction of a water barrier with a zero net charge and steric hindrance to keep the bacterial cell at a distance from the surface.[61,77] As a consequence of this, polymer brushes (polymer chains tethered from a surface) have been shown to lead to the best resistance to biofilm formation.[78–80] In particular, brushes of *N*-(2-hydroxypropyl)methacrylamide and carboxybetaine acrylamide have resisted biofilm formation by repelling the bacteria.[78]

4.1.6 Effect of Flow Regime in Blood–Surface Interactions

The fluid dynamics at the surface affect the formation of thrombi and the fibrin network. Flow can dictate the rates of transport of cells to the surface, as well as change the level of receptor expression on platelets and the composition of thrombi. These effects are widely studied on platelets and, to a lesser extent, other cells. High shear stress results in higher platelet deposition and lower fibrin deposition, while, at low shear stress, the inverse is true.[81]

Flow also influences the coagulation cascade. However, the understanding of these effects is still limited to a few factors. For instance, activation of factor X to Xa has been shown to increase with the flow rate.

Red blood cells are usually considered as passive participants in the process of hemostasis and thrombosis. Red blood cells rarely adhere to surfaces even of artificial materials.[82] At venous flows (low shear stress), red blood cells comprise an important proportion of the thrombus, which is red colored (red thrombus). On the other hand, in arterial flows (higher shear stress), the thrombus formed is mostly off-white (white thrombus) due to the large proportion of platelets in the plug.

4.1.7 Tumorigenesis

In humans, sarcomas and neoplasms at the site of implant are rarer than on animal models. Cancers associated with implants may occur at any post-operative time interval, but are usually observed after many years. Both the physical and the chemical nature of the implant may play a role in tumorigenesis.

4.2 Blood-Contacting Devices

4.2.1 Vascular Stents

Cardiovascular diseases are the number one cause of death world-wide. In this category, in particular, coronary artery disease (CAD) still remains as the main contributor to mortality. Advances in the management of CAD have led to enormous improvement in its outcome, thanks to the introduction of new treatments. The condition is characterized by insufficient blood irrigation to the tissue of the cardiac muscle (*myocardium*), which often leads to myocardial infarctions (heart attack) and death of the tissue. The cause for the decreased perfusion of the myocardium is the narrowing of the lumen of the coronary arteries, generally due to formation of plaque, which eventually restricts the blood flow in the artery. The narrowing of the blood vessel is called *stenosis*. Besides pharmacological therapy, the procedures employed for the treatment of CAD are aimed at achieving *revascularization* of the heart tissue.

Percutaneous coronary interventions consist on accessing the stenotic vessel *via* a catheter inserted through another vessel such as the femoral artery. In an angioplasty, a balloon-tipped catheter is inserted in the lumen of the narrowed section of the artery and inflated in order to widen it by forcing it open. However, after the balloon angioplasty, the blood vessel often suffers from *restenosis* due to the contraction of the smooth muscle tissue in the arterial wall. An improvement introduced to this procedure consists in including a metallic mesh scaffold –called *stent*– lining the balloon, which is expanded and left behind to support the vessel in the open position. In this way the rates of occurrence of restenosis are markedly reduced.[83,84] On the other hand, even after the placement of a stent, restenosis of the vessel can still occur by a distinct pathophysiological process, leading to what is called *in-stent restenosis*. The interactions between the biomaterial surface of the stent, the tissues of the blood vessel, and the blood itself play a determinant role in the origin of this problem. The pathophysiological aspects leading to in-stent restenosis and the strategies to overcome it have been reviewed recently.[85] A bare metal stent presents the uncoated metallic surface to the interior of the blood vessel (typically stainless steel 316L, nickel–titanium, or cobalt–chromium alloys). The implantation of the stent itself causes physical damage to the surrounding tissue, as it deforms the layers of tissue that make up the wall of the blood vessel. After the balloon is removed following angioplasty, the blood vessel tissue responds by contracting the vascular smooth muscle cells that make up the medial tissue layer of the vessel wall, resulting in elastic recoil. While the introduction of the stent prevents the elastic recoil from taking place, the inflammation caused on the surrounding tissue is increased. Moreover, the permanent insertion of a foreign object into the body's contacting tissue and blood leads to the rapid non-specific adsorption of proteins. These in turn mediate the activation of the different protease pathways (intrinsic coagulation cascade, kinin–kallikrein system, and complement system) and the activation of platelets, neutrophils, and monocytes. Furthermore, the damage to the vascular endothelium lining the *tunica intima* (innermost layer of arterial tissue) and even the disruption of deeper levels of vessel tissue activate the tissue factor pathway of the coagulation cascade. As the stent ruptures the vascular endothelium, the sites of injury expose a multitude of adhesion motifs from the extracellular matrix proteins, which further promote inflammation as they trigger thrombus formation. Leukocytes are attracted to the site of injury by chemotaxis and are activated to produce a variety of growth factors and cytokines. These stimulate the fibroblasts and smooth muscle cells present in the tissue of the blood vessel to generate the extracellular matrix and to proliferate. The tissue growth causes the thickening of the artery wall. Eventually, the tunica intima invades the lumen of the blood vessel beyond the structure of the stent, leading to *neointimal hyperplasia* (NIH). The resulting restenosis is the main complication of the use of bare metal stents.

To overcome this issue, drug-eluting stents (DES) have been introduced. The current generation of DES present a polymer coating embedded with a pharmacological agent that is released slowly, thereby maintaining a local concentration without systemic effects. The drugs applied to DES are targeted to prevent the proliferation of the smooth muscle cells in the intima and, in this way, suppress NIH. The two main substances employed for this purposes are sirolimus (an immunosupressor) and paclitaxel (an antiproliferative, commonly used in cancer therapy). The use of DES is associated with a marked reduction in restenosis rates.[86,87] However, while NIH is effectively prevented, it has been established that the implantation of DES is associated with increased rates of late[88] stent thrombosis events (*i.e.* occurring over one month after the stent is in place).[89–92] Nevertheless, studies have often cited the benefits of the reduced need for revascularization to outweigh the side effect of thrombosis events. The consequences of thrombosis complications are severe, leading to myocardial infarction and death, while prolonged antiplatelet therapy needs to be used for prevention. The mechanism behind this adverse effect is explained by a delay of the healing process of the artery and the inhibition of re-endothelialization of the vessel wall. Furthermore, the coverage of the vascular endothelium growing onto the struts of implanted DES has been found in autopsies to be reduced in comparison to bare metal stents.[88]

In light of these findings, the focus of stent-related research has shifted in recent years from the prevention of NIH to achieving improved biocompatibility of the material surface and minimizing the damage induced by the introduction of this foreign device. The polymer coatings employed to modulate the release kinetics of stents have been associated with hypersensitivity reactions, possibly contributing to the reported impaired healing and thrombosis. Therefore, a newer generation of DES, taking this observation into consideration, has been designed to eliminate the polymer component or replace it with a biodegradable polymer which can be absorbed by the body after the elution of the drug.[93]

As a further alternative, fully biodegradable stents have been proposed. The main advantage lies in that they are expected to be resorbed by the tissue after their function is complete. Both polymeric and metal degradable stents present interesting developments, holding the potential to minimize late stent thrombosis.[94,95] Biodegradable drug-eluting devices containing the same types of pharmacological agents employed in regular DES have also been introduced.[96,97] Some models are already undergoing clinical trials.[85]

In an effort to decrease the adverse responses elicited by the stent surface, a variety of stable surface coatings have been employed with an expected increase in biocompatibility. These should enable prompt healing and re-endothelialization of the affected artery and stent structure. Among the strategies employed, inorganic coatings consisting either of metals or ceramics have been tested, including gold, copper, platinum, titanium, carbon, silicon carbide, iridium oxide, and titanium nitride oxide (TiNOX). On the other hand, organic coatings consist of polymers that can be classified broadly as biostable, biodegradable, or biocompatible/bioinspired. The first category includes polymers commonly employed in the fabrication of biomedical devices (including poly(ethylene terephthalate), polyurethanes, polydimethylsiloxanes, and expanded polytetrafluoroethylene (ePTFE) among others). Their success in preventing in-stent restenosis has varied, while significant inflammatory response and NIH were observed in animal models in some cases.[85] Biodegradable polymers, such as poly(caprolactone), poly(glycolic acid)/poly(lactic acid), and other polyesters, have been applied as drug-releasing coatings in a new generation of DES. They are expected not to elicit the inflammatory response and hypersensitivity associated with previous generation DES. Polymers emulating the properties of some natural substances have been labeled as biocompatible and include heparin, fibrin, hyaluronic acid, and phosphorylcholine. However, these polymers have not led to improved outcomes in human clinical trials.[85]

The complexity of the reactions elicited in the body by a biomaterial is mediated by the non-specific adsorption of proteins. Clearly, in order to be able to prevent the deleterious responses elicited in the body by the introduction of the stent, several factors need to be taken into account. The damage to the vascular tissue induced upon introduction of the stent during angioplasty leads to an exaggerated healing response that causes NIH, which, so far, only DES have managed to effectively mitigate. At the same time, the impairment of cell proliferation associated with these drugs leads to a delay in the healing of the artery while the artificial materials (metals and polymers) remain exposed to the lumen of the blood vessel. Hence, while DES have succeeded in partially modulating the response of the smooth muscle cell layer, problems arising from undesired blood–material interactions are in fact a greater cause for concern than for bare metal stents.

While fully biodegradable polymeric or metallic DES appear as a highly promising alternative to limit the potential for negative responses to the material, to date, they have not managed to achieve a satisfactory performance. Therefore, if a stable material is to be implanted into an artery without eliciting adverse responses, it should additionally support enhanced re-endothelialization to avoid prolonged direct contact with blood, especially after discontinuation of antiplatelet therapy. Strategies to achieve this goal have been presented based on the seeding of the scaffold with endothelial cells.[85] However, damage of the endothelial cells during the stent implantation procedure and loss of cell adhesion in the flow regime of the artery have limited the success of this approach. In order to circumvent these problems, endothelial cells are transferred locally to the stent after implantation. While enhanced coverage of the stent surface has been achieved with this approach, the ability of these endothelial cells to restore the functional tissue has been found to be very limited. An alternative strategy to facilitate the formation of vascular endothelium consists in taking advantage of the body's own circulating cells known to repair endothelial tissue, namely endothelial progenitor cells (EPCs). This has been achieved by coating the stent surface with an antibody against CD34, which is able to specifically capture endothelial progenitor cells from blood. Thus, a stent coated with anti-CD34 is expected to facilitate the formation of a layer of endothelium. Nevertheless, while metal stents coated with anti-CD34 have been found to reduce restenosis, they do so less effectively than standard DES owing to their decreased ability to prevent NIH. This highlights the complexity involved in simultaneously controlling all possible interactions of biomaterials after implantation. Taking these concerns into account, stents combining both antiproliferative drug-eluting capabilities and the capture of circulating endothelial progenitor cells by introduction of anti-CD34 antibodies were introduced.[98–100] In particular, one of such devices (COMBO Dual Therapy Stent, OrbusNeich, Fort Lauderdale, Florida) has been designed taking into consideration the distinct responses required from the side of the scaffold facing the vessel wall, where proliferation of smooth muscle cells and inflammatory responses should be minimized, and the arterial lumen, where the attachment and proliferation of endothelial progenitor cells should be supported (Figure 4.7). Further trials of this promising technology are needed to evaluate its potential for supporting the healing of the artery after angioplasty.

Regarding the support of the proliferation of specific cells, it is clear that the morphology of the stent surface plays a central role. An ideal stent surface should favor the adhesion and proliferation of endothelial cells, associated with the healing process and prevention of thrombosis, while suppressing the

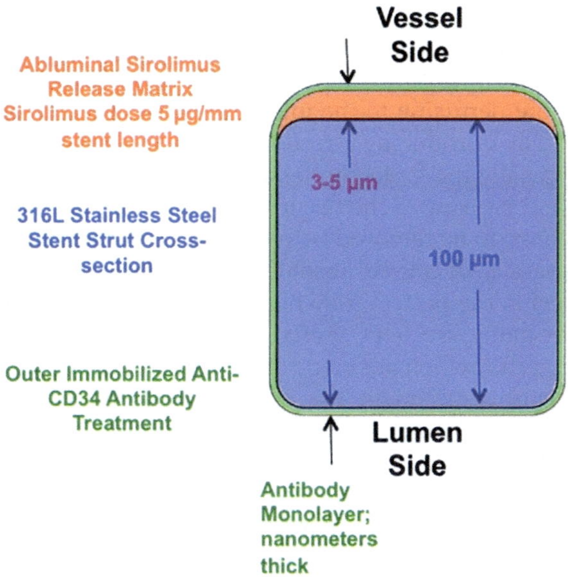

Figure 4.7 Cross-section of the strut of a Combo Dual Therapy Stent (OrbusNeich, Fort Lauderdale, Florida), showing the different functionalization applied on the sides facing the vessel wall and the lumen of the artery Reprinted from M. Haude, S. W.L. Lee, S. G. Worthley, S. Silber, S. Verheye, S. Erbs, M. A. Rosli, R. Botelho, I. Meredith, K. H. Sim, P. R. Stella, H.-C. Tan, R. Whitbourn, S. Thambar, A. Abizaid, T. H. Koh *et al.*, The REMEDEE Trial A Randomized Comparison of a Combination Sirolimus-Eluting Endothelial Progenitor Cell Capture Stent With a Paclitaxel-Eluting Stent, *JACC: Cardiovasc. Interventions*, **6**, 334–343, Copyright 2013 with permission from Elsevier.

excessive proliferation of fibroblast and smooth muscle cell, both of which are behind NIH and restenosis. The influence of substrate morphology has been reviewed in the previous chapter, but it should be noted that the precise mechanisms behind its impact on cell behavior are not clear yet. Thus, it is expected that substantial improvement in the control of the fate of cells can be achieved by elucidating the factors participating in this process. Furthermore, the interplay of morphology, surface-immobilized molecular cell-adhesion cues, and the release of soluble substances can lead to strategies to promote the healing of the endothelium while minimizing restenosis.

4.2.2 Vascular Grafts

The treatment of localized vascular disease often requires the repair of a portion of vein or artery. In cases such as dissection of a vessel or a localized enlargement of an artery that risks bursting (aneurysm), the threatened section needs to be substituted or bypassed using a *vascular graft*. In principle, either another vein from the same person's body (autologous graft) or a donor (allograft) can be used, or the graft can be synthetic. The surgical treatment of CAD consists of bypassing the affected coronary artery (or arteries) in what is called *coronary artery bypass grafting*. For this purpose, either another artery close to the heart is diverted to bypass the affected vessel and restore perfusion to the myocardium or, alternatively, the bypass can be achieved by introducing a section of a peripheral vein harvested from the patient (typically the great saphenous vein, from the leg). Even though the structure and layering of the tissue is different in veins and arteries (and venous valves need to be removed before implantation), the biocompatibility of veins for bypass is, of course, absolute, as the native vascular endothelium lining the vessel is the perfect fully hemocompatible material.

On the other hand, synthetic grafts suffer from a number of limitations. Since vascular grafts, which are made of artificial materials, will be subjected to permanent implantation in the endovascular environment, they must display appropriate characteristics with regard not only to their hemocompatibility, but also their mechanical properties, which must also be compatible with the natural vascular tissue surrounding it. This includes the ability to be joined surgically to vessels by suturing, forming *anastomoses* at both ends of the graft. If the material is too rigid, the different compliance of the graft and the vessel at the anastomosis might lead to bleeding. Vascular grafts used medicine consist of finely woven or knitted fabrics of ePTFE or Dacron (poly(ethylene terephthalate)).

A *stent graft* is basically a covered stent. An expandable metallic scaffold, very similar to the ones employed for the treatment of coronary stenosis, sheathed in a Dacron or ePTFE fabric, is used for the treatment of aneurysms. As the stent is expandable, the graft is inserted *endovascularly*, accessing the vessel through a catheter without the need for a major surgical intervention. In spite of the woven material being porous and, therefore, not preventing hemorrhage, the rapid clotting on the fiber surface can effectively prevent the leakage of blood out of the vessel.[101] Additionally, coatings of collagen added on the graft surface can help to induce rapid clotting, which is desired to achieve a sealing effect, even on a highly porous fabric. On the other hand, the porosity of the fabric supports the formed clot in place, favoring its adherence to the graft. The clotted layer, consisting mainly of fibrin, platelets, and white blood cells, covers the graft surface forming a *pseudointima*. In contrast, the healing of the vessel from the anastomosis sites extends to cover part of the graft surface (up to 1–2 cm) and forms a new layer of structured tissue called *neointima*.[102] As the neointima is in contact with blood through a layer of vascular endothelium, its surface is non-thrombogenic and it is considered beneficial in comparison with the pseudointima. In the internal mid-portion of the graft, where the ingrowth of the vessel does not reach, the healing and neointimal formation is possible either by advancing growth from the endothelium of the capillaries through the pores and into the graft, or by adhesion of endothelial progenitor cells circulating in blood. However, as the external side of the graft is not subjected to a constant flow of blood, it elicits a usual foreign body response. This leads to the progressive formation of a fibrous capsule around the implant.

The characteristics of the pseudointima and its mediation in the blood–implant interaction pose several risks and limitations on the application of vascular grafts. The formation of non- or poorly adherent thrombi in the blood stream is a source of complications. In particular, at the anastomoses, points with impaired blood flow can be a site of thrombus formation. Especially, distal anastomosis can give rise to thrombus formation if there are imperfections in the joint with the vessel. In addition, the thrombosis on the inner surface is particularly challenging in prosthetic grafts of small diameter (less than 6 mm). The occlusion of the lumen of small diameter grafts by formation of thrombus is a significant challenge for their use in the repair of very small blood vessels. Moreover, as explained above (see Section 4.1), the implant surface can offer a favorable location for the attachment of bacteria, leading to infection. Occlusion of the implant can also develop as a consequence of hyperplasia at the anastomoses, resulting from the implant injury, in a similar way as explained for stents after balloon angioplasty.

Coatings have also been employed to reduce the thrombogenicity of the graft surface and are commercially available.[103] *Pyrolitic carbon* coatings have been employed in long-term blood-contacting devices to provide resistance to platelet adhesion and thrombus formation.[104] Even though this material is highly hydrophobic and it is known to elicit rapid protein adsorption and conformational changes, it has been postulated that its surface supports the formation of a strongly bound layer of adsorbed proteins which are not displaced by other proteins and effectively achieve passivation. Carbon-impregnated ePTFE vascular grafts have been reported to reduce the incidence of complications during the first two years after implantation.[105] Heparin coatings have also been proposed to reduce platelet adhesion and thrombogenicity, thereby enabling the use of small-diameter prosthetic grafts with positive results in an animal model.[106] Clinical evaluation of heparin-coated grafts also indicated improved outcomes.[107]

Current research is focused on improving the endothelialization of the graft surface. The main approach consists of employing tissue engineering to promote the attachment of endothelial progenitor cells. The ultimate goal is to replace the vascular graft with an artificial autologous vessel, *i.e.* grown from cells harvested from the patient himself. This requires the development of appropriate biodegradable tissue engineering scaffolds, the seeding of cells, and the delivery of both surface-immobilized and soluble molecular cues. While the challenges involved in achieving artificially grown autologous blood vessels are considerable, the expected advantages include non-thrombogenicity and the ability to create tailor-made adaptable vessels of the exact shape and size required for a particular procedure.

4.2.3 Prosthetic Heart Valves

Valvular heart disease is a group of conditions in which one or more of the heart valves are unable to properly control the flow of blood. While this type of disease can be congenital, the loss of valve function is mostly associated with increasing age. Two types of valve failure can be distinguished: *stenosis* (narrowing of the opening leading to decreased flow) and *regurgitation* (failure to prevent retrograde blood flow due to incomplete closing). Surgical implantation of prosthetic heart valves to replace the function of the diseased one has been developed as a treatment with great success, leading to considerable improvement in patient outcomes. The success achieved with the use of these devices has been considered one of the most valuable improvements in healthcare facilitated by implantable biomaterials.[108] Nevertheless, while the design and construction of prosthetic heart valves has made important advances, valve replacement does not provide a definitive cure to the underlying disease as it brings risks of additional complications that need to be carefully treated and managed, in what has been dubbed "prosthetic valve disease".[109] The potential complications arise from a number of factors and depend on the type and design of the valve. Two basic types of prosthetic heart valves are used, which will be discussed in the following.

Artificial mechanical valves consist of a fixed base ring and a rigid moving element (called occluder), which responds to the change in blood flow direction by shutting the opening and preventing reverse flow. The base is surrounded by an ePTFE or Dacron fabric which can be sutured in the final position. The vast majority of mechanical valves employ occluders made of pyrolytic carbon due to its wear resistance, rigidity, and very low thrombogenicity in comparison with other artificial materials. However, one device employing a rigid synthetic polymer has also been used.[110] The valve ring is made of pyrolytic carbon or metal. In early designs, the occluder mechanism was comprised of a ball held in a wire cage, which would move to block the flow in response to the change in flow direction. Nowadays the occluding mechanisms used consist of either one tilting disc held by struts (single leaflet) or two tilting semilunar disc pieces (bileaflet design). Modern mechanical valves are extremely durable and mechanical failure of their components is highly exceptional. However, as with the previous examples, the introduction of artificial materials in permanent contact with blood is associated with thrombogenicity. Mechanical heart valves also disrupt the blood flow regime, creating pressure gradients and regions of stagnant flow, which dramatically raises the risk of thromboembolic complications. Moreover, regions of high shear stress caused by impaired flow characteristics can cause platelet activation, leading to thrombosis, and also rupture of red blood cells, called *hemolysis*. Thrombosis can lead to progressive obstruction of the valve itself. Additionally, release of the thrombus into the blood stream results in embolism, with potentially catastrophic consequences. For this reason, patients receiving a mechanical heart valve implant must undergo constant anticoagulation therapy (typically with warfarin) for the rest of their lives. Thus, they are also subjected to increased risk of bleeding brought about by the medication itself. Keeping that in mind, the anticoagulant dosage needs to be regularly monitored and adjusted.

A second type of heart valve used in implants is termed *bioprosthetic* (or also biological) heart valve. The fundamental difference is that this type of implant is mainly derived from processed natural biological tissue. The animal-derived tissue is chemically treated to stabilize its structure and inactivate any immunogenic cues remaining by crosslinking. As the pig's heart is anatomically similar to the human heart, porcine aortic valves are harvested and treated to be employed as bioprosthesis. They are mounted on a Dacron or ePTFE fabric sewing ring for implantation or on an expandable stent. Bioprosthetic valves are also fashioned from harvested bovine pericardial tissue, which is given the desired shape and subsequently mounted on a sewing ring or stent. In contrast to mechanical valves, bioprosthetic valves allow a much more natural flow regime, closely resembling normal physiological conditions. The elimination of high shear forces results in a marked reduction of thrombogenicity, which becomes negligible. Hence, prolonged use of anticoagulation medication is not necessary after implantation of a bioprosthetic heart valve, which eliminates the extra risk of bleeding. On the other hand, the structural durability of bioprosthetic valves is much lower than that of mechanical valves.[111] After implantation, they are subjected to various *in vivo* degradation mechanisms.[112] Firstly, mechanical deterioration occurs gradually due to repeated cycling and it cannot be repaired as the leaflets of bioprosthetic heart valves contain no viable cells. The extracellular matrix components, while chemically crosslinked, are still subjected to the action of enzymes in the body.[113] Moreover, *calcification* of the valve leads to the slow accumulation of hydroxyapatite crystals, which eventually degrade its performance and cause failure. The deterioration of the valve is highly dependent on the age of the patient, being dramatically accelerated in younger patients. Thus, they are generally reserved for patients older than 65 years.[109] Nevertheless, bioprosthetic heart valves, not containing any rigid parts and mounted on expandable stents, have the distinct advantage of allowing implantation by means of a catheter introduced *via* the femoral artery or the apex of the heart. In this way, the trauma associated with open heart surgery is eliminated, enabling the implantation to be performed on patients who are older or present comorbidities that make replacement surgery too risky.

A significant complication of heart valve replacement surgery is infection of the implant site. Both types of valves present a foreign surface which facilitates the attachment of microorganisms. This leads to *endocarditis*, which is potentially fatal and generally requires immediate replacement surgery.

As an alternative to bioprosthetic and rigid mechanical devices, artificial heart valves consisting of leaflets of flexible polymeric materials have been investigated for many decades. A number of materials have been employed to tackle the demanding challenges of mimicking natural flow conditions and retaining resistance to thrombus formation and mechanical integrity, while allowing minimal invasive implantation owing to the device flexibility and collapsability. However, so far, this approach has not met with significant success.[108] Similarly as with vascular grafts, tissue engineering is also envisioned as a promising strategy. A prosthetic valve comprising living autologous cells in a tissue engineering scaffold is also expected to have additional advantages, including the ability to self-repair and adapt. In particular, for pediatric patients who receive the implant before their hearts are fully developed, a tissue-engineered heart valve could potentially grow with the patient, eliminating the need of subsequent valve replacement to adjust for the growth of the heart.

4.2.4 Extracorporeal Blood Circulation

While the advances in cardiovascular implants have undoubtedly revolutionized the treatment options for many pathologies such as the ones described above, it must be mentioned that they have been accompanied by the development of novel surgical techniques in the second half of the last century. A relevant example explained above (Section 4.2.2) is coronary artery bypass grafting for the surgical treatment of CAD. Together with this procedure, the surgical correction of large cardiac structural defects, the replacement of valves, and heart transplants are among the types of surgery that require the blood flow to be diverted from the heart and the patient's blood circulation and oxygenation to be supported artificially. Nevertheless, nowadays, less invasive methods have been developed for some cases. The development of an artificial machine that can temporarily take over the functions of the heart and the lungs has enabled the evolution of *cardiopulmonary bypass* (CPB).[114] The bypass can be maintained for several hours, allowing the surgeon to carry out complex operations on the heart without obstruction of the operating field with blood.

The bypass is achieved by connecting cannulas to the central arteries and veins (aorta and vena cava) and clamping the aorta. The "heart–lung" machine consists of a blood reservoir, heat exchanger, pump, and oxygenator as the central pieces, as well as sensors to monitor the temperature, pressure, and flow. Additionally, an arterial filter is placed in-line at the exit of the circuit to eliminate gas bubbles, platelet and other cell aggregates, and any other debris that could cause embolisms if injected back to the patient.

The heat exchanger is employed to lower the patient's core body temperature inducing hypothermia, which lowers the body's oxygen requirements. The two main types of pumps employed in heart–lung machines are roller pumps and centrifugal pumps. Roller pumps are a type of positive displacement pump, in which the blood is pushed through a flexible tube by peristaltic action. On the other hand centrifugal pumps work by transferring momentum to the blood by a rotating impeller. However, roller pumps create regions of high shear stress and abrupt pressure gradients in the points where the roller compresses the tubing, leading to hemolysis (damage of the red blood cells). Therefore, modern cardiopulmonary bypass machines employ pumps of the centrifugal type.

The oxygenator is a device that enables the blood being circulated to release carbon dioxide and capture oxygen by contact with oxygen-rich air.

Original CPB machines have achieved this by bubbling oxygen gas through the blood, thereby producing a large contact area between the gas and blood for the exchange to take place directly. On the other hand, this process is associated with activation of the complement system, increased hemolysis, and formation of microscopic gas bubbles leading to microemboli.[115] Thus, in modern devices, membrane oxygenators for CPB are employed, in which the blood is generally not contacted with the gas directly, but through a thin polymeric membrane that allows the diffusion of gaseous components. Additionally, membrane oxygenators reduce the risk of infection as they prevent direct contact of the blood with airborne microorganisms.

A major breakthrough that made the use extracorporeal heart–lung machines possible was the discovery of the potent anticoagulant drug heparin. Adequate anticoagulation must be achieved before connecting the patient's circulation to the bypass machine and it has to be maintained throughout the procedure. Once the bypass is disconnected, protamine is administered, which acts to neutralize the anticoagulant effect of heparin. Failure to properly anticoagulate the blood would result in extensive thrombus formation as the blood is exposed to an enormous area of artificial materials. An acute *post-perfusion syndrome* is observed in patients who have undergone CPB. The symptoms can include coagulation abnormalities, stroke, coma, seizures, sepsis, and neurological problems such as memory loss and short- and long-term cognitive decline.[116–118] A multitude of factors are thought to play a role in this process, of which the main agent is the contact of the blood with the vast area of foreign surface.[119] Others include tissue ischemia and hypothermia, the formation of microemboli, and the administration of important amounts of blood products. A systemic inflammatory response is triggered as the extensive contact sets in motion the complement system *via* the alternative pathway, the coagulation cascade, and the kinin–kallikrein system, as well as the activation of platelets and leukocytes.[41] This leads, on one hand, to the release of mediators of the protease cascades (tissue factor and complement mediators) and cytokines and effectors of inflammation. On the other hand, the massive activation of cellular defense systems causes transient depletion of coagulation factors from platelets, with the concomitant coagulation alterations after the CPB is reversed, and leukocytes and complement molecules.[120] Moreover, the formation of microemboli is considered an important cause of neurological disorders, as they can cause mini-strokes giving rise to ischemia in the brain.[121]

Thus, while the introduction of cardiopulmonary bypass has contributed enormously to the development of heart surgery, several questions remain about its side-effects, and considerable effort has been and continues to be invested in improving the hemocompatibility of bypass systems.[122] Heparin-coating of the artificial surfaces has been introduced in an effort to reduce the requirements of heparin infusion and minimize its adverse effects. Studies have shown a relative decrease in the markers of inflammation as well as length of hospital stays in patients who have undergone CPB surgery with heparin-coated devices.[123,124] Phosphorylcholine has also been employed in coatings in an attempt to mimic the chemical motifs present on the surface of cells.[125,126] A novel approach recently proposed involves replacing the commonly used heparin anticoagulant with an inhibitor of factor XIIa. In this way, a specific inhibition of the contact activation pathway is expected to be achieved, without increasing the risk of hemostatic complications.[32] However, other authors have observed little or no correlation between factor XII and thrombin generation *in vivo* during CPB, so the exact pathway by which coagulation is initiated remains a subject of study.[127] This is further complicated by the extensive interactions of the different systems activated during blood–surface contact, especially the release of tissue factor by monocytes or the procoagulant activity of complement in the presence of platelets.

Many other devices operate on the principle of temporarily removing part of the blood from the body for processing in an extracorporeal circuit. These include ventricular assist devices, which are used to support the blood circulation of patients with heart failure (in the short or long term) while the patient's heart is healing or until a transplant becomes available.[128] It has been found that reduction of the load to which the heart is subjected can induce regeneration and proliferation of the cells of the cardiac muscle.[129,130] Moreover, extracorporeal membrane oxygenators (ECMO) are used to support the life of patients even for several days. Working on a similar principle as CPB they face the same challenges.

A major application of extracorporeal circulation is hemodialysis, the most common procedure used to replace kidney function in patients suffering from renal insufficiency. The dialysis process involves extensive contact of the blood with the membrane and other artificial surfaces composing the extracorporeal circuit. Thus, heparin is employed to maintain anticoagulation. Nevertheless, chronic kidney disease and dialysis have been associated with cognitive impairment, cerebrovascular lesions, and high rates of stroke.[131] While multiple factors are probably contributing to these problems (including coexisting cardiovascular disease and metabolic disturbances), it has been argued that the dialysis process itself may be leading to complications, as ischemic strokes are common 30 min or less after a dialysis session.[132] Chronic systemic inflammation has been proposed as a contributor to cardiovascular risk in patients undergoing prolonged dialysis treatment, as the dialysis membranes are activators of the complement system.[41] Moreover, dialysis membranes also cause activation of the coagulation cascade pathways (although controversy remains on the importance of factor XII) and platelet activation and adhesion.[5,133] In light of these findings, also for dialysis applications, several coatings have been presented to prevent unwanted interactions and improve the hemocompatibility at the blood–membrane interface.[126]

4.2.5 Fouling and Label-free Biosensing

The continuous development of bioanalytical techniques is a cornerstone of advances in biomedical research, on the one hand, and on the evolution of clinical practice, on the other hand. Biosensors, in particular, offer significant possibilities in this regard. Techniques such as surface plasmon resonance (SPR) and quartz-crystal microbalance are routinely used in molecular biology to elucidate and quantify biomolecular affinity interactions, thus unraveling the chemical mechanisms at work in living systems.

With regard to the direct application of biosensors in medicine, possibly the largest impact has been achieved by the introduction of the glucose biosensor. The self-monitoring of the blood glucose concentration in patients with diabetes mellitus allows them to take the measures to avoid prolonged hyperglycemia and maintain tight glycemic control, thereby avoiding long-term complications, and prevent life-threatening episodes of hypoglycemia.[134] The device relies on the action of the enzyme *glucose oxidase* to catalyze the oxidation reaction of glucose.[135] The electrons exchanged in the redox process are employed to reduce either oxygen to hydrogen peroxide or a mediator such as a metal complex. The change in the concentration of oxygen, hydrogen peroxide, or oxidation state of the mediator is detected on a nearby electrode, generally amperometrically. Other novel designs are based on the direct transfer of the exchanged electrons between the enzyme and the electrode.[136] The selectivity for the analyte is provided by the enzyme, as the biological element in the biosensor.

Even with the success that this biosensor has already enjoyed, glucose biosensing remains an exciting, incredibly active area of research. Currently, emphasis is placed in developing systems that allow *continuous glucose monitoring*.[137] The goal is to track the blood glucose concentration in real time quasi-continuously *in vivo* (dozens of measurements per day). This is essential for the application of insulin pumps in a closed feedback loop, to enable diabetic patients to maintain normal glucose levels with concomitantly significantly improved outcomes.[138] However, indwelling devices for the long-term *in vivo* measurement of glucose concentration are subjected to the same types of biological response as any other implant, besides changes in time inherent to the sensor design and stability.[139] Additionally, any biosensor designed to operate *in vivo* is constrained with regards to the selection of components and materials to avoid any leaching of toxic substances.

Ultimately, the inflammation of the tissue and the foreign body response experienced by different types of indwelling sensing probes limits their durability, as the medium in which they are inserted changes. Direct measurement of the glucose concentration in blood *in vivo* requires the permanent placement of the sensor in a blood vessel. The risk of thrombus formation limits this approach to very short periods of time, such as the duration of a surgical procedure. The most commonly used alternative to gain access to the blood glucose concentration is to place the sensor in contact with interstitial fluid, for example, subcutaneously, as the concentration in this medium only lags the concentration in blood slightly. However, as foreign body response advances, the device becomes increasingly shielded from the medium it is meant to sense, necessitates frequent recalibration and, in the worst case, sensor replacement. Coating strategies to minimize these undesired effects are also being pursued, mostly based on the passivation of the outer layer of the sensor.[140] This is just one area of medical biosensing, which requires the understanding and control of the interactions between the device and the biological environment. Thus, improved therapeutic outcomes, reduced patient discomfort, and decreased need for device maintenance and replacement are expected from advances in the prevention of biosensor-induced inflammation and foreign body response.

Modern health care increasingly involves diagnostic methods based on monitoring biomarkers in body fluids. This development can allow the early identification of pathologies, even before any symptoms can be detected. Monitoring the concentration of some biomarkers is also a tool to determine the predisposition to a disease or its progression. Timely discovery of risk factors can allow their correction and management to minimize their potential health effects. Furthermore, the precise simultaneous detection of many biomarkers is expected to lead to a paradigm shift in the treatment of disease, taking advantage of the advances of molecular biology.

Direct detection of biological analytes through their intrinsic physico-chemical properties, such as mass, size, electrical impedance, or dielectric permittivity, has proved very difficult. Therefore, biological research has historically relied upon attachment of a "label" to the molecules or cells under study. The label is designed to be easily detected and acts as a surrogate to indirectly indicate the presence of the analyte. Typical labels include molecular fluorescent probes, nanoparticles, quantum dots, enzymes, and radioisotopes. Currently, most routine methods for the determination of biomarkers in body fluids include enzyme-linked immunosorbent assay (ELISA), chemiluminescence, fluorescence, radioimmunoassay, immunoprecipitation, and immunoagglutination.[38] However, these methods are very laborious and require expensive labeled reagents and highly trained staff, while some of them offer limited possibilities for automation and integration.[141–143] In practice, label-based assays also require a high degree of development to confirm that the label does not block an important active site on the tagged molecule or negatively modify its conformation resulting in unreliable results. Additionally, the use of tagged reagents makes it impossible to study multi-step sequential processes with these techniques. This is a serious drawback for research in molecular biology, where most reactions occur as a cascade of events and require real-time monitoring. In addition, emerging key disciplines such as proteomics will greatly benefit from novel technologies capable of performing rapid and multiplexed analysis of biological samples in a highly specific and sensitive manner.[141,144–146] As previously stated, label-based detection techniques are generally not able to adapt to these changes. Thus, methods offering reduced costs and complexity, while providing more quantitative information with high throughput, are required.

Label-free biosensor technology has emerged as the most promising candidate to meet the challenges of modern diagnostics and medicine. There are now various methods allowing direct detection of biological analytes without labels. *Affinity biosensors* are devices incorporating a biologically active component – a bioreceptor that is capable of interacting with the selected analyte – and a physical transducer. The sensor functions as a transducer that can convert changes in a physical property after a biorecognition event on the sensor's active surface into a quantifiable signal (Figure 4.8). In this way, label-free detection avoids the experimental uncertainties induced by the effect of the label on the molecular conformation, the blocking of active binding epitopes, or the inaccessibility of the labeling site. Furthermore, label-free detection methods greatly simplify the time and effort required for assay development, while minimizing experimental artefacts from quenching, shelf life, and background fluorescence, generally found on label-based assays.

Various physical principles have been used to design biosensors, being the most sensitive the sensors based on optical waves, electrochemical signals, and the piezoelectric effect. In particular, optical biosensors offer potential advantages as the measurement does not require any physical or electrical contact between the sensor and the detection instrument (*e.g.* spectrophotometer), simplifying the preparation of "multi-sensing" surfaces. A single multi-sensing surface has the potential to assess a huge number of analytes by illuminating different areas of the sensor decorated with an array of biorecognition elements in discrete positions. This allows the inexpensive preparation of devices for high-throughput measurement.[147–149] Detection

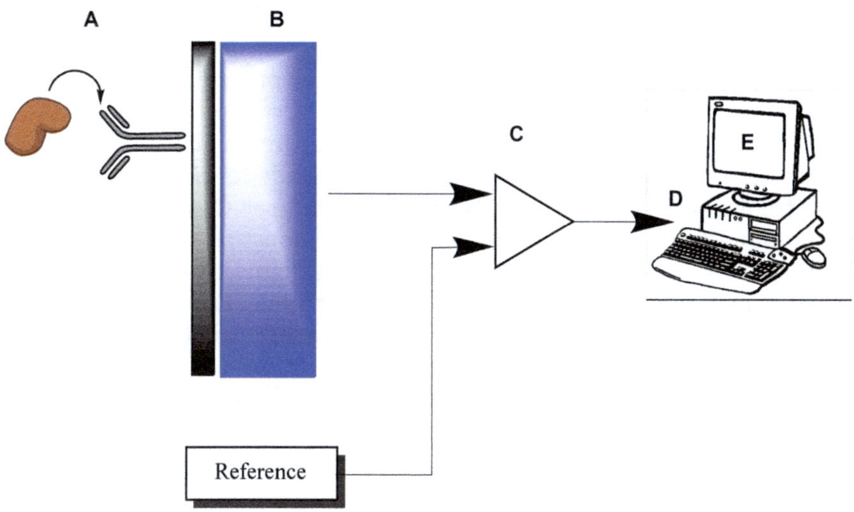

Figure 4.8 Scheme of a label-free biosensor. The binding event occurs at the biorecognition element (A). This reaction is converted into an output by the transducer (B) which is amplified (C) and processed (D) leading to a read-out (E).

of biomarkers for various diseases, pathogens, and monitoring biomolecular interactions in complex matrices is of key importance in the analytical field. Remarkably, label-free affinity biosensors have the potential to reduce the time of analysis while having excellent sensitivity and selectivity.[150–153]

However, contrary to other multi-step analytical methods or biosensors based on labeled reagents, the direct detection in biological fluids is generally subjected to interference by *non-specific adsorption* phenomena. When quantifying an analyte in a complex biological matrix, fouling on the transducer surface can also give rise to a signal.[154,155] As the sensor cannot differentiate between the specific response due to the capture of analytes by the immobilized bioreceptors and the non-specific response caused by fouling of other components, it is imperative to suppress fouling on the transducer surface. The transducer surface must be able to *selectively* capture the desired analyte from the sample while preventing fouling of other biomolecules. Analytes of biomedical interest are present in body fluids, especially blood. In particular, blood plasma and serum allow easier manipulation and sample storage than whole blood, and, in general, are preferred. Thus, prevention of fouling from blood serum and plasma is critical for the success of label-free biosensing applications.

A vast number of examples of sensing platforms have been introduced for a plethora of highly relevant analytes. However, only very few reports have tackled the direct detection of analytes in blood plasma or serum. The rapid protein fouling commonly observed on most artificial materials poses serious restrictions to the surfaces which can be used for real applications.

Some authors have proposed the use of a reference channel to separate the specific from the non-specific signal by subtracting the reference to the sensing channel. Two strategies have been applied: (i) injection of the same sample in a channel modified with "dummy" bioreceptors (*i.e.* not able to bind the analyte), and (ii) injection of blood plasma or serum from a different donor, not containing the analyte of interest. The first approach, in general, is not fully effective due to the impossibility to prepare an absolutely identical dummy reference surface. The second strategy leads to bigger errors, since blood from different donors can be very different (see Chapter 2) and their fouling can vary dramatically.[57,156] Figure 4.9 demonstrates the variability of fouling from blood plasma. The right panel highlights that on specific surfaces the fouling can vary up to an order of magnitude. Effective suppression of fouling from all donors, as displayed by poly(carboxybetaine acrylamide) and poly[*N*-(2-hydroxypropyl) methacrylamide], is the only possibility for eliminating the influence of fouling on biosensing measurements.

The general principle behind the design of biosensor targeting measurements in complex bodily (especially blood-derived) fluids is illustrated in Figure 4.10. The surface coating strategies employed to achieve the required antifouling effect will be reviewed in the next chapter. At the same time, appropriate *biofunctionalization* techniques must be applied for the attachment of bioreceptors, which should be immobilized in sufficient surface density and proper orientation. The chemical functionalization techniques

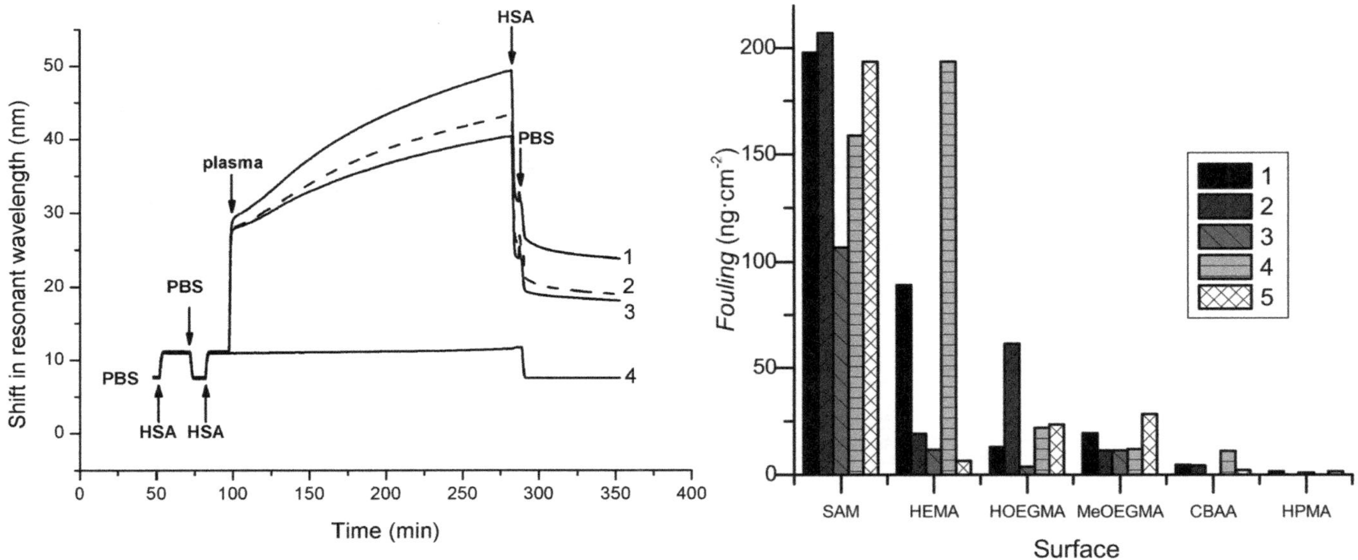

Figure 4.9 Difference in fouling observed in samples of different donors. Left: surface plasmon resonance sensogram showing the interaction of blood plasma collected from three different donors with a carboxy-terminated poly(ethylene glycol) coating. HSA: human serum albumin, 5 mg mL^{-1} in phosphate-buffered saline (PBS, pH 7.4). Reprinted with permission from ref. 57. Copyright 2009 American Chemical Society. Right: Fouling from blood plasma from 5 different donors on several antifouling surfaces as measured by surface plasmon resonance during 15 min of contact. SAM, self-assembled monolayer of oligo(ethylene glycol) alkanethiols; HEMA, poly(2-hydroxyethyl methacrylate); HOEGMA, poly[oligo(ethylene glycol methacrylate)]; MeOEGMA, poly[oligo(ethylene glycol methacrylate) methyl ether]; CBAA, poly(carboxybetaine acrylamide); and HPMA, poly[*N*-(2-hydroxypropyl)methacrylamide]. Reproduced from Ref. 156 with permission from The Royal Society of Chemistry.

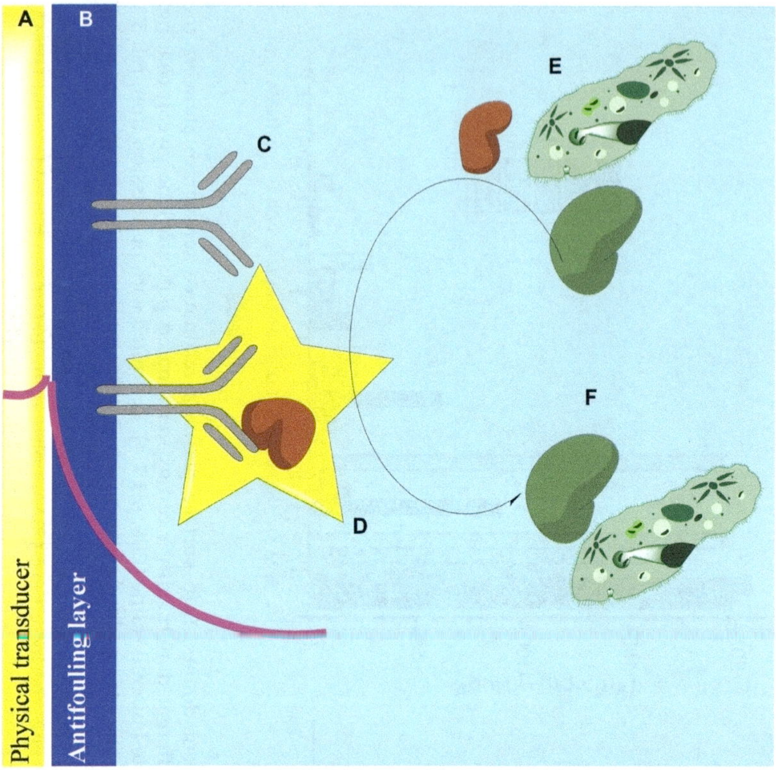

Figure 4.10 Scheme of an affinity biosensor: the physical transducer (A) produces a signal from the binding event (D) between the bioreceptor (C) and the analyte. An antifouling layer (B) prevents the non-specific adsorption of the rest of the components (F) present in the sample (E).

employed must also be mild enough to, on the one hand, prevent damage or changes in the conformation of the bioreceptors and, on the other hand, not impair the antifouling performance of the underlying layer.[157] Attaining functionalization methods that fulfill these requirements is one of the challenges of current research, often taking advantage of novel chemical ligation protocols and control of the surface architecture.

The combination of antifouling layers and the careful immobilization of bioreceptors has enabled the sensitive detection of analytes of interest in complex biological media. Recently, a label-free optical affinity biosensor capable of detecting the antibodies against a common virus at their natural concentration in real clinical samples has been reported.[158] The results obtained with the biosensor corresponded well with the established ELISA protocol for the same samples.

Thus, the enormous advances made in novel physical sensing principles must be accompanied by (i) the understanding of the mechanisms of fouling from biological fluids, and (ii) the development of new and perfected surface chemistries to suppress it. The next chapter will describe the progress made in this regard, which is expected to bring biosensor technology to a wider range of applications that will greatly benefit from it.

References

1. D. F. Williams, *Biomaterials*, 2009, **30**, 5897–5909.
2. S. Franz, S. Rammelt, D. Scharnweber and J. C. Simon, *Biomaterials*, 2011, **32**, 6692–6709.
3. J. M. Anderson, A. Rodriguez and D. T. Chang, *Semin. Immunol.*, 2008, **20**, 86–100.
4. F. De Somer, A. Van Landschoot, G. Van Nooten and J. Delanghe, *Perfusion*, 2008, **23**, 215–221.
5. M. B. Gorbet and M. V. Sefton, *Biomaterials*, 2004, **25**, 5681–5703.
6. B. Sivaraman and R. A. Latour, *Biomaterials*, 2010, **31**, 1036–1044.
7. C. Sperling, M. Fischer, M. F. Maitz and C. Werner, *Biomaterials*, 2009, **30**, 4447–4456.
8. L. Vroman and A. L. Adams, *J. Biomed. Mater. Res.*, 1969, **3**, 43–67.
9. L. Vroman and A. L. Adams, *J. Colloid Interface Sci.*, 1985, **111**, 391–402.
10. L. Vroman, *Materials*, 2009, **2**, 1547–1557.
11. T. Riedel, Z. Riedelova-Reicheltova, P. Majek, C. Rodriguez-Emmenegger, M. Houska, J. E. Dyr and E. Brynda, *Langmuir*, 2013, **29**, 3388–3397.
12. A. H. Schmaier, L. Silver, A. L. Adams, G. C. Fischer, P. C. Munoz, L. Vroman and R. W. Colman, *Thromb. Res.*, 1984, **33**, 51–67.
13. C. Rodriguez-Emmenegger, E. Brynda, T. Riedel, M. Houska, V. Šubr, A. B. Alles, E. Hasan, J. E. Gautrot and W. T. S. Huck, *Macromol. Rapid Commun.*, 2011, **32**, 952–957.
14. A. E. Engberg, P. H. Nilsson, S. Huang, K. Fromell, O. A. Hamad, T. E. Mollnes, J. P. Rosengren-Holmberg, K. Sandholm, Y. Teramura, I. A. Nicholls, B. Nilsson and K. N. Ekdahl, *Biomaterials*, 2015, **36**, 55–65.
15. E. Bastida, G. Escolar, A. Ordinas and J. J. Sixma, *Blood*, 1987, **70**, 1437–1442.
16. B. Sivaraman and R. A. Latour, *Biomaterials*, 2011, **32**, 5365–5370.
17. C. Blaszykowski, S. Sheikh and M. Thompson, *Trends Biotechnol.*, 2014, **32**, 61–62.
18. K. N. Ekdahl, J. Hong, O. A. Hamad, R. Larsson and B. Nilsson, *Adv. Exp. Med. Biol.*, 2013, **735**, 257–270.
19. B. Sivaraman and R. A. Latour, *Biomaterials*, 2010, **31**, 832–839.
20. L. C. Xu, J. W. Bauer and C. A. Siedlecki, *Colloids Surf., B*, 2014, **124**, 49–68.
21. M. Gawaz, F.-J. Neumann, T. Dickfeld, A. Reininger, H. Adelsberger, A. Gebhardt and A. Schömig, *Circulation*, 1997, **96**, 1809–1818.
22. R. Zhuo, C. A. Siedlecki and E. A. Vogler, *Biomaterials*, 2006, **27**, 4325–4332.

23. J. Ding, Z. Chen, S. Niu, J. Zhang, N. K. Mondal, B. P. Griffith and Z. J. Wu, *Artif. Organs*, 2015, **39**, 576–583.
24. B. Sivaraman and R. A. Latour, *Biomaterials*, 2010, **31**, 1036–1044.
25. *Molecular Hematology*, ed. D. Provan and J. G. Gribben, Blackwell Publishing Ltd, 2005.
26. *Hemostasis and Thrombosis Basic Principles and Clinical Practice*, ed. V. J. Marder, W. C. Aird, J. Bennett, S. Schulman and G. C. White, Lippincott Williams & Wilkins, 2013.
27. M. Huang, A. C. Rigby, X. Morelli, M. A. Grant, G. Huang, B. Furie, B. Seaton and B. C. Furie, *Nat. Struct. Mol. Biol.*, 2003, **10**, 751–756.
28. K. Mann, M. Nesheim, W. Church, P. Haley and S. Krishnaswamy, *Blood*, 1990, **76**, 1–16.
29. M. Sunnerhagen, T. Drakenberg, S. Forsen and J. Stenflo, *Haemostasis*, 1996, **26**, 45–53.
30. *Hemostasis and Thrombosis*, ed. T. G. DeLoughery, Springer 2015.
31. E. W. Davie and O. D. Ratnoff, *Science*, 1964, **145**, 1310–1312.
32. M. Larsson, V. Rayzman, M. W. Nolte, K. F. Nickel, J. Bjorkqvist, A. Jamsa, M. P. Hardy, M. Fries, S. Schmidbauer, P. Hedenqvist, M. Broome, I. Pragst, G. Dickneite, M. J. Wilson, A. D. Nash, C. Panousis and T. Renne, *Sci. Transl. Med.*, 2014, **6**, 222ra17.
33. E. A. Vogler and C. A. Siedlecki, *Biomaterials*, 2009, **30**, 1857–1869.
34. J. M. Anderson, *J. Mater. Sci.: Mater. Med.*, 2015, **26**, 121.
35. J. M. Anderson, *Annu. Rev. Mater. Res.*, 2001, **31**, 81–110.
36. G. Asimakopoulos and K. M. Taylor, *Ann. Thorac. Surg.*, 1998, **66**, 2135–2144.
37. S. M. Albelda, C. W. Smith and P. A. Ward, *FASEB J.*, 1994, **8**, 504–512.
38. K. Murphy, P. Travers and M. Walport, *Janeway's Immunobiology*, Taylor & Francis, New York, 2008.
39. R. B. Johnston Jr., *N. Engl. J. Med.*, 1988, **318**, 747–752.
40. M. Fukutomi, S. Kobayashi, K. Niwaya, Y. Hamada and S. Kitamura, *Artif. Organs*, 1996, **20**, 767–776.
41. B. Nilsson, K. N. Ekdahl, T. E. Mollnes and J. D. Lambris, *Mol. Immunol.*, 2007, **44**, 82–94.
42. M. M. Markiewski, B. Nilsson, K. N. Ekdahl, T. E. Mollnes and J. D. Lambris, *Trends Immunol.*, 2007, **28**, 184–192.
43. U. Amara, D. Rittirsch, M. Flierl, U. Bruckner, A. Klos, F. Gebhard, J. D. Lambris and M. Huber-Lang, 2008, **632**, 68–76.
44. P. J. Sims and T. Wiedmer, *Immunol. Today*, 1991, **12**, 338–342.
45. C. V. Gould, C. A. Umscheid, R. K. Agarwal, G. Kuntz, D. A. Pegues and C. Healthcare, Infection Control Practices Advisory, *Infect. Control Hosp. Epidemiol.*, 2010, **31**, 319–326.
46. H. J. Busscher and H. C. van der Mei, *PLoS Pathog.*, 2012, **8**, e1002440.
47. J. W. Costerton, G. G. Geesey and K.-J. Cheng, *Sci. Am.*, 1978, **238**, 86.
48. J. W. Costerton, P. S. Stewart and E. P. Greenberg, *Science*, 1999, **284**, 1318–1322.

49. M. Y. Wu, V. Sendamangalam, Z. Xue and Y. Seo, *Biofouling*, 2012, **28**, 1119–1128.
50. B. Schachter, *Nat. Biotechnol.*, 2003, **21**, 361–365.
51. H. Ceri, M. E. Olson, C. Stremick, R. R. Read, D. Morck and A. Buret, *J. Clin. Microbiol.*, 1999, **37**, 1771–1776.
52. W. L. Cochran, G. A. McFeters and P. S. Stewart, *J. Appl. Microbiol.*, 2000, **88**, 22–30.
53. M. E. Davey and G. A. O'Toole, *Microbiol. Mol. Biol. Rev.*, 2000, **64**, 847–867.
54. R. Firoozabadi, T. Alton and J. Wenke, *J. Am. Acad. Orthop. Surg.*, 2015, **23**, 443–451.
55. O. Rzhepishevska, S. Hakobyan, R. Ruhal, J. Gautrot, D. Barbero and M. Ramstedt, *Biomater. Sci.*, 2013, **1**, 589–602.
56. C. Rodriguez-Emmenegger, M. Houska, A. B. Alles and E. Brynda, *Macromol. Biosci.*, 2012, **12**, 1413–1422.
57. C. Rodriguez-Emmenegger, E. Brynda, T. Riedel, Z. Sedlakova, M. Houska and A. B. Alles, *Langmuir*, 2009, **25**, 6328–6333.
58. A. Clements, F. Gaboriaud, J. F. Duval, J. L. Farn, A. W. Jenney, T. Lithgow, O. L. Wijburg, E. L. Hartland and R. A. Strugnell, *PLoS One*, 2008, **3**, e3817.
59. R. Bos, H. C. van der Mei and H. J. Busscher, *FEMS Microbiol. Rev.*, 1999, **23**, 179–230.
60. Y. Chen, A. K. Harapanahalli, H. J. Busscher, W. Norde and H. C. van der Mei, *Appl. Environ. Microbiol.*, 2014, **80**, 637–643.
61. A. Roosjen, H. J. Busscher, W. Norde and H. C. van der Mei, *Microbiology*, 2006, **152**, 2673–2682.
62. X. Laloyaux, E. Fautré, T. Blin, V. Purohit, J. Leprince, T. Jouenne, A. M. Jonas and K. Glinel, *Adv. Mater.*, 2010, **22**, 5024–5028.
63. K. Glinel, A. M. Jonas, T. Jouenne, J. r. m. Leprince, L. Galas and W. T. S. Huck, *Bioconjugate Chem.*, 2008, **20**, 71–77.
64. S. H. Hakobyan, O. Rzhepishevska, R. Ruhal and M. Ramstedt, *European Cells and Materials*, 2013, **26**, 37.
65. M. Krishnamoorthy, S. Hakobyan, M. Ramstedt and J. E. Gautrot, *Chem. Rev.*, 2014, **114**, 10976–11026.
66. F. Hui and C. Debiemme-Chouvy, *Biomacromolecules*, 2013, **14**, 585–601.
67. J. Hasan, R. J. Crawford and E. P. Ivanova, *Trends Biotechnol.*, 2013, **31**, 295–304.
68. W. J. Yang, T. Cai, K. G. Neoh, E. T. Kang, G. H. Dickinson, S. L. Teo and D. Rittschof, *Langmuir*, 2011, **27**, 7065–7076.
69. R. Wang, K. G. Neoh, Z. Shi, E. T. Kang, P. A. Tambyah and E. Chiong, *Biotechnol. Bioeng.*, 2012, **109**, 336–345.
70. R. Hu, G. Li, Y. Jiang, Y. Zhang, J.-J. Zou, L. Wang and X. Zhang, *Langmuir*, 2013, **29**, 3773–3779.
71. Y. Wang, E. Y. Chi, D. O. Natvig, K. S. Schanze and D. G. Whitten, *ACS Appl. Mater. Interfaces*, 2013, **5**, 4555–4561.

72. G. Gao, D. Lange, K. Hilpert, J. Kindrachuk, Y. Zou, J. T. J. Cheng, M. Kazemzadeh-Narbat, K. Yu, R. Wang, S. K. Straus, D. E. Brooks, B. H. Chew, R. E. W. Hancock and J. N. Kizhakkedathu, *Biomaterials*, 2011, **32**, 3899–3909.
73. M. Ramstedt, N. Cheng, O. Azzaroni, D. Mossialos, H. J. Mathieu and W. T. S. Huck, *Langmuir*, 2007, **23**, 3314–3321.
74. Q. Cui, W. M. Mihalko, J. S. Shields, M. Ries and K. J. Saleh, *J. Bone Jt. Surg., Am. Vol.*, 2007, **89**, 871–882.
75. G. Cheng, H. Xue, Z. Zhang, S. Chen and S. Jiang, *Angew. Chem., Int. Ed.*, 2008, **47**, 8831–8834.
76. M. Ramstedt, B. Ekstrand-Hammarstrom, A. V. Shchukarev, A. Bucht, L. Osterlund, M. Welch and W. T. S. Huck, *Biomaterials*, 2009, **30**, 1524–1531.
77. C. J. Van Oss, M. K. Chaudhury and R. J. Good, *Chem. Rev.*, 1988, **88**, 927–941.
78. C. Rodriguez-Emmenegger, S. Janel, A. de los Santos Pereira, M. Bruns and F. Lafont, , *Polym. Chem*, 2015, **6**, 5740–5751.
79. C. Rodriguez-Emmenegger, A. Decker, F. Surman, C. M. Preuss, Z. Sedláková, N. Zydziak, C. Barner-Kowollik, T. Schwartz and L. Barner, *RSC Adv.*, 2014, **4**, 64781–64790.
80. K. Glinel, A. M. Jonas, T. Jouenne, J. r. m. Leprince, L. Galas and W. T. S. Huck, *Bioconjugate Chem.*, 2008, **20**, 71–77.
81. H. R. Baumgartner, V. Turitto and H. J. Weiss, *J. Lab. Clin. Med.*, 1980, **95**, 208–221.
82. F. Surman, T. Riedel, M. Bruns, N. Y. Kostina, Z. Sedlakova and C. Rodriguez-Emmenegger, *Macromol. Biosci.*, 2015, **15**, 636–646.
83. H. C. Lowe, S. N. Oesterle and L. M. Khachigian, *J. Am. Coll. Cardiol.*, 2002, **39**, 183–193.
84. E. D. Grech, *BMJ [Br. Med. J.]*, 2003, **326**, 1080–1082.
85. F. Nazneen, G. Herzog, D. W. Arrigan, N. Caplice, P. Benvenuto, P. Galvin and M. Thompson, *J. Biomed. Mater. Res., Part B*, 2012, **100**, 1989–2014.
86. M. C. Morice, P. W. Serruys, J. E. Sousa, J. Fajadet, E. Ban Hayashi, M. Perin, A. Colombo, G. Schuler, P. Barragan, G. Guagliumi, F. Molnar and R. Falotico, *N. Engl. J. Med.*, 2002, **346**, 1773–1780.
87. L. Pendyala, R. Jabara, T. Shinke, N. Chronos, K. Robinson, J. Li and D. Hou, *Cardiovasc. Hematol. Agents Med. Chem.*, 2008, **6**, 105–115.
88. M. E. Pfisterer, *Circulation*, 2008, **118**, 1117–1119.
89. D. J. Malenka, A. V. Kaplan, F. L. Lucas, S. M. Sharp and J. S. Skinner, *JAMA*, 2008, **299**, 2868–2876.
90. M. Joner, A. V. Finn, A. Farb, E. K. Mont, F. D. Kolodgie, E. Ladich, R. Kutys, K. Skorija, H. K. Gold and R. Virmani, *J. Am. Coll. Cardiol.*, 2006, **48**, 193–202.
91. P. K. Kuchulakanti, W. W. Chu, R. Torguson, P. Ohlmann, S. W. Rha, L. C. Clavijo, S. W. Kim, A. Bui, N. Gevorkian, Z. Xue, K. Smith, J. Fournadjieva, W. O. Suddath, L. F. Satler, A. D. Pichard, K. M. Kent and R. Waksman, *Circulation*, 2006, **113**, 1108–1113.

92. J. Daemen and P. W. Serruys, *Circulation*, 2007, **116**, 316–328.
93. N. Nikam, T. B. Steinberg and D. H. Steinberg, *Med. Devices*, 2014, **7**, 165–178.
94. J. A. Ormiston and P. W. Serruys, *Circ.: Cardiovasc. Interventions*, 2009, **2**, 255–260.
95. H. Hermawan, D. Dube and D. Mantovani, *Acta Biomater.*, 2010, **6**, 1693–1697.
96. F. Vogt, A. Stein, G. Rettemeier, N. Krott, R. Hoffmann, J. vom Dahl, A. K. Bosserhoff, W. Michaeli, P. Hanrath, C. Weber and R. Blindt, *Eur. Heart J.*, 2004, **25**, 1330–1340.
97. J. P. Lane, L. E. Perkins, A. J. Sheehy, E. J. Pacheco, M. P. Frie, B. J. Lambert, R. J. Rapoza and R. Virmani, *JACC Cardiovasc. Interventions*, 2014, **7**, 688–695.
98. G. Nakazawa, J. F. Granada, C. L. Alviar, A. Tellez, G. L. Kaluza, M. Y. Guilhermier, S. Parker, S. M. Rowland, F. D. Kolodgie, M. B. Leon and R. Virmani, *JACC Cardiovasc. Interventions*, 2010, **3**, 68–75.
99. F. Chen, F. Yang, Q. Zhao, S. Feng, W. Li, Y. Bi, S. Zhang, Y. Wang and B. O. Feng, *J. Interventions Cardiol.*, 2015, **28**, 257–263.
100. M. Haude, S. W. Lee, S. G. Worthley, S. Silber, S. Verheye, S. Erbs, M. A. Rosli, R. Botelho, I. Meredith, K. H. Sim, P. R. Stella, H. C. Tan, R. Whitbourn, S. Thambar, A. Abizaid, T. H. Koh, P. Den Heijer, H. Parise, E. Cristea, A. Maehara and R. Mehran, *JACC Cardiovasc. Interventions*, 2013, **6**, 334–343.
101. F. J. Schoen and R. F. Padera, in *Biomaterials Science*, ed. B. D. Ratner, A. S. Hoffman, F. J. Schoen and J. E. Lemons, 3rd edn, 2013, pp. 771–784.
102. G. C. Hunter, K. J. Woodside and J. J. Naoum, in *Comprehensive Vascular and Endovascular Surgery*, ed. J. W. Hallett, J. L. Mills, J. J. Earnshaw, J. A. Reekers and T. W. Rooke, Elsevier, Philadelphia, Second edn, 2009, pp. 665–687.
103. A. I. Cassady, N. M. Hidzir and L. Grøndahl, *J. Appl. Polym. Sci.*, 2014, **131**, 40533.
104. L. Feng and J. D. Andrade, *J. Biomater. Sci., Polym. Ed.*, 1996, **7**, 439–452.
105. R. Y. Kannan, H. J. Salacinski, P. E. Butler, G. Hamilton and A. M. Seifalian, *J. Biomed. Mater. Res., B Part B*, 2005, **74**, 570–581.
106. P. H. Lin, C. Chen, R. L. Bush, Q. Yao, A. B. Lumsden and S. R. Hanson, *J. Vasc. Surg.*, 2004, **39**, 1322–1328.
107. A. B. Lumsden and N. J. Morrissey, *J. Vasc. Surg.*, 2015, **61**, 703–712.e1.
108. D. Bezuidenhout, D. F. Williams and P. Zilla, *Biomaterials*, 2015, **36**, 6–25.
109. P. Pibarot and J. G. Dumesnil, *Circulation*, 2009, **119**, 1034–1048.
110. S. Muralidharan, V. Muthubaskeran and P. Chandrasekar, *Indian J. Thorac. Cardiovasc. Surg.*, 2011, **27**, 24–27.
111. F. J. Schoen and R. J. Levy, *J. Biomed. Mater. Res.*, 1999, **47**, 439–465.
112. P. Zilla, J. Brink, P. Human and D. Bezuidenhout, *Biomaterials*, 2008, **29**, 385–406.

113. M. Moczar, J. P. Mazzucotelli, P. Bertrand, M. Ginat, J. Leandri and D. Loisance, *ASAIO J.*, 1994, **40**, M697–M701.
114. S. P. Nolan, R. Zacour and J. F. Dammann, *Ann. Thorac. Surg.*, 1997, **64**, 1540–1543.
115. J. R. Utley, *J. Cardiac Surg.*, 1990, **5**, 177–189.
116. S. Uysal and D. L. Reich, *J. Cardiothorac. Vasc. Anesth.*, 2013, **27**, 958–971.
117. A. Silverstein and H. P. Krieger, *Arch. Neurol.*, 1960, **3**, 601–605.
118. G. J. Murphy and G. D. Angelini, *J. Cardiac Surg.*, 2004, **19**, 481–488.
119. D. Royston, *J. Cardiothorac. Vasc. Anesth.*, 1997, **11**, 341–354.
120. L. A. Harker, T. W. Malpass, H. E. Branson, E. A. Hessel 2nd and S. J. Slichter, *Blood*, 1980, **56**, 824–834.
121. N. Stroobant, G. Van Nooten, Y. Van Belleghem and G. Vingerhoets, *Acta Cardiol.*, 2010, **65**, 557–564.
122. J. E. Marcoux and T. R. Mycyk, *Perfusion*, 2013, **28**, 433–439.
123. H. P. Wendel and G. Ziemer, *Eur. J. Cardio-Thorac. Surg.*, 1999, **16**, 342–350.
124. A. K. Zimmermann, N. Weber, H. Aebert, G. Ziemer and H. P. Wendel, *J. Biomed. Mater. Res., Part B*, 2007, **80**, 433–439.
125. F. Pappalardo, P. Della Valle, G. Crescenzi, C. Corno, A. Franco, L. Torracca, O. Alfieri, L. Galli, A. Zangrillo and A. D'Angelo, *Ann. Thorac. Surg.*, 2006, **81**, 886–891.
126. C. Werner, M. F. Maitz and C. Sperling, *J. Mater. Chem.*, 2007, **17**, 3376–3384.
127. J. F. Burman, H. I. Chung, J. C. R. Lincoln, D. A. Lane, H. Philippou and A. Adami, *Lancet*, 1994, **344**, 1192–1193.
128. L. W. Miller, F. D. Pagani, S. D. Russell, R. John, A. J. Boyle, K. D. Aaronson, J. V. Conte, Y. Naka, D. Mancini, R. M. Delgado, T. E. MacGillivray, D. J. Farrar and O. H. Frazier, *N. Engl. J. Med.*, 2007, **357**, 885–896.
129. D. C. Canseco, W. Kimura, S. Garg, S. Mukherjee, S. Bhattacharya, S. Abdisalaam, S. Das, A. Asaithamby, P. P. Mammen and H. A. Sadek, *J. Am. Coll. Cardiol.*, 2015, **65**, 892–900.
130. A. A. Gupte, D. J. Hamilton, A. M. Cordero-Reyes, K. A. Youker, Z. Yin, J. D. Estep, R. D. Stevens, B. Wenner, O. Ilkayeva, M. Loebe, L. E. Peterson, C. J. Lyon, S. T. C. Wong, C. B. Newgard, G. Torre-Amione, H. Taegtmeyer and W. A. Hsueh, *Circ.: Cardiovasc. Genet.*, 2014, **7**, 266–276.
131. A. M. Murray, *Adv. Chronic Kidney Dis.*, 2008, **15**, 123–132.
132. K. Toyoda, K. Fujii, S. Fujimi, Y. Kumai, H. Tsuchimochi, S. Ibayashi and M. Iida, *Am. J. Kidney Dis.*, 2005, **45**, 1058–1066.
133. T. Renne, M. Pozgajova, S. Gruner, K. Schuh, H. U. Pauer, P. Burfeind, D. Gailani and B. Nieswandt, *J. Exp. Med.*, 2005, **202**, 271–281.
134. The Diabetes Control and Complications Trial Research Group, *N. Engl. J. Med.*, 1993, **329**, 977–986.
135. R. Wilson and A. P. F. Turner, *Biosens. Bioelectron.*, 1992, **7**, 165–185.

136. J. Wang, *Chem. Rev.*, 2008, **108**, 814–825.
137. A. Liebl, H. R. Henrichs, L. Heinemann, G. Freckmann, E. Biermann and A. Thomas, *J. Diabetes Sci. Technol.*, 2013, **7**, 500–519.
138. A. Heller, *AlChE J.*, 2005, **51**, 1054–1066.
139. S. Vaddiraju, D. J. Burgess, I. Tomazos, F. C. Jain and F. Papadimitrakopoulos, *J. Diabetes Sci. Technol.*, 2010, **4**, 1540–1562.
140. S. P. Nichols, A. Koh, W. L. Storm, J. H. Shin and M. H. Schoenfisch, *Chem. Rev.*, 2013, **113**, 2528–2549.
141. *Optical Biosensors*, ed. F. S. Ligler and C. R. Taitt, Elsevier, Amsterdam, Second edn, 2008.
142. *Label-Free Biosensors. Techniques and Applications*, ed. M. A. Cooper, Cambridge University Press, Cambridge, 2009.
143. *Surface Plasmon Resonance Based Sensors*, ed. J. Homola, Springer-Verlag, Berlin, 2006.
144. *Principles of Bacterial Detection: Biosensors, Recognition Receptors and Microsystems*, ed. M. Zourob, S. Elwary and A. P. F. Turner, Springer, 2008.
145. A. Sadana, *Binding and Dissociation Kinetics for Different Biosensor Applications Using Fractals*, Elsevier, 2006.
146. *Affinity Biosensors. Techniques and Protocols*, ed. K. R. Rogers and A. Mulchandani, Humana Press Inc, Totowa, New Jersey, 1998.
147. G. MacBeath and S. L. Schreiber, *Science*, 2000, **289**, 1760–1763.
148. *Protein Arrays, Biochips and Proteomics. The next Phase of Genomic Discovery*, ed. J. S. Albala and I. Humphery-Smith, Marcel Dekker, Inc., New York, 2003.
149. P. Jonkheijm, D. Weinrich, H. Schröder, C. M. Niemeyer and H. Waldmann, *Angew. Chem., Int. Ed.*, 2008, **47**, 9618–9647.
150. K. De Vos, J. Girones, S. Popelka, E. Schacht, R. Baets and P. Bienstman, *Biosens. Bioelectron.*, 2009, **24**, 2528–2533.
151. J. Homola, *Anal. Bioanal. Chem.*, 2003, **377**, 528–539.
152. J. Homola, *Chem. Rev.*, 2008, **108**, 462–493.
153. L. M. Lechuga, K. Zinoviev, L. Fernandez, J. Elizalde, O. E. Hidalgo and C. Dominguez, in *Proc. SPIE 7220, Silicon Photonics IV*, ed. J. A. Kubby and G. T. Reed, 2009, p. 72200L.
154. C. Rodriguez-Emmenegger, O. A. Avramenko, E. Brynda, J. Skvor and A. B. Alles, *Biosens. Bioelectron*, 2011, **26**, 4545–4551.
155. H. Vaisocherova, W. Yang, Z. Zhang, Z. Cao, G. Cheng, M. Piliarik, J. Homola and S. Jiang, *Anal. Chem.*, 2008, **80**, 7894–7901.
156. A. de los Santos Pereira, C. Rodriguez-Emmenegger, F. Surman, T. Riedel, A. Bologna and E. Brynda, *RSC Adv.*, 2014, **4**, 2318–2321.
157. H. Vaisocherova, V. Sevcu, P. Adam, B. Spackova, K. Hegnerova, A. de los Santos Pereira, C. Rodriguez-Emmenegger, T. Riedel, M. Houska, E. Brynda and J. Homola, *Biosens. Bioelectron*, 2014, **51**, 150–157.
158. T. Riedel, C. Rodriguez-Emmenegger, A. de los Santos Pereira, A. Bědajánková, P. Jinoch, P. M. Boltovets and E. Brynda, *Biosens. Bioelectron.*, 2014, **55**, 278–284.

CHAPTER 5

Antifouling Surface Chemistries to Minimize Signal Interference from Biological Matrices in Biosensor Technology

5.1 The Issue of Signal Interference from the Biological Matrix in Biosensor Technology

The year 2015 marked the 40th anniversary of the first commercial biosensor.[1–3] Originally, the worldwide market for such a novel technology remained somewhat limited, only reaching an estimated US$ 5 million per year a decade later (Figure 5.1).[4] Sales modestly but steadily increased in the ensuing 1990s before experiencing, at the turn of the new millenium, a genuine exponential growth (Figure 5.1). The situation today is such that biosensor technology has become a very lucrative, hence fiercely competitive field worth well over US$ 10 billion per year, and one that still is promised a bright commercial future.[1,3–6] Such a staggering number may be misleading however since, admittedly, biosensors clearly are far from being ubiquitous in our society and, apart from occasional advertising campaigns, largely remain a curiosity for the general population. A closer look at the realm of commercial biosensors reveals a much different reality indeed, the market being in actuality dominated by only one type of such devices,[1,3,5,6] those healthcare-oriented that are designed to monitor the level of glucose in the blood of individuals afflicted with diabetes[2,3] (pregnancy tests also represent a substantial share of the market[5]). The following question then naturally

RSC Detection Science Series No. 8
Biological Fluid–Surface Interactions in Detection and Medical Devices
By Michael Thompson, Christophe Blaszykowski, Sonia Sheikh, Cesar Rodriguez-Emmenegger and Andres de los Santos Pereira

Published by the Royal Society of Chemistry, www.rsc.org

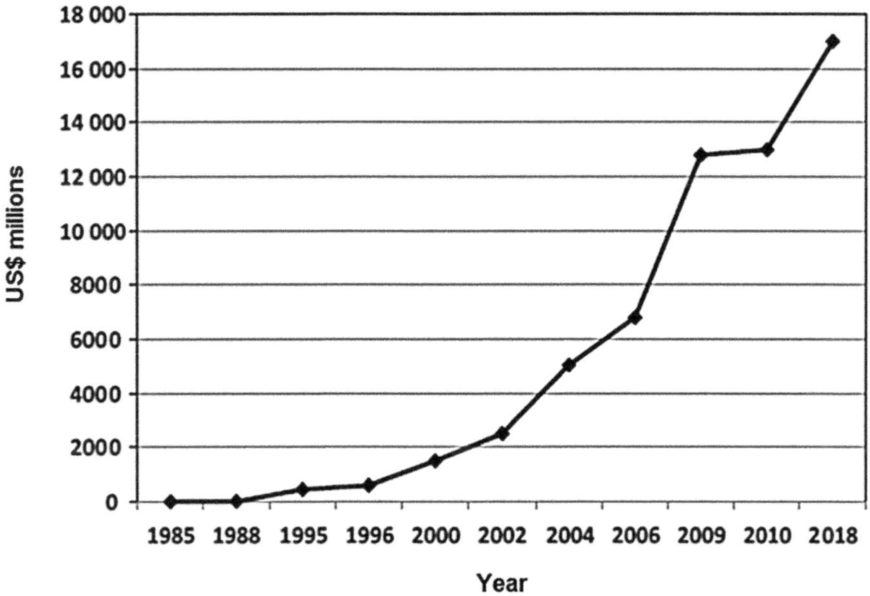

Figure 5.1 30+ years of commercial biosensor technology: estimated past, present, and future world market (in US$ millions).
Reprinted from A. Turner, Biosensors: then and now, *Trends Biotechnol.*, **31**, 119–120, Copyright 2013 with permission from Elsevier.

arises: Why have commercial biosensors only matured in niche applications?[1,6] More fundamentally, why has biosensor technology – after all these years of dedicated research – still not found a prominent place as screening/diagnostic alternative despite attractive features notably in terms of convenience and cost of operation, speed of analysis, and miniaturization capability for decentralized, point-of-care testing?[6] Beyond the usual, business-driven financial imperatives in terms of upfront R&D capital investment *vs.* projected commercial profitability (which the large-scale 'hyperglycemia' market justifies and meets),[1,3,5] the major hurdle for biosensor development is technical in nature and relates, to a large extent, to the pivotal issue of signal interference from the biological matrix where detection, more often than not, is intended to be performed.[6–8]

Conceptually, biosensors are analytical devices designed to convert biomolecular interactions occurring at the sensor–liquid boundary into measurable signals (Figure 5.2).[1,7] Composed of a biorecognition element (made of biomolecular receptor probes) intimately interfaced with a transducing component,[6] biosensors are engineered in practice to selectively and sensitively detect in a real-time and ideally label-free manner target analytes that can range in size and complexity from small molecular species to larger biological entities. However, for a biosensor to be fully functional and the resulting bioassay to aspire for commercialization/marketability, a number of critical parameters and requirements need to be

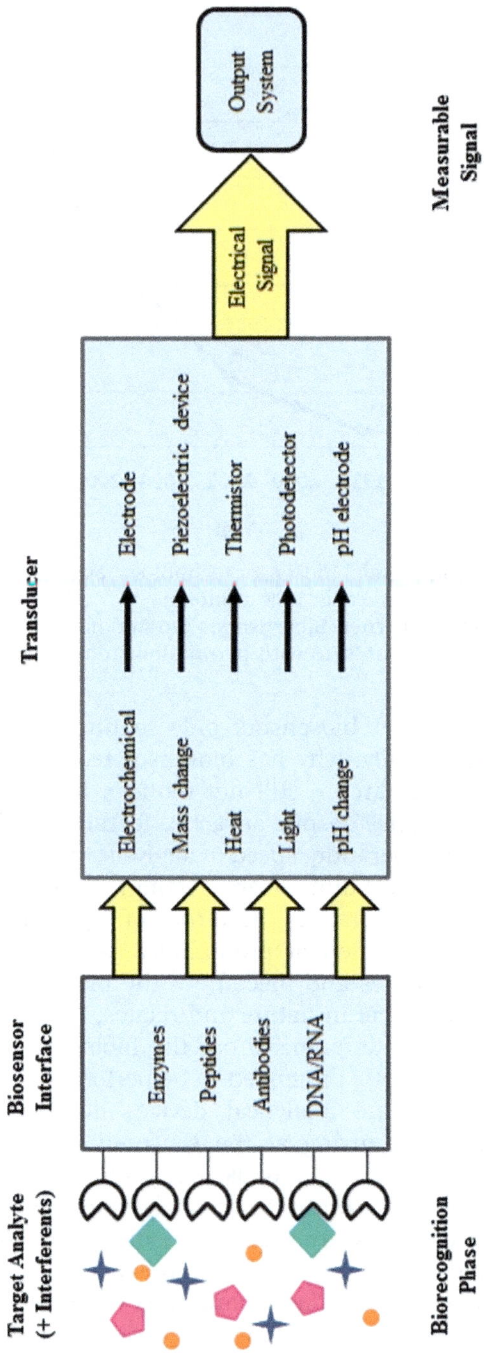

Figure 5.2 Schematic representation of the generic architecture of biosensors featuring (left to right) the biosensing interface whereupon the biorecognition of target analytes occurs *via* specific bioreceptors, the intimately interfaced transducing component for conversion of biomolecular interactions into an analytical signal and the output system. Adapted from ref. 7 with permission from the Royal Society of Chemistry.

controlled/fulfilled.[6,7,9–11] Architecturally, the biorecognition element must be securely attached to the transducer – with proper spatial orientation and loading density, as well as homogeneous distribution, to ensure optimal analyte binding capability.[7] Furthermore, molecular affinity must be maximally retained upon immobilization in order for the specific analyte to effectively interact with surface-attached bioreceptors.[7] Analytically, a biosensor device should ideally display both high selectivity and sensitivity, as well as provide reliable and reproducible measurements – even in the presence of potentially interfering species.[3,7,9]

This latter issue of 'non-specific adsorption' (NSA) – which constitutes a particular case of the larger, ubiquitous adsorption phenomenon known as 'fouling' – is recurrent in biosensor technology[6] and particularly relevant to those devices intended to perform detection in complex biological matrices, whose components have the notorious tendency to spontaneously adsorb and non-specifically accumulate onto sensing platforms made of exogenous synthetic materials (Figure 5.3).[7] The undesirable outcome is the generation of often overwhelming interference signals that prevent the detection, not to mention the quantification, of target analytes present at considerably lower concentration (several orders of magnitude).[12] Unfortunately, indeed, NSA species – most notably (but not only) proteins[13] – also produce physicochemical stimuli that are indiscriminately detected by biosensors and interfere with the specific response of target analytes (Figure 5.3). From an analytical point of view, a consequence is the possibility for the occurrence of 'false positive' results; that is, biosensor responses incorrectly interpreted as originating from genuine binding events between analytes and biorecognition elements. Conversely, erroneous 'false negative' readings might also result from the hindrance of analyte binding. Understandably, both these worrisome situations pose serious dangers, most particularly in clinical diagnostics[9] for detection in biosamples possessing highly interfering matrices (Figure 5.3) – such as blood or, more commonly, its cell-cleared plasma/serum derivatives.[7,12] In addition to this issue of signal interference also adds – for implanted biosensors for continuous *in vivo* monitoring – that of device bio-incompatibility,[2,14,15] a concern that is linked to the possibility for biological processes/responses (such as thrombosis and the 'foreign body reaction') with potentially harmful effects both for the implant recipient and device to be surface-activated[16] upon biofluid–artificial material contact.[17–19]

The literature on biosensors is extensive, encompassing tens of thousands of articles, the vast majority of these accounts describing detection capability in buffer solutions with 'analyte-only' samples. Although such undertaking is legitimate in early stages of development, the relevant objective for biosensors in bioanalytical screening/diagnostic applications – where the greatest technical difficulty arguably resides – still remains however the ability to move beyond analysis in these ideal research laboratory samples and to distinguish target analytes from potentially interfering species present in complex, real-life biosamples. In a similar vein, the literature on antifouling surface chemistry strategies is overwhelmingly composed of

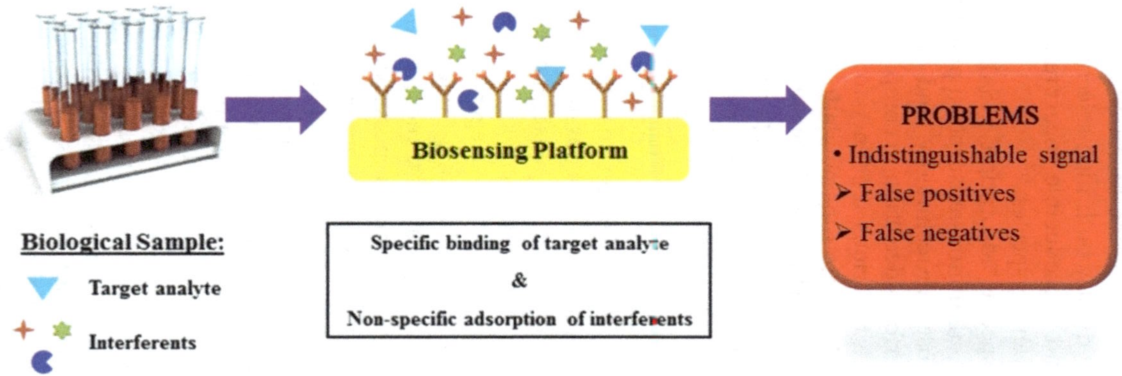

Figure 5.3 The recurrent issue of non-specific adsorption of interfering species (*vs.* the specific binding of target analytes) in biosensor technology, which is especially relevant to complex biological matrices and may be responsible for the occurrence of erroneous readings.
Adapted from ref. 7 with permission from the Royal Society of Chemistry.

publications praising the outstanding, often nearly perfect performance of coatings against buffered solutions lacking practical relevance as they are merely spiked with a single type, or an extremely simple mixture, of model proteins. It is in this peculiar context that the present chapter sets out to first inventory the state-of-the-art antifouling surface chemistries developed to date (January 2015) to minimize fouling from real-life biosamples (whether of human or animal origin, Figure 5.4) – which are indeed uncomparatively more challenging but the sole ones to offer the stringent conditions of a real-world testing ground – and then to describe the actual implementation of these stealth artificial coatings into biosensor applications or the like. In order to be practically relevant to most, this chapter – at the exception of some key references[20] – further restricts the discussion to biosample solutions containing ≥10 vol.% of biofluid, and to a fouling surface coverage $\Gamma < 30\ \text{ng cm}^{-2}$ (quantified on flat, non-particulate surfaces with a suitable technique[21] – most usually Surface Plasmon Resonance (SPR)[22,23]). Indeed, more dilute biosamples may certainly yield negligible fouling/NSA levels;[11] however, the risk with excessive dilution is to produce undetectable concentrations of analyte and therefore preclude analysis all together. Considering that the early detection in clinical diagnostics of trace amounts of biomarkers may dramatically improve the management and prognosis of associated diseases/conditions,[24] one may then easily understand why biosample dilution, although tempting, should not automatically be considered as an option to alleviate fouling/NSA – a problem, which should rather be

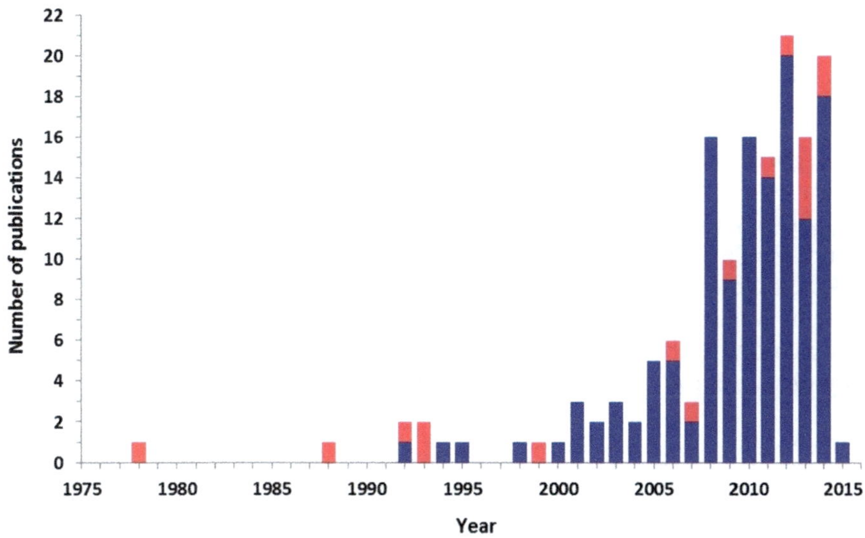

Figure 5.4 Forty years of antifouling surface chemistry against human or animal (blue) blood plasma/serum and (pink) cell lysate, cerebrospinal fluid, egg, milk, saliva, or urine (from Scifinder® database – last updated January 2015).
Adapted from ref. 25 with permission from the Royal Society of Chemistry.

chosen to be tackled through use of pronouncedly antifouling surface chemistry. Finally, another matter of concern also discussed in this chapter – which is rarely touched upon but of utmost importance – is that of the dependence of the antifouling performance of coatings on the compositional variability of biosamples.

5.2 State-of-the-art Antifouling Surface Chemistries

5.2.1 Against Blood Plasma/Serum

Blood is the biofluid of Life in which bathe both cellular and non-cellular species that are essential in maintaining the organism healthy and functional as they provide a means of delivering vital nutrients and gases to cells (or conversely disposing of metabolic waste products), of regulating the internal environment (*e.g.* pH, osmosis), of protecting the organism against foreign bodies (through the action of the immune system), or of preventing blood loss (*via* haemostasis and coagulation).[26–29] Whole blood is essentially composed of three parts, which easily separate upon centrifugation and are, from the densest to the lightest fraction, the red blood cells, the 'buffy coat' (contains platelets and white blood cells), and 'blood plasma' (gathers the non-cellular components of blood, most notably proteins).[26–31] Blood serum used in 'serology' is obtained upon coagulation then centrifugation of whole blood.[26] In terms of fouling potency, plasma and serum, even cleared of the cellular components of blood, still constitute highly complex matrices that feature thousands[30] of adhesive proteins at high total concentration ($\sim$60–80 g L^{-1}).[32]

5.2.1.1 *'Bioinspired' Constructs: Single Amino Acids, Peptides and Peptoids*

We begin our review of antifouling surface chemistries against blood plasma/serum with the 1995 work of Lundström and co-workers, who used *in situ* ellipsometry to study the antifouling properties of L-cysteine (Figure 5.5A) monolayers on gold.[33] After 10 min exposure to 10% human plasma (diluted in PBS) followed by successive PBS/water rinsings, the amount of adsorbed proteins was estimated to be $\Gamma \sim 200$ ng cm^{-2} – a value slightly lower than that recorded on bare gold. A real significant difference ($\Gamma < 50$ ng cm^{-2}) was only observed for gold coated with a monolayer of glutathione tripeptide

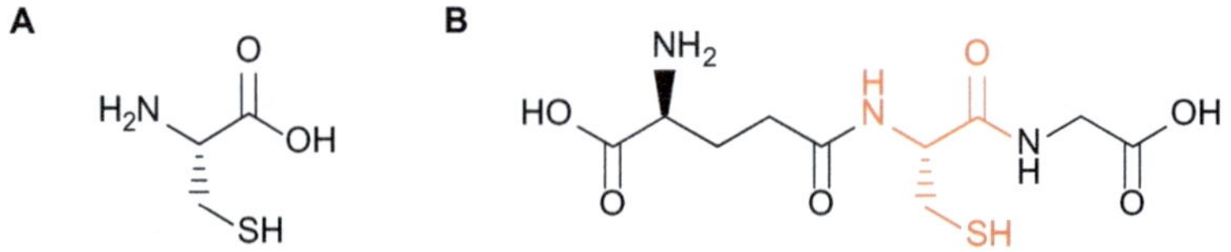

Figure 5.5 Molecular structure of (A) L-cysteine amino acid, and (B) glutathione tripeptide.
Adapted from ref. 25 with permission from the Royal Society of Chemistry.

(Figure 5.5B). This work was an extension of previous studies by the investigators, wherein total and individual protein adsorption from 10% human plasma/serum on these nanometer-thick monolayers had been probed using ellipsometry combined with antisera binding experiments.[34,35]

Further investigations on the antifouling properties of single amino acid-based monolayers – this time indirectly tethered to gold substrates through *N*-3-mercaptopropionic acid (3-MPA) graft residues – were conducted by Masson's group in 2008.[36] Fouling upon contact with undiluted bovine serum (76 mg protein mL^{-1}) for 20 min was quantified using SPR. Among the 19 natural amino acids investigated, the lowest fouling levels ($\Gamma \sim 418$ ng cm^{-2}) were monitored for self-assembled monolayers (SAM) made of polar (L-Asn, L-Ser) or ionic (L-Asp) amino acids (Figure 5.6A). Conversely, highest fouling ($\Gamma \sim 808$ ng cm^{-2}) was recorded for coatings

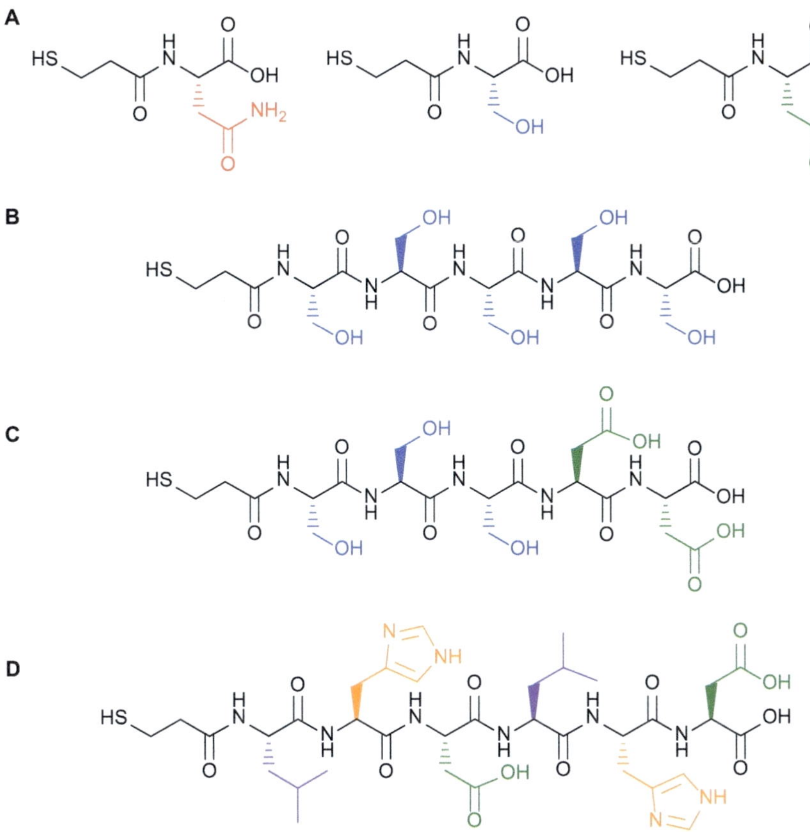

Figure 5.6 Molecular structure of (A) 3-MPA-Asn-OH, 3-MPA-Ser-OH, and 3-MPA-Asp-OH single amino acids, (B) 3-MPA-Ser_5-OH homopentapeptide, (C) 3-MPA-Ser_3-Asp_2-OH binary-patterned pentapeptide, and (D) 3-MPA-(Leu-His-$Asp)_2$-OH ternary-patterned hexapeptide.
Adapted from ref. 25 with permission from the Royal Society of Chemistry.

based on hydrophobic L-tyrosine. The following year, Masson and co-workers were able to substantially decrease fouling by expanding the scope of their study to short 3-MPA-homopeptide-based SAMs.[37] Fouling surface coverage was the lowest ($\Gamma = 132 \pm 33$ ng cm^{-2}) for pentapeptide SAMs featuring L-serine subunits (Figure 5.6B). The team achieved further tremendous progress using 3-MPA-pentapeptides presenting binary sequences of amino acids.[38,39] For the nine different peptide designs under investigation, the level of fouling was consistently lower than $\Gamma = 80$ ng cm^{-2}, the highest resistance to serum adsorption belonging to 3-MPA-Ser$_3$-Asp$_2$-OH SAMs with $\Gamma = 23 \pm 10$ ng cm^{-2} (Figure 5.6C). At last, the best antifouling performance was reached in 2011 with a third-generation of ternary-patterned hexapeptide constructs, of which 3-MPA-(Leu-His-Asp)$_2$-OH (Figure 5.6D) SAMs were the most effective at reducing fouling with $\Gamma = 12 \pm 11$ ng cm^{-2}.[40]

Unlike Lundström in 1995[33] and Masson in 2008,[36] Booksh was just recently able to prepare a single amino acid-based coating with pronounced antifouling properties.[41] In this 2015 account, anchorage to gold substrates did not rely on thiol chemistry, however,[33,36] but on an electrografting reaction involving a diazonium salt of phenylalanine (Figure 5.7).[41] Fouling surface coverage by crude bovine serum (undiluted, 75 mg protein mL^{-1}), measured by SPR upon 20 min exposure, was relatively good at $\Gamma = 42 \pm 4$ ng cm^{-2}. In comparison, the level of adsorption on bare gold was reported to be $\Gamma = 96 \pm 2$ ng cm^{-2}.

In 2002, Picart and co-workers reported their findings on the antifouling properties of several polyelectrolyte multilayered films made of alternating strata of poly(L-lysine) (PLL) and poly(L-glutamic acid) (PGA) polypeptides shown in Figure 5.8.[42] Using 10% serum (diluted in MES-Tris buffer) and

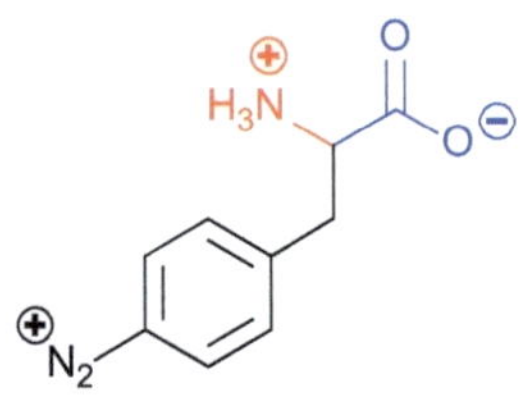

Figure 5.7 Molecular structure of 4-diazonium-DL-phenylalanine.

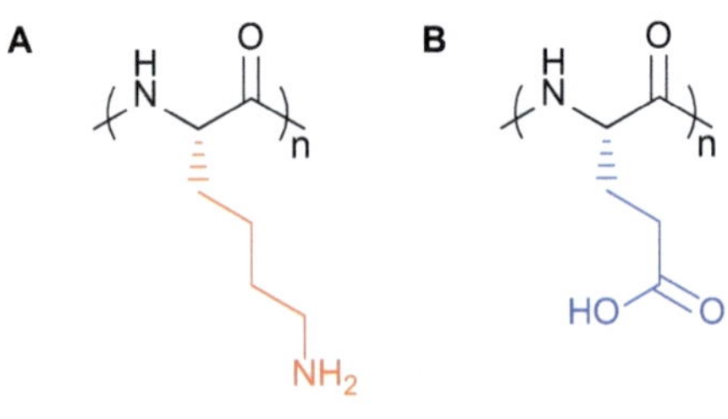

Figure 5.8 Molecular structure of (A) poly(L-lysine) PLL, and (B) poly(L-glutamic acid) PGA.
Adapted from ref. 25 with permission from the Royal Society of Chemistry.

optical waveguide lightmode spectroscopy (OWLS) as the method of detection, the authors observed that PLL-terminated coatings invariably adsorbed a significant amount of proteins ($\Gamma > 300$ ng cm^{-2}, 60 min contact) – regardless of film thickness; that is, irrespectively of the number of alternating layers beneath. Conversely, serum adsorption on films exposing a terminal PGA layer was undetectable, at least up to six strata of polyelectrolytes for ~16 nm-thick (PLL/PGA)$_3$ coatings.

More recently, in 2013, Liu and co-workers prepared poly(L-serine) zwitterionic brushes on gold through an environmentally-friendly (catalyst-free), 'grafting from' photochemical process – known as 'surface-initiated photoiniferter-mediated polymerization' (SI-PIMP) – that involved a pre-coated thiol SAM of dithiocarbamoyl photoiniferter (Figure 5.9A) and L-serine methacrylate monomer (Figure 5.9B).[43] Considering film thickness (and uniformity) could be easily controlled through simple variation of the UV-photoirradiation time, the investigators were able to determine the relationship existing between this parameter and fouling by 100% human blood plasma/serum. Interestingly, results showed that serum fouling was minimal ($\Gamma = 9.2 \pm 1.3$ ng cm^{-2}) at the intermediate thickness of ~37 nm, as measured by SPR upon 10 min exposure. Both thinner and thicker brushes (~5–55 nm range) were indeed less resistant to serum adsorption with $10 < \Gamma < 50$ ng cm^{-2}. Gratifyingly, fouling by full human plasma at optimal

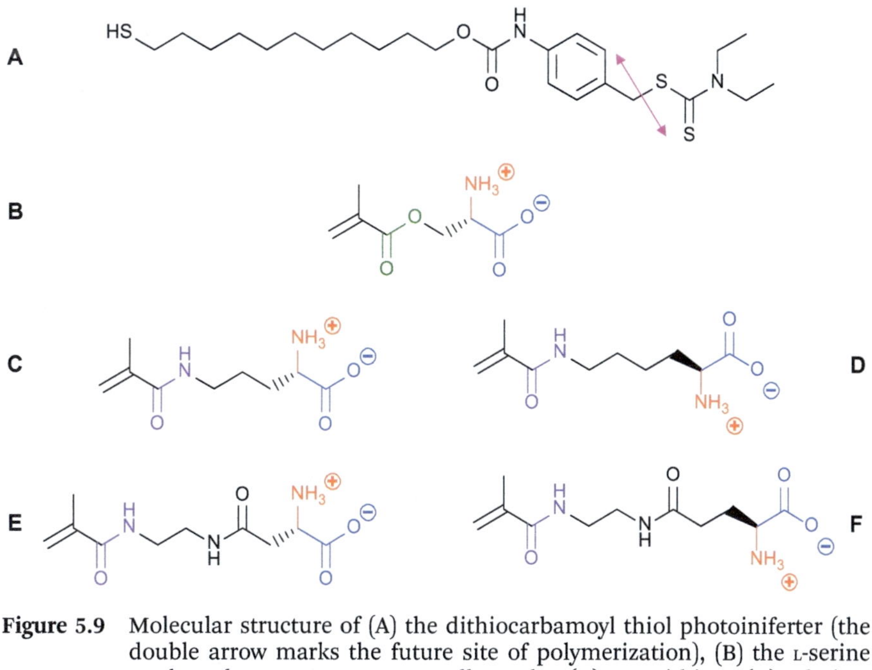

Figure 5.9 Molecular structure of (A) the dithiocarbamoyl thiol photoiniferter (the double arrow marks the future site of polymerization), (B) the L-serine methacrylate monomer; as well as the (C) L-ornithine, (D) L-lysine, (E) L-aspartic acid-based, and (F) L-glutamic acid-based methacrylamide monomers.

film thickness was also limited ($\Gamma = 12.9 \pm 2.1$ ng cm^{-2}). In contrast, both bare gold and photoiniferter precursor SAM surfaces exhibited substantial fouling by both these fluids ($\Gamma > 150$ ng cm^{-2}).[43–45] In a series of follow-up papers published in 2014, Liu's group expanded the scope of poly(L-amino acid) zwitterionic antifouling brushes grown on gold through the versatile SI-PIMP process.[44–46] Another novelty, beside the chemical structure of the amino acid side-chain (Figure 5.9C–F), was the 'amide' nature (*vs.* 'ester') of the newly polymerized monomers (Figure 5.9C–F *vs.* B). Methacrylamide-based brushes were invariably more antifouling than the methacrylate ester variety, whether against 100% human plasma or 100% human serum (10 min contact).[44–46] Experimentally, although there was essentially no statistical difference in terms of antifouling properties within the amide series, the best performance against full plasma ($\Gamma = 3.2 \pm 2.3$ ng cm^{-2}) was recorded for ~15 nm-thick 'ornithine' brushes (Figure 5.9C).[44,45] With respect to full serum, ~12 nm-thick aspartic acid-based brushes (Figure 5.9E) offered superior protection against fouling with $\Gamma = 0.8 \pm 0.8$ ng cm^{-2}.[45,46]

In 2005, Messersmith and co-workers reported the synthesis of a novel peptidomimetic polymer with which they would modify Ti surfaces for antifouling purposes.[47] This hybrid macromolecule features two distinct domains: (1) a short, biomimetic substrate-anchoring pentapeptide made of alternating L-DOPA and L-lysine residues, itself *N*-coupled to (2) a *N*-(2-methoxyethyl)-substituted 20-mer polypeptoid (Figure 5.10A). Excellent resistance to protein adsorption upon exposure to whole human serum for 20 min was demonstrated by means of OWLS ($\Gamma \sim 4$ ng cm^{-2} as opposed to $\Gamma \sim 435$ ng cm^{-2} for unmodified Ti substrates). The project was later expanded to explore the influence on surface antifouling of the peptoid side-chain's nature. In the new study, three different side-chains would be employed (2-methoxyethyl, 2-hydroxyethyl, and 2-hydroxypropyl), the adhesion pentapeptide sequence being kept identical (Figure 5.10A–C).[48] Although mixed success was encountered among polymers, all three coatings (~4.3 nm-thick) displayed substantially superior antifouling properties against whole human serum than the unmodified TiO_2 substrates of the study ($\Gamma = 15$–83 *vs.* 342 ± 21 ng cm^{-2}), when exposed for 20 min at 37 °C. Varying the repeat length of *N*-(2-methoxyethyl)-substituted peptoid segments was also shown soon after to have statistically no significant effect on protein adsorption ($\Gamma = 15$–53 ng cm^{-2}).[49]

The most recent report on peptoid-based antifouling coatings was published in 2011 by Liu and Jia.[50] In this study, the authors introduced a new type of peptoid polymer featuring *N*-ethyl-β-alanine subunits (Figure 5.11). These poly(β-peptoid)s were grafted to gold SPR sensor chips through C-terminal cysteamine residues. Fouling was assessed upon exposure for 10 min to dilute (10% in PBS) as well as full human plasma and serum. Gold substrates derivatized with $PEtA_{20}$ – a 20-mer of *N*-ethyl-β-alanine (Figure 5.11, $n = 20$) – exhibited excellent antifouling properties, whether against 10% serum ($\Gamma = 10.7 \pm 3.6$ ng cm^{-2}) or 10% plasma ($\Gamma = 5.2 \pm 6.2$ ng cm^{-2}). Gratifyingly, comparable levels of performance were

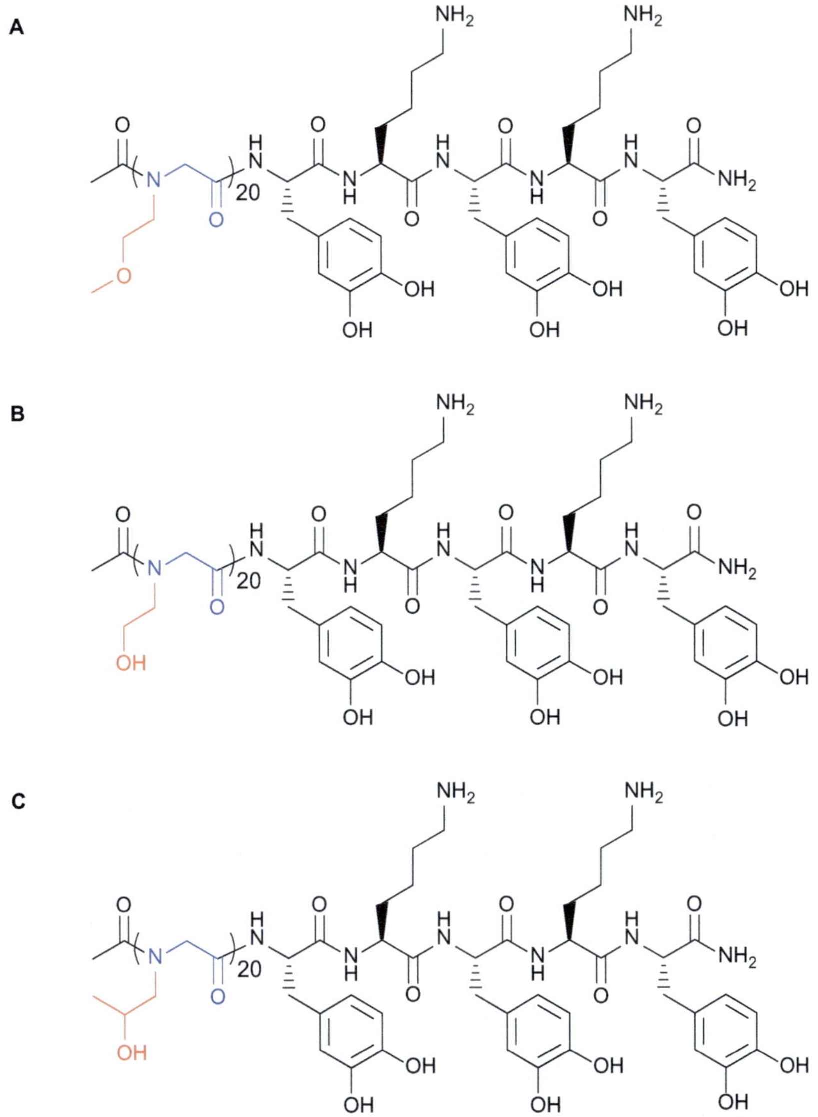

Figure 5.10 Molecular structure of the (A) *N*-(2-methoxyethyl)-, (B) *N*-(2-hydroxyethyl)-, and (C) *N*-(2-hydroxypropyl)-substituted 20-mer polypeptoids, all *C*-coupled to a substrate-anchoring pentapeptide sequence made of alternating L-DOPA and L-lysine residues.
Adapted from ref. 25 with permission from the Royal Society of Chemistry.

achieved both for 100% serum ($\Gamma = 10.8 \pm 14.8$ ng cm^{-2}) and 100% plasma ($\Gamma = 9.8 \pm 12.7$ ng cm^{-2}). Better yet, it turned out that PEtA$_{20}$ coatings are quasi-systematically outperformed, both in terms of performance and reproducibility of measurement, by adlayers constructed with the longer

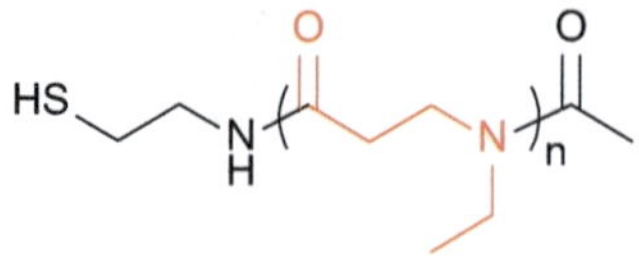

Figure 5.11 Molecular structure of $PEtA_{20}$ ($n = 20$) and $PEtA_{40}$ ($n = 40$) poly(*N*-ethyl-β-alanine) poly(β-peptoid)s.
Adapted from ref. 25 with permission from the Royal Society of Chemistry.

$PEtA_{40}$ poly(β-peptoid)s shown in Figure 5.11 ($n = 40$). In fact, fouling coverages on $PEtA_{40}$ coatings from 10% serum and 10% plasma were respectively $\Gamma = 3.9 \pm 1.2$ and 5.7 ± 1.5 ng cm^{-2}, while those for more challenging 100% serum and 100% plasma were $\Gamma = 5.9 \pm 3.4$ and 9.7 ± 3.2 ng cm^{-2}.

5.2.1.2 *Oligo- and Polyethylene Oxides/Glycols*

We now turn our attention towards oligo- and polyethylene oxide (OEO and PEO) constructs, and their derivatives, which all together constitute historically the most prevalent coating materials employed for antifouling surface chemistry purposes.[51–54]

We begin with the work of Horbett and co-workers, who in 2007 explored with SPR the antifouling properties of tetraglyme (Figure 5.12) coatings on gold against human plasma.[55] The study concluded that these typically 100 nm-thick films strongly resist protein adsorption, even at low plasma dilution in PBS. Fouling from 10, 50, and 100% human plasma (10–20 min exposure) was calculated to be $\Gamma = 4.8$, 17.3, and 24.1 ng cm^{-2}, respectively. In comparison, bare gold adsorbed a considerably greater amount of proteins ($\Gamma \sim 244$ ng cm^{-2}), even from 1% plasma.

Later (in 2010), this short/light tetrameric OEG chain was integrated by Chang and co-workers into a linear, OEG_4-terminated alkylthiol molecule (Figure 5.13A) with which OEG_4-SAMs were formed on gold SPR sensor chips.[56–58] These adlayers exhibited relatively good antifouling properties ($\Gamma = 38.3$ ng cm^{-2}) against 20% human platelet-poor plasma (PPP), as measured at 37 °C upon ~15 min exposure. Two years earlier, Jiang's group had not been as successful ($\Gamma \sim 200$ ng cm^{-2}) using 10% human plasma in this case.[59] Gratifyingly, 10% and even 100% human serum triggered significantly less fouling with $\Gamma \sim 15$ and 85 ng cm^{-2}, respectively.[59–61] Recently (in 2012), Rodriguez-Emmenegger and co-workers reported that ultrathin OEG_6-SAMs (9 Å-thick), prepared with a slightly longer thiol molecule (Figure 5.13B), performed reasonably well against undiluted foetal bovine

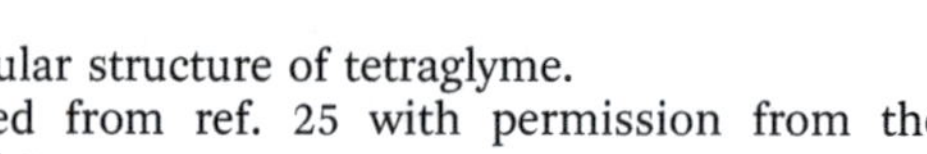

Figure 5.12 Molecular structure of tetraglyme.
Adapted from ref. 25 with permission from the Royal Society of Chemistry.

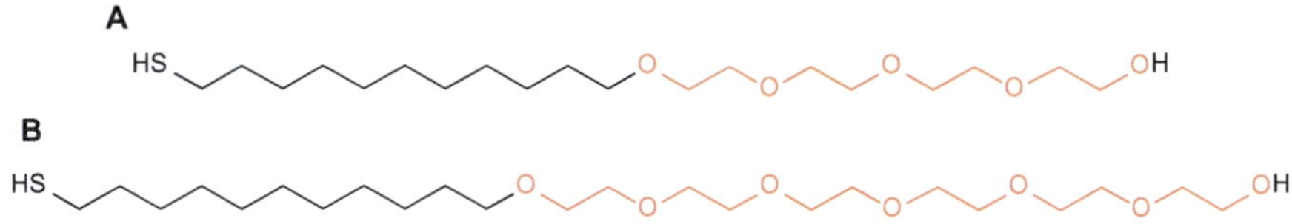

Figure 5.13 Molecular structure of the OEG_n-terminated alkylthiol building block molecules (A: $n=4$, B: $n=6$) for antifouling SAM formation on gold. Figure 5.13A adapted from ref. 25 with permission from the Royal Society of Chemistry.

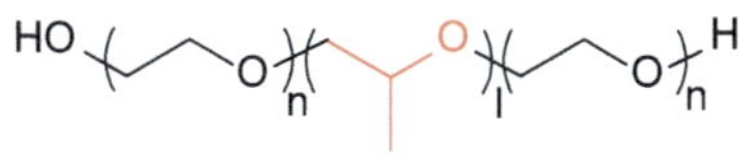

Figure 5.14 Molecular structure of Pluronic™ F108 triblock copolymer ($n=133$, $l=50$).
Adapted from ref. 25 with permission from the Royal Society of Chemistry.

serum (FBS), with $\Gamma=26.1\pm2.7$ ng cm^{-2} after contact for 15 min.[62,63] Fouling from an undiluted human biofluid (blood plasma) was again an issue ($\Gamma=71.0\pm8.0$ ng cm^{-2}).

Higher molecular weight PEG molecules (2–20 kDa range),[64] grafted onto gold SPR sensor chips through sulfhydryl moieties as well, were also used to form coatings (~1 to 5 nm-thick) with good antifouling properties.[65] Resistance to protein adsorption – against 10% FBS for 30 min at 25 °C – was indeed measured to be $\Gamma=6$–10 ng cm^{-2}, the best performance belonging to intermediately dense/thick PEG films formed with 5 kDa polymer molecules. With respect to human blood-based fluids, another study reported in 2012 PEG_{20kDa} coatings (8.1 nm-thick) able to minimize fouling from 100% human blood plasma to $\Gamma\sim25$ ng cm^{-2}, as measured by SPR upon 20 min exposure.[66] In comparison, bare gold was heavily fouled with $\Gamma=293\pm70$ ng cm^{-2}.

The aforementioned experiment by Chang and co-workers with OEG_4-SAMs constituted a control reference to a study primarily dedicated to explore the resistance to protein adsorption of Pluronic™ coatings on gold.[56] Pluronics™ – or poloxamers – are amphiphilic PEO-*bl*-PPO-*bl*-PEO triblock copolymers consisting of a central hydrophobic polypropylene oxide (PPO) segment flanked by two hydrophilic PEO chains. Poloxamer adlayers were prepared upon physisorption of the copolymers' middle PPO block onto pre-coated hydrophobic SAMs of 1-undecane-thiol, an arrangement which compels hydrophilic PEO side-chains to project outwards in the liquid environment. Interestingly, it was shown for Pluronic™ F108 coatings (Figure 5.14) that PEO surface coverage plays a critical role on plasma protein adsorption. High-density surfaces (11 PEO chains per 100 nm^2) adsorbed the least amount of proteins from 20% PPP ($\Gamma<1.5$ ng cm^{-2}). In contrast, low-density surfaces (3 PEO chains per 100 nm^2) adsorbed a substantially greater amount of proteins ($\Gamma=61.5$ ng cm^{-2}), a value even higher than that recorded for the OEG_4 reference SAMs discussed earlier ($\Gamma=38.3$ ng cm^{-2}).[56]

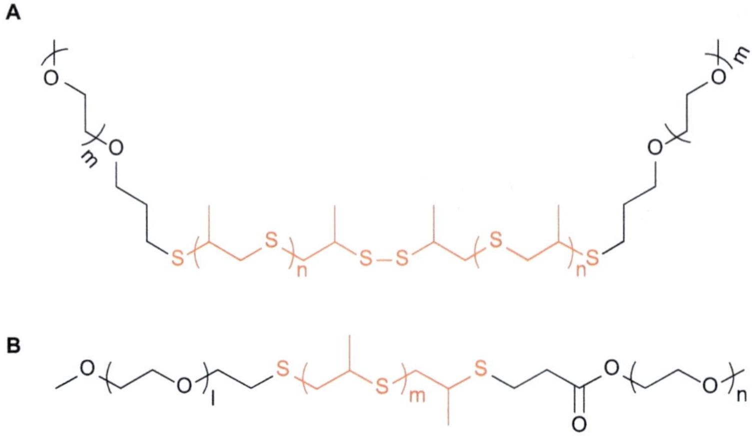

Figure 5.15 Molecular structure of two variants of sulfur-based poloxamer analogues (PEG-*bl*-PPS-*bl*-PEG triblock copolymers).
Adapted from ref. 25 with permission from the Royal Society of Chemistry.

In 2005, Tosatti and co-workers had explored closely related triblock copolymer structures (Figure 5.15A), in which the two external PEG chains were now sandwiching a central polypropylene sulfide (PPS) block.[67] Unlike the preceding Pluronics™ F108 coatings that were physisorbed onto pre-coated alkanethiol SAMs through hydrophobic–hydrophobic interactions,[56] these PEG-*bl*-PPS-*bl*-PEG adlayers (~34 Å-thick) were directly anchored to the gold surface through multisite polysulfide chemisorption of their central PPS segment.[67,68] The impact of these PEG films on fouling from whole human serum, measured by means of SPR (30 min exposure), was also considerable, protein adsorption being reduced to $\Gamma = 16 \pm 5$ ng cm^{-2} – from $\Gamma = 380 \pm 14$ ng cm^{-2} for bare gold and $\Gamma = 198 \pm 81$ ng cm^{-2} for PPS-only coatings.[67] Similar antifouling performance ($\Gamma = 25 \pm 17$ ng cm^{-2}) was recorded for adlayers constructed with another variety of PEG-*bl*-PPS-*bl*-PEG triblock copolymer (Figure 5.15B).[68] These latter coatings were also shown to display great stability for many weeks under ambient conditions.

Another type of multidentate copolymeric adsorbate with pendant PEG chains was introduced in 2010 by Robin and co-workers for the surface modification of titanium oxide surfaces.[69,70] This macromolecule – a terpolymer featuring undecylphosphonate, PEG, and *n*-butyl side-chains branching out a central methacrylate backbone in a 1 : 1 : 8 ratio (Figure 5.16) – displayed the ability to self-organize onto TiO_2 substrates and impart the latter with excellent antifouling properties. In fact, protein adsorption on this 30 Å-thick adlayer was $\Gamma = 4 \pm 1$ ng cm^{-2}, as measured with OWLS upon 15 min exposure to full human serum.[69]

In 2009, Schaaf and Mésini had also described poly(acrylic acid) (PAA)-based coatings derivatized with short OEO_3 side-chains (Figure 5.17) at different grafting ratios (GR).[71] These polymer constructs were immobilized

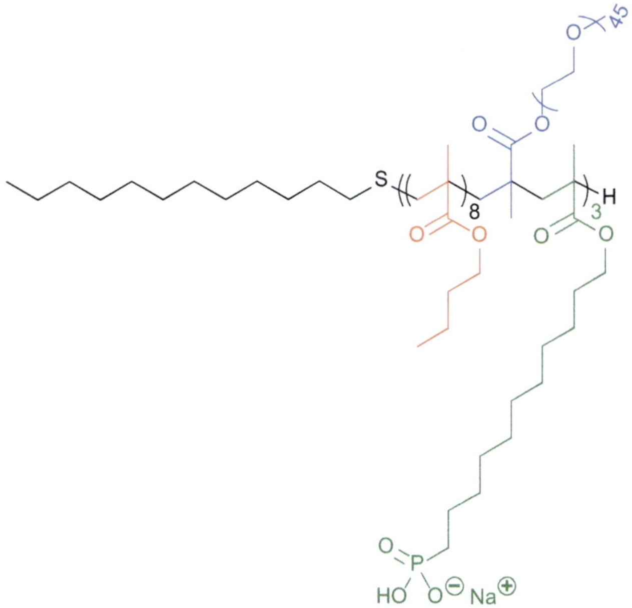

Figure 5.16 Molecular structure of the PEG-polyalkyl phosphonate terpolymer. Adapted from ref. 25 with permission from the Royal Society of Chemistry.

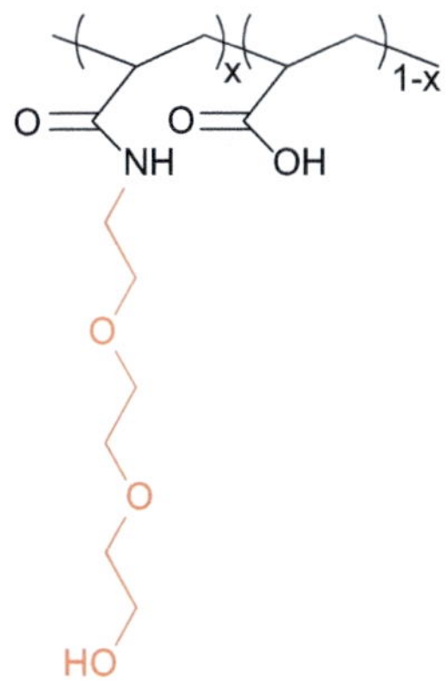

Figure 5.17 Molecular structure of PAA-*g*-OEO_3 polyelectrolyte adsorbates.

on silicon/titanium oxide substrates through a precursor film made of alternating poly(styrene sulfonate)/poly(allylamine hydrochloride) polyelectrolyte layers – this stratified structure being itself anchored to the substrate *via* a poly(ethylene imine) adhesion coating. PAA-*g*-OEO_3 films with GR < 20% were reasonably antifouling with $\Gamma = 18$–$51\ ng\ cm^{-2}$, as measured with OWLS upon exposure to 10% FBS. It was only when GR reached 24% (and up to 42%) that the imposed surface chemistry was truly antifouling

($\Gamma = 1\ \text{ng cm}^{-2}$). In stark contrast, non-derivatized PAA coatings were heavily fouled with $\Gamma = 297\ \text{ng cm}^{-2}$.[71,72]

Soon after (in 2011), Textor and co-workers reported OEG adsorbates with a more complex dendritic architecture that were able to self-assemble onto TiO_2 surfaces.[73] These macromolecules featured OEG dendrimer arms conjugated to substrate-anchoring L-DOPA/dopamine multidentate oligopeptides (Figure 5.18). Resistance to protein adsorption was measured using ellipsometry, upon exposure to full blood serum for 20 min, and shown to strongly depend on dendron surface coverage. Lowest fouling, actually below the limit of detection (LOD) of the technique, was obtained for monomolecular adlayers at saturation coverage. *In situ* OWLS measurements with an ~2 ng cm^{-2} LOD confirmed this observation.

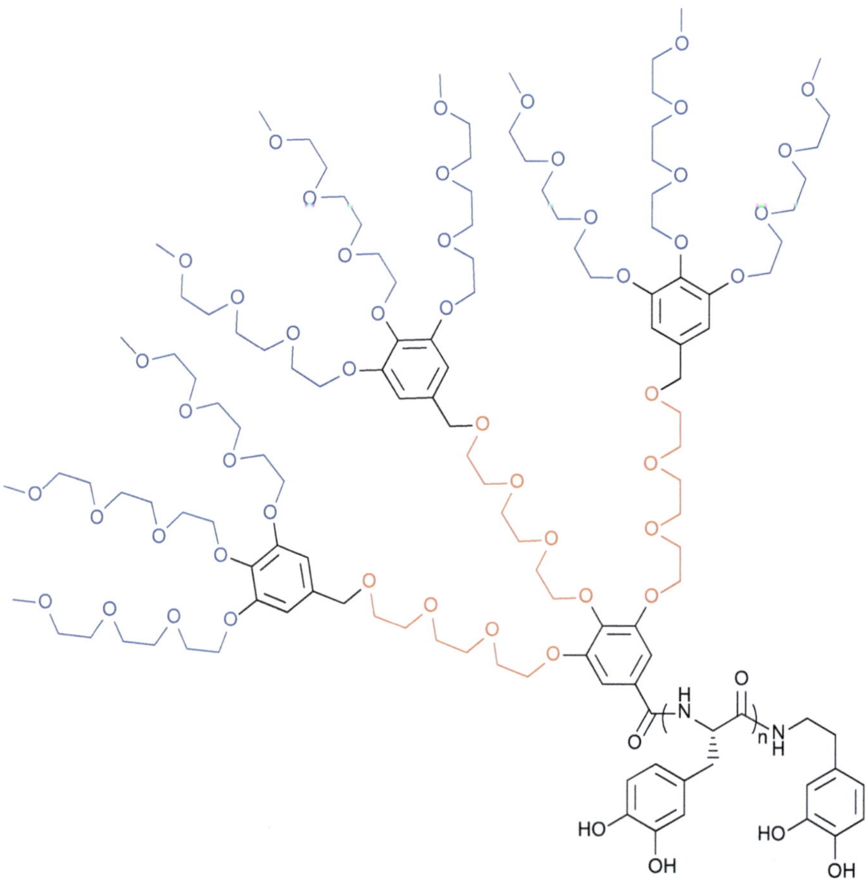

Figure 5.18 Molecular structure of OEG dendritic multisorbates featuring substrate-anchoring L-DOPA and dopamine catechol-based residues in an oligopeptide sequence.
Adapted from ref. 25 with permission from the Royal Society of Chemistry.

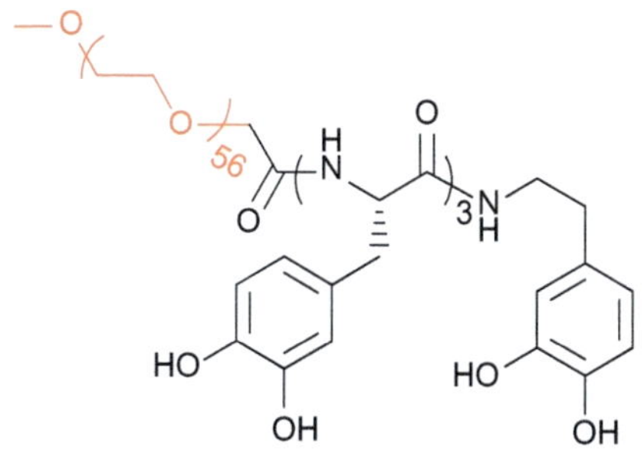

Figure 5.19 Molecular structure of the linear PEG multisorbate featuring substrate-anchoring L-DOPA and dopamine catechol-based residues in an oligopeptide sequence.
Adapted from ref. 25 with permission from the Royal Society of Chemistry.

Coatings prepared with a linear PEG multisorbate of similar molecular weight (Figure 5.19) also displayed such remarkable antifouling properties.[73] This work constituted the latest addition to several previous studies, wherein linear PEG adsorbates featuring various L-DOPA-, dopamine- or derivative-based anchors had also been successfully tested for their ability to form robust and stable adlayers with high fouling resistance against full human serum/plasma.[74–80]

Later in 2011, Huck and co-workers published the findings of a more systematic structural study, wherein the influence on protein adsorption of OEO polymer brush architecture was investigated.[81] Various brushes with regular, linear, or dendritic side-chain substructures were prepared *in situ* through atom transfer radical polymerization (ATRP) of mono/oligoglycerol (meth)acrylate monomers (Figure 5.20A–C) – a 'grafting from' process surface-initiated by precursor thiol SAMs of ω-mercapto-undecyl α-bromoisobutyrate (Figure 5.20D) pre-coated on gold SPR sensor chips. The highest resistance to non-diluted serum adsorption ($\Gamma = 20 \pm 10$ ng cm^{-2}) was recorded for 17 nm-thick first-generation dendritic brushes (5 min exposure at 25 °C). In contrast, their second-generation counterparts performed poorly, as did the regular[82] and linear methoxylated (R = Me) brushes. All these less well performing coatings indeed adsorbed greater amounts of serum proteins ($\Gamma > 100$ ng cm^{-2}), and this regardless of brush thickness (3 to 22 nm). Only the linear hydroxylated 'bottle' brushes (R = H, 6 nm-thick) approached such a level of antifouling performance with $\Gamma = 53 \pm 12$ ng cm^{-2}. Similar architectural dependence and trends of protein adsorption were observed for experiments run with full plasma, the highest resistance to fouling ($\Gamma = 8 \pm 6$ ng cm^{-2}) still being the feat of first-generation dendritic brushes.[81] Irrespectively, all polymer brushes did display superior antifouling performance, against both human serum and plasma, compared to bare gold ($\Gamma = 217 \pm 5$ and 385 ± 5 ng cm^{-2}, respectively) and the ATRP initiator SAM ($\Gamma = 207 \pm 11$ and 203 ± 1 ng cm^{-2}).

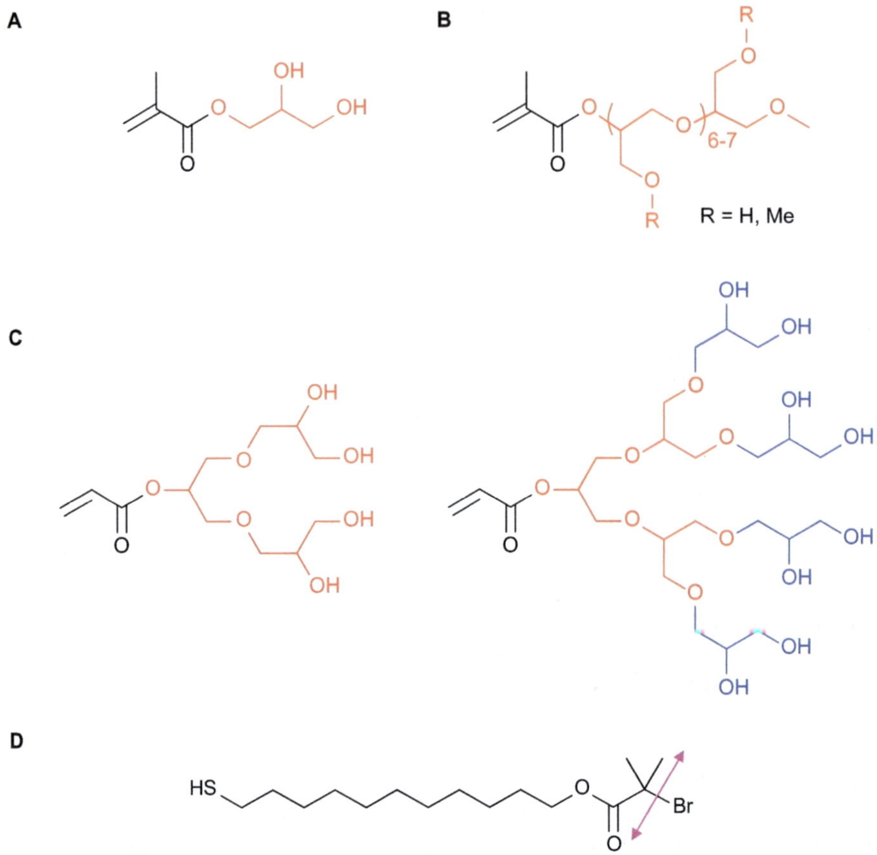

Figure 5.20 Molecular structure of the (A) regular, monoglycerol- and (B) linear, oligoglycerol-based methacrylate monomers, as well as the (C) first- and second-generation (left/right) dendritic oligoglycerol acrylate monomers. (D) Molecular structure of the ω-mercapto-undecyl α-bromoisobutyrate ATRP building block initiator (the double arrow marks the future site of polymerization).
Adapted from ref. 25 with permission from the Royal Society of Chemistry.

Other types of OEO bottle brushes – also prepared on gold *via* this 'grafting from' polymerization strategy using the linear, hydroxy- (HOEGMA) or methoxy-terminated (MeOEGMA) OEG methacrylate monomers drawn in Figure 5.21 – have also been shown on several occasions to possess remarkable antifouling properties. In 2008 for instance, a collaboration between Horbett, Ratner, and Jiang resulted in the preparation of polyMeOEGMA brushes (20–25 nm-thick) able to limit fouling from 100% human plasma to $\Gamma = 9.2 \pm 6.5$ ng cm^{-2}, as measured by SPR.[83] In separate investigations published in 2011 and 2012, Rodriguez-Emmenegger *et al.* reported similar low-fouling performance for both polyHOEGMA and polyMeOEGMA brushes with

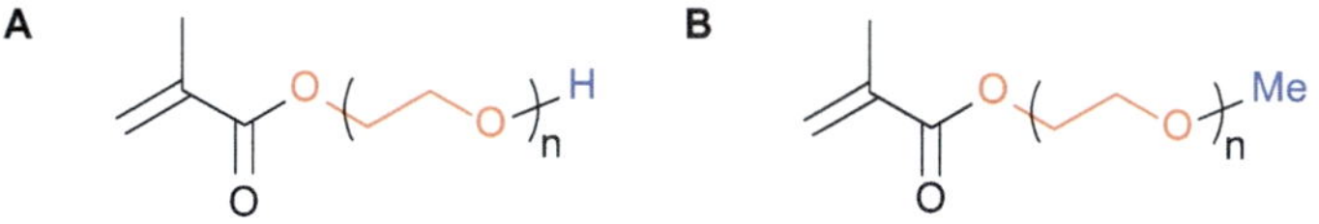

Figure 5.21 Molecular structure of (A) HOEGMA, and (B) MeOEGMA methacrylate monomers.
Adapted from ref. 25 with permission from the Royal Society of Chemistry.

$\Gamma = 16.3$ and 18.9 ng cm^{-2}, respectively.[62,84–86] In 2014, these results were confirmed on several occasions by the investigators for polyMeOEGMA[87–89] and polyHOEGMA[87,89] coatings – as well as diblock polyMeOEGMA-*bl*-poly-HOEGMA copolymer structures of different thickness[87] – upon 15 min exposure at 25 °C. The same year, the authors described similarly good antifouling properties against undiluted human serum, for ~32 nm-thick polyHOEGMA brushes exposed for 30 min at 25 °C.[90] The antifouling performance against 10% human serum, of ~31 nm-thick polyMeOEGMA brushes (15 min contact), was excellent at $\Gamma = 4.3$ ng cm^{-2}.[91]

Earlier (in 2004), Chilkoti's group had already reported that fouling from 100% FBS on ~15 nm-thick polyMeOEGMA brushes was below the ~1 ng cm^{-2} LOD of their SPR instrument (20 min exposure).[92] This remarkable achievement was later reproduced (and extended to human serum and plasma) by Jiang in 2008,[59] then recently confirmed in 2011–2012 by Rodriguez-Emmenegger for 30–40 nm-thick brushes ($\Gamma < 0.6$ ng cm^{-2}),[62,84,86,93] and Huck in 2010 ($\Gamma < 2$ ng cm^{-2} for 20–30 nm-thick films).[94] Interestingly however, this latter study revealed that whole sera obtained from a different species (horse) or at a different stage of maturity (adult *vs.* foetal bovine serum) did foul polyMeOEGMA coatings. Finally, Rodriguez-Emmenegger *et al.* also reported (in 2014) that ~31 nm-thick polyMeOEGMA brushes offer excellent antifouling protection against 10% FBS ($\Gamma < 0.3$ ng cm^{-2}, 10 h contact).[91]

Several research groups also demonstrated that polyHOEGMA and poly-MeOEGMA brushes exhibiting excellent antifouling properties could as well be ATRP-grown onto a variety of other substrate materials than gold. In 2006 for instance, Chilkoti and co-workers polymerized 'non-fouling' poly-MeOEGMA brushes onto silicon oxide.[95] Fouling from undiluted FBS (60 min exposure) was indeed below the LOD (1 Å) of the ellipsometer apparatus used by the authors for polymer brushes thicker than ~95 Å. The following year, García in collaboration with Collard prepared ~135 Å-thick polyHOEGMA brushes on titanium oxide that were able to reduce fouling below $\Gamma = 20$ ng cm^{-2}, as measured by SPR against 10% FBS.[96] More recently (in 2011), Rodriguez-Emmenegger *et al.* described the growth of both types of polyOEGMA brush on nylon-6/6 adhesion films, themselves sputtered onto various inorganic (gold, silicon, TiAlV alloy) or organic (polypropylene) substrates.[84] Fouling from undiluted FBS on polyHOEGMA brushes polymerized on nylon/Au surfaces was measured by SPR to be $\Gamma = 9.7$ ng cm^{-2} (15 min exposure). Their polyMeOEGMA counterparts, on the other hand, were not fouled by 100% FBS ($\Gamma < 0.03$ ng cm^{-2}). With respect to undiluted human plasma, fouling surface coverages were certainly higher (respectively $\Gamma = 28.3$ and 19.5 ng cm^{-2}), but otherwise much lower than those recorded on the control surfaces constituted by bare gold and pristine nylon ($\Gamma = 307$ and 177 ng cm^{-2}). Also demonstrated by the investigators were the excellent antifouling properties against human plasma ($\Gamma < 15$ ng cm^{-2}, as measured by FTIR-ATR upon 15 min exposure at 37 °C) displayed by polyMeOEGMA brushes grown on nylon-coated polypropylene sheets.

Moving away from gold as the substrate material meant that a new type of initiator adlayer – other than that based on the thiol SAM surface chemistry of ω-mercapto-undecyl α-bromoisobutyrate (drawn in Figure 5.20D) – needed to be developed. As discussed earlier, this was successfully accomplished by Chilkoti[95] and García/Collard[96] (chlorosilane chemistry on SiO_2 or TiO_2), as well as Rodriguez-Emmenegger[84] (acylation of amino-rich nylon). The latter research team recently (in 2013) continued their effort in a proof-of-concept study, whose objective was to propose a versatile, substrate-independent surface modification route.[66,97] The approach was to find its universality in the fact that the chosen adhesion surface chemistry – that of dopamine–melanin polymer films featuring bioinspired catechol anchoring moieties – is known for exhibiting high affinity toward virtually any substrate, and for offering subsequent functionalizability.[66,79,97] This latter characteristics was taken advantage of to covalently install α-bromoisobutyryl initiation residues in a second step, from which polyOEGMA brushes were finally ATRP-grown. Demonstration was made on the model surface of gold SPR sensor chips that fouling from undiluted human plasma (15 min exposure) could be minimized to $\Gamma = 2.7 \pm 1.2$ ng cm^{-2} for ~20 nm-thick polyHOEGMA brushes, or even nearly suppressed for ~18 nm-thick polyMeOEGMA ones ($\Gamma < 0.3$ ng cm^{-2}).[97] Another two types of polymer brush that will be introduced later on, polyCBAA[98] and polyHEMA (both ~18 nm-thick), were also shown to display pronounced antifouling properties with $\Gamma = 1.5 \pm 0.2$ and 10.9 ± 1.1 ng cm^{-2}, respectively.[97] In comparison, the bare control surface of gold was heavily fouled ($\Gamma = 293 \pm 70$ ng cm^{-2}), and so was that of the ~13 nm-thick dopamine–melanin adhesion adlayer ($\Gamma = 331 \pm 38$ ng cm^{-2}).[66,97]

5.2.1.3 *Zwitterionic Sulfo- and Carboxybetaines*[99]

Zwitterionic monomers of diverse nature have also been employed for the 'grafting from' formation of antifouling polymer brushes onto gold substrates overcoated with a precursor thiol SAM of the ω-mercapto-undecyl α-bromoisobutyrate ATRP initiator (Figure 5.20D). These include the sulfobetaine and carboxybetaine methacrylates (SBMA and CBMA – Figure 5.22A and B), as well as the carboxybetaine acrylamide variant (CBAA – Figure 5.22D). Together with the polyOEGMA brushes described earlier, these recently established polySBMA, polyCBMA, and polyCBAA zwitterionic brushes rank among the coatings with highest known antifouling performance against blood plasma/serum.

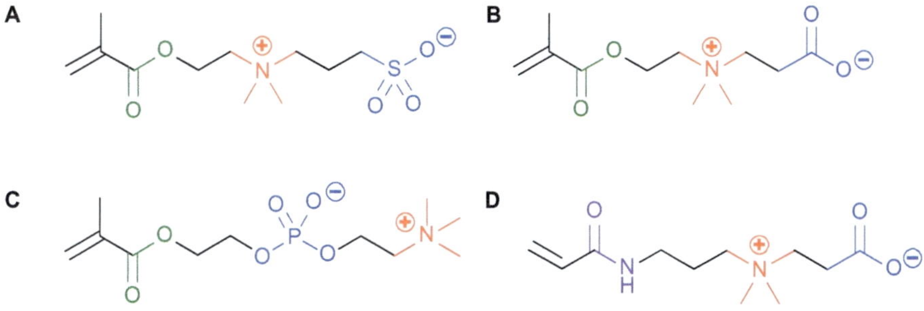

Figure 5.22 Molecular structure of the (A–C) SBMA, CBMA, and PCMA methacrylate, as well as (D) CBAA acrylamide zwitterionic monomers.
Adapted from ref. 25 with permission from the Royal Society of Chemistry.

A fourth kind of zwitterionic brush, built from bioinspired phosphorylcholine methacrylate monomer (PCMA – Figure 5.22C), has also been described but shown not to offer great protection against fouling from human plasma,[100] even if the latter was diluted to 33% in PBS[101] ($\Gamma > 345$ ng cm^{-2}, as measured by SPR upon 30 min exposure). [PC-terminated thiol SAMs challenged by 100% human plasma performed as poorly ($\Gamma > 300$ ng cm^{-2}).[83]] Acceptable antifouling performance was only achieved against 10% FBS ($\Gamma = 13$–45 ng cm^{-2} at 25 °C).[65] Unlike the previous two examples however,[100,101] these latter PCMA-based antifouling surfaces (~0.5 to 2 nm-thick) were prepared using a 'grafting to' method, wherein polymers (5–25 kDa) were first ATRP-grown in solution then deposited onto gold SPR sensor chips.[102,103] In this account as well, a linear relationship was observed between fouling surface coverage Γ and film thickness/polymer MW, thickest coatings prepared with heaviest molecules being the most efficient.[65]

5.2.1.3.1 Sulfobetaine Methacrylate (SBMA). As part of a 2008 study that aimed at correlating platelet adhesion/activation with protein adsorption, Chen and co-workers reported polySBMA brushes only ~7 nm-thick capable of limiting protein adsorption from 20% PPP to $\Gamma = 1.65$ ng cm^{-2}, as measured by SPR upon ~15 min exposure at 37 °C.[57,58] The same year, a collaboration by Horbett, Ratner, and Jiang challenged slightly thicker polySBMA brushes (10–15 nm) with undiluted human plasma, and determined the level of protein adsorption to be $\Gamma = 9.1 \pm 0.6$ ng cm^{-2} using SPR detection.[83] [This value was recently lowered to $\Gamma = 3.4$ ng cm^{-2} by Zheng's team with ~20 nm-thick brushes.[104]] Still in 2008, Jiang and co-workers published a more systematic study, wherein polySBMA brushes of various thicknesses (15–90 nm range) were screened by means of SPR for their ability to withstand protein adsorption from 100% human serum (10 min exposure).[105] The interesting conclusion of that work was that fouling was minimal ($\Gamma = 6.1$ ng cm^{-2}) at the intermediate thickness of 62 nm. Both thinner and thicker brushes indeed adsorbed a greater amount of serum proteins, always below $\Gamma \sim 20$ ng cm^{-2} however. It was only when zwitterionic brushes (~20 nm-thick) were built from an amide (*vs.* ester) variant of the SBMA monomer that 'ultralow' fouling against both 100% human plasma and serum was reached ($\Gamma < 0.3$ ng cm^{-2}, 10 min exposure).[104]

Other types of α-bromoisobutyryl building block molecules – other than the ω-mercapto-undecyl-based one shown in Figure 5.20D – for the surface pre-modification of gold or other underlying substrate materials with ATRP initiator SAMs have been studied as well. For instance, Jiang's team reported in 2008 that polySBMA brushes with excellent resistance to protein adsorption could also be polymerized from an α-bromoisobutyramide initiator underlayer grafted on gold SPR sensor chips through bioinspired dopamine residues (Figure 5.23A).[106] Fouling from 10% human serum was low at $\Gamma \sim 12 \pm 3$ ng cm^{-2} (15 min contact). Even lower adsorption ($\Gamma = 1.1 \pm 2.0$ ng cm^{-2}) was achieved for polySBMA brushes grown from ATRP initiator residues immobilized onto pre-coated, NH_2-terminated alkylthiol SAMs. These particular coatings maintained good performance ($\Gamma = 14.9 \pm 6.0$ ng cm^{-2}) against more challenging undiluted human serum. With respect to 100% human plasma, fouling was nearly undetectable ($\Gamma = 0.8 \pm 0.7$ ng cm^{-2}). Remarkably, these antifouling properties were preserved – against both undiluted fluids – upon storage in PBS for 42 days under ambient conditions of temperature. In a simultaneous separate publication, the authors observed comparable antifouling performance when the polySBMA polymer chains were grown in solution from free ATRP initiator molecules (Figure 5.23B), then deposited onto NH_2-terminated SAMs ('grafting to' method).[107] Indeed, fouling surface coverages from 100% human serum and plasma respectively were $\Gamma = 22.5 \pm 7.5$ and 1.6 ± 7.3 ng cm^{-2}, as measured by SPR upon 15 min exposure.

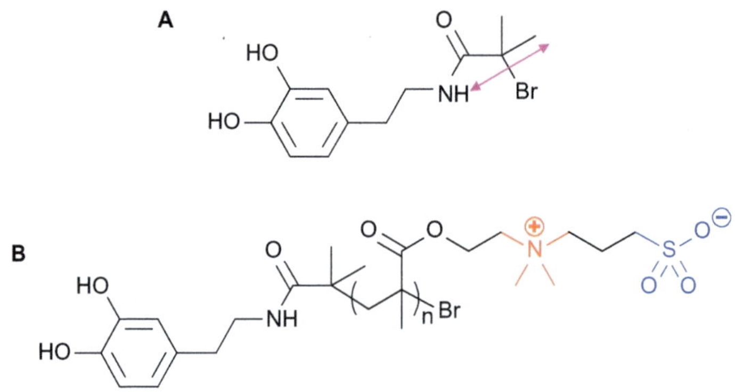

Figure 5.23 Molecular structure of (A) the α-bromoisobutyramide ATRP initiator (the double arrow marks the future site of polymerization), and (B) the 'grafting to' polySBMA polymer grown in solution – both bearing a bioinspired, catechol-based dopamine residue for surface adhesion. Adapted from ref. 25 with permission from the Royal Society of Chemistry.

With respect to other substrate materials (than gold), Bailey's team reported in 2011 – in the manner of Chilkoti in 2006 for polyMeOEGMA brushes[95] – that polySBMA brushes with good antifouling properties against undiluted FBS ($\Gamma = 26$ ng cm^{-2}) could also be ATRP-grown onto the silicon (oxide) surface of photonic microring resonators through precursor siloxane SAMs of undecyltrichlorosilane α-bromoisobutyrate initiator.[108] In comparison, fouling on bare oxide-passivated silicon surfaces was evaluated to be $\Gamma = 300$ ng cm^{-2}.

Finally, there is also the approach by Chang *et al.*, who devised hybrid copolymer coatings combining the antifouling characteristics of zwitterionic polySBMA with the thermoresponsive properties of non-ionic poly(*N*-isopropylacrylamide) (polyNIPAAm).[109] In that 2009 study, the authors prepared a series of statistical poly(SBMA-*co*-NIPAAm) copolymers, whose chemical structure is shown in Figure 5.24A, that were next physisorbed through hydrophobic–hydrophobic interactions onto CH_3-terminated SAMs pre-coated on gold ('grafting to' method). Measured by SPR upon exposure to 10% PPP at 37 °C, the amount of adsorbed plasma proteins was found to be invariably below $\Gamma = 5$ ng cm^{-2}, even for copolymer films containing as little as 15 mol% of SBMA monomer. In comparison, CH_3-terminated conditioning SAMs adsorbed a considerably greater amount of plasma proteins ($\Gamma \sim 150$ ng cm^{-2}), and so did polyNIPAAm homopolymer coatings ($\Gamma \sim 90$ ng cm^{-2}). Conversely, polySBMA homopolymer brushes offered much greater resistance against fouling ($\Gamma < 5$ ng cm^{-2}). To note, Leckband's group did recently succeed in preparing polyNIPAAm homopolymer films with pronounced antifouling properties, upon growth from mixed thiol initiator SAMs on gold.[110] Coatings with high polymer grafting density (650 Å^2 per chain) were indeed able to effectively resist fouling from 10% FBS with $\Gamma < \sim 20$ ng cm^{-2}, as measured with ellipsometry upon 20 h exposure at 25 °C. On the other hand, polyNIPAAm coatings with lower grafting density (1450 Å^2 per chain) were not as antifouling with $\Gamma \sim 25$–200 ng cm^{-2}.

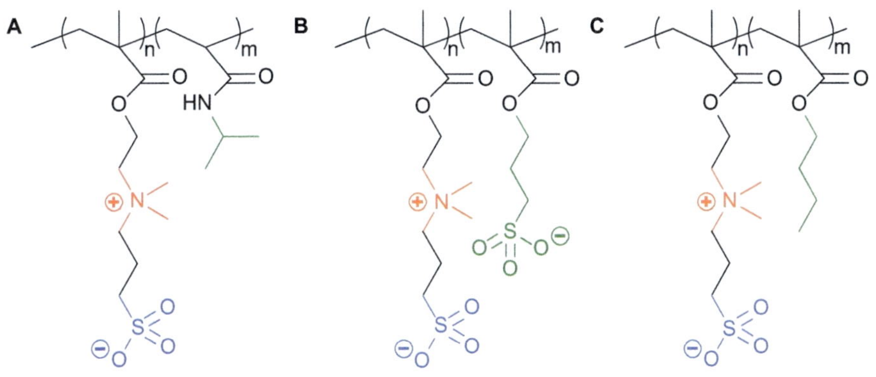

Figure 5.24 Molecular structure of the (A) poly(SBMA-*co*-NIPAAm), (B) polySBMA-*bl*-polySA, and (C) poly(SBMA-*co*-BMA) copolymers.
Figure 5.24A adapted from ref. 25 with permission from the Royal Society of Chemistry.

The same year, in 2013, Chang and Jiang described SBMA-containing diblock methacrylate copolymers incorporating polysulfonate (SA) subunits (Figure 5.24B) – for ion-pairing, interpenetrating physisorption on polycationic, trimethylammonium-based methacrylate brushes, themselves ATRP-grown on gold SPR sensor chips *via* precursor SAMs of ω-mercaptoundecyl α-bromoisobutyrate initiator (shown in Figure 5.20D).[111] Electrostatically 'grafted to' polySBMA-*bl*-polySA assemblies presenting longer polysulfonate anchorage chains (that is, polySBMA$_{40}$-*bl*-polySA$_{40}$ *vs.* polySBMA$_{40}$-*bl*-polySA$_{20}$) were shown to invariably exhibit superior antifouling properties, both against 100% serum ($\Gamma \sim 9$ *vs.* 37 ng cm^{-2}) and 100% plasma ($\Gamma \sim 9$ *vs.* 44 ng cm^{-2}), as measured upon 20 min exposure at 37 °C. Not surprisingly, fouling from diluted fluids (20% and 50% in PBS) was lower. Whereas trimethylammonium-based 'control' adhesion layers were shown to be heavily fouled both by serum and plasma, even if diluted to 20% ($\Gamma > 210$ ng cm^{-2}), polySBMA homopolymer 'reference' brushes exhibited their habitual pronounced antifouling properties against both full fluids[83,105] with $\Gamma < 10$ ng cm^{-2}.[111]

At last, another variety of antifouling, SBMA-based zwitterionic methacrylate copolymer brush was described in 2014 by Liu's group.[112] These constructs were the result of the random copolymerization of *n*-butyl methacrylate (BMA) and SBMA monomers (Figure 5.24C). The authors showed that, among a series of ~40 nm-thick poly(SBMA-*co*-BMA) brushes deposited onto gold SPR sensor chips, only those with a minimal 10 mol% SBMA content possessed pronounced antifouling properties – the best performance ($\Gamma = 15 \pm 4$ ng cm^{-2} for 100% human plasma, and $\Gamma = 17 \pm 3$ ng cm^{-2} for 100% human serum, 10 min exposure) belonging to the brush with the highest SBMA content studied, poly(SBMA$_{20}$-*co*-BMA$_{80}$). In contrast, brushes with a SBMA content below 10% were heavily fouled ($\Gamma > 70$ ng cm^{-2}), regardless of the nature of the challenging fluid. The same held true for bare gold ($\Gamma > 150$ ng cm^{-2}).

5.2.1.3.2 Carboxybetaine Methacrylate (CBMA). Along with polySBMA coatings, Horbett, Ratner and Jiang also described in 2008 polyCBMA brushes (10–15 nm-thick) capable of reducing protein adsorption from 100% human plasma to as little as $\Gamma = 0.4 \pm 0.9$ ng cm^{-2}, as measured by SPR.[83] Such an 'ultralow' fouling level had already been reported shortly before by Jiang's group for both human serum and plasma.[59] In 2009, the latter team investigated using SPR the influence of incubation temperature on protein adsorption from undiluted human plasma (~10 min exposure).[113] This work revealed that 29 nm-thick polyCBMA brushes exhibit 'ultralow' fouling properties ($\Gamma < 0.3$ ng cm^{-2}), at all examined temperatures (25, 30, and 37 °C). The same year, Rodriguez-Emmenegger *et al.* also described polyCBMA brushes capable of efficaciously resisting protein adsorption ($\Gamma < 0.6$ ng cm^{-2}); however, only from 33% human plasma (30 min exposure).[101] In comparison, fouling on uncoated gold substrates was considerably much greater ($\Gamma = 400 \pm 98$ ng cm^{-2}).

Interestingly, and in stark contrast with the polyCBMA system, polySBMA brushes prepared for the purpose of the study adsorbed a substantial amount of plasma proteins ($\Gamma = 320 \pm 78$ ng cm^{-2}) – a quantity nearly equivalent to that observed for unmodified, control gold surfaces. This issue was alleviated when polyCBMA chains were sequentially grown over polySBMAs in a diblock layered configuration ($\Gamma = 24 \pm 10$ ng cm^{-2}), but still very much present when CBMA and SBMA monomers were randomly copolymerized ($\Gamma = 180 \pm 26$ ng cm^{-2}). Tremendous improvement was finally recently achieved by the investigators (in 2012) when fouling on 20 nm-thick polyCBMA brushes from now undiluted plasma (*vs.* 33% in PBS) was found to be simply undetectable by SPR ($\Gamma < 0.03$ ng cm^{-2}, 30 min exposure).[100]

Extending their own work with SBMA (see Figure 5.23B),[107] Jiang and coworkers designed in 2010 polyCBMA adsorbates functionalized with bioinspired L-DOPA adhesion residues (Figure 5.25) – for the direct, 'grafting to' surface modification of gold SPR sensor chips.[114] Fouling from 100% human serum and plasma on films estimated to be 9–10 nm-thick was low at $\Gamma = 11.0 \pm 5.0$ and 8.9 ± 3.4 ng cm^{-2} (10 min exposure), respectively. Comparable levels of performance were monitored when surface chemistry was implemented to derivatize silicon oxide substrates.[115–117] Interestingly however, when the polymer conjugate was reduced to a single L-DOPA/polyCBMA arm, fouling dramatically increased to $\Gamma \sim 64$ and 68 ng cm^{-2}.[114]

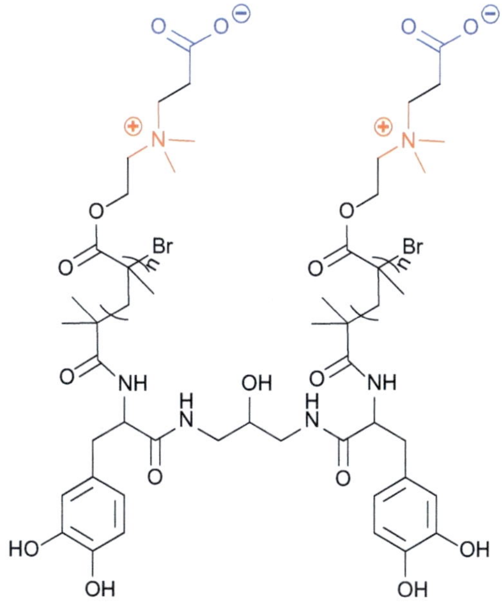

Figure 5.25 Molecular structure of the L-DOPA/polyCBMA polymer conjugate with bioinspired catechol residues for substrate anchorage.
Adapted from ref. 25 with permission from the Royal Society of Chemistry.

5.2.1.3.3 Carboxybetaine Acrylamides (CBAA). The last variant of zwitterionic brush to be discussed – based on the CBAA monomer (Figure 5.22D) – has been the object of multiple studies. One such work was conducted by Jiang's team, who reported in 2008 polyCBAA brushes (15–20 nm-thick) exhibiting 'ultralow' fouling performance below/near the LOD of their SPR sensor (0.2 ng cm^{-2}) upon contact with either 10, 50, or 100% human plasma for 7 min under ambient temperature conditions.[118,119] The following year, the investigators published a more comprehensive SPR study, wherein the effects of film thickness (10–55 nm range) and incubation temperature (25 or 37 °C) on fouling were determined for both full human serum and plasma.[120] As it was the case for the polySBMA brushes described earlier, there also seemed to exist an intermediate, optimal thickness (~21 nm) at which fouling was minimal (*i.e.* undetectable), and this regardless of the incubation temperature or the nature of the contacting fluid (10 min exposure). To note, the range in thickness for which $\Gamma < 5$ ng cm^{-2} was 15–26 nm, at 25 °C. Gratifyingly, these polyCBAA brushes were capable of maintaining their high resistance to protein adsorption at the more relevant body temperature of 37 °C, for an even wider range of thicknesses (15–40 nm). Soon after, Jiang's team extended the scope of polyCBAA surface chemistry by polymerizing monomer molecules of various spacing length between the inner ammonium and terminal carboxylate charged moieties.[121] Polymer brushes 15–20 nm-thick, prepared either with regular-ethylene CBAA-2 (the structure actually drawn in Figure 5.22D) or shorter-methylene CBAA-1 monomers, decreased protein adsorption from full human plasma and serum below $\Gamma = 5$ ng cm^{-2}, as measured by SPR upon 10 min exposure. In contrast, polymerization of CBAA-3 monomers featuring a longer propylene spacer arm yielded slightly thicker brushes (15–25 nm) on which fouling from undiluted fluids was higher, especially for serum ($\Gamma \sim 70$ ng cm^{-2}). Recently, the refractive index (RI) – a material characteristic that can be correlated to film density – was demonstrated by Jiang's group to also constitute an important parameter to take into consideration for the identification and development of antifouling polymer coatings.[122] In that 2012 investigation, fouling from undiluted human serum (measured by SPR after contact for 10 min at 25 °C) was shown to systematically be lower than $\Gamma = 5$ ng cm^{-2} only for polyCBAA brushes with RI values $> \sim 1.5$, and this regardless of dry film thickness (~9–40 nm). Below this threshold in dry RI, fouling was substantially higher (up to 29 ng cm^{-2}). In 2011, Rodriguez-Emmenegger *et al.* also reported polyCBAA brushes capable of resisting fouling, from 100% FBS for 15 min ($\Gamma < 0.03$ ng cm^{-2}).[84] Unfortunately, with respectively $\Gamma = 15$ and 16 ng cm^{-2}, SPR measurements with undiluted human plasma and serum biosamples were not as satisfactory. Soon after however, the investigators finally reported polyCBAA brushes (18 nm-thick) able to nearly suppress fouling from undiluted human plasma ($\Gamma < 0.03$ ng cm^{-2}).[62,85,100,123] Similar resistance to adsorption was observed with undiluted foetal bovine and calf sera.[62] [To note, fouling from undiluted food media – *i.e.* raw orange, tomato, or cucumber blend juices, as well as fresh half-fat milk – was also investigated ($\Gamma < 5$ ng cm^{-2}, 10 min exposure at 25 °C).[123]] In comparison, plasma proteins avidly adsorbed on bare gold ($\Gamma = 307$ ng cm^{-2}).[62,85,100] Lastly, the authors also showed that fouling on these polyCBAA brushes was still undetectable after longer exposure (120 *vs.* 15 min periods).[85]

In 2011–2012, Jiang's team additionally demonstrated that polyCBAA brushes could also be polymerized photochemically through an environmentally-friendly, catalyst-free SI-PIMP process[43] that allowed for film thickness and uniformity to be readily controlled.[124] Measured by SPR after contact for 10 min at 25 °C, fouling surface coverage from undiluted human plasma was very low ($\Gamma < 5$ ng cm^{-2}),[124–126] and comparable to that recorded for regular, ATRP-grown polyCBAA brushes of similar thickness.[125]

5.2.1.4 Hybrid, Derivative and Biomimetic Materials

The remaining surface-modifying (macro)molecules to be discussed in this chapter do not belong to any of the three families presented thus far, but rather are combinations thereof (*i.e.* hybrids), derivatives, or totally unrelated materials. Nonetheless, these 'outliers' are equally as important and form, as a matter of fact, some of the best antifouling coatings known against blood plasma/serum.

5.2.1.4.1 PLL-*graft*-PEG Hybrids and PLL-*graft*-PMOXA Analogues. In a series of studies published between 2000 and 2012, several research groups described hybrid copolymeric surface-modifiers having for generic design a PLL polypeptide backbone partially grafted with PEG side-chains through primary amine residues (Figure 5.26).[127–140] At pH 7.4 (PBS or HEPES), these PLL-*g*-PEG macromolecules were shown to be able to spontaneously self-assemble onto a variety of metal oxide substrates made of niobium,[127–129,136,137,139,140] titanium,[127,128,130–133,135,138] silicon/titanium,[127] or tantalum.[128] The adsorption mechanism involves multiple electrostatic interactions between the residual, positively-charged amine moieties of PLL side-chains and the negatively-charged oxide surfaces.[128,130–132,134,136] As a result, the polypeptide backbone lies parallel to the substrate and the PEG side-chains stretch perpendicularly into the liquid environment, in a comb-like overall architecture.[128–132,134,139,140] For optimized copolymer structures, protein adsorption from human serum (15–60 min exposure) was near/lower than the 2 ng cm^{-2} LOD of the OWLS sensor.[127–130,132–134,136,137,139,140] This result is to be contrasted with that obtained with bare, control oxide surfaces for which fouling surface coverage ranged between $\Gamma = 223$ and 596 ng cm^{-2}.[127–129,137,139,141] Interestingly, it was also shown in one example that resistance to fouling is maintained whether PLL-*g*-PEG coatings are exposed to multiple successive injections of serum over several hours, or stored dry for several months.[127]

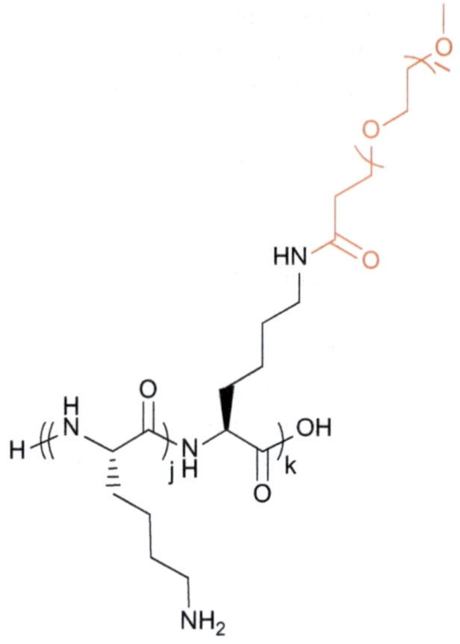

Figure 5.26 Molecular structure of the PLL-*g*-PEG copolymers.
Adapted from ref. 25 with permission from the Royal Society of Chemistry.

Finally, the demonstration was also made in 2005 that oxidized poly(dimethylsiloxane) (PDMS) substrates can be derivatized with antifouling PLL-*g*-PEG surface chemistry as well.[142] Using OWLS, the resulting coatings were shown to be capable of preventing protein adsorption from serum with $\Gamma \sim \pm 1.0$ ng cm^{-2}. This remarkable performance very much contrasts with the heavy fouling situation observed for bare PDMS ($\Gamma = 385 \pm 18$ ng cm^{-2}). Another key finding was the long-term resistance exhibited by such surface chemistry (up to 12 weeks).

A PLL-graft copolymer analogue featuring poly(2-methyl-2-oxazoline)[143] PMOXA side-chains (Figure 5.27A) was also reported by Konradi *et al.* in 2008.[144,145] Likewise, PLL-*g*-PMOXA coatings exhibited excellent antifouling properties against full human serum ($\Gamma < 2$ ng cm^{-2}), even after repeated 15 min contact (as measured by OWLS).[143–145] A PMOXA copolymer variant introduced soon after (Figure 5.27B) performed as efficiently,[137,146] even following prolonged exposure (up to 2 weeks) to simulated physiological conditions of ionic strength or/and an oxidative environment.[137,143] These PLL-*g*-PMOXA films were also shown to be significantly more stable in such situations than the PLL-*g*-PEG parent coatings. A very recent 2014 account revealed that PLL-*g*-PEG adlayers are also sensitive to trypsin (a protease that cleaves polypeptide chains at lysine (and arginine) residues[147]); the extent of degradation, assessed by OWLS as a decrease in adsorbed mass upon 1.5 h exposure at 37 °C, depending on the temperature (37 or 80 °C) at which the coatings were deposited.[147] The higher this temperature was, the more stable the PLL-*g*-PEG assemblies turned out to be. [To note,

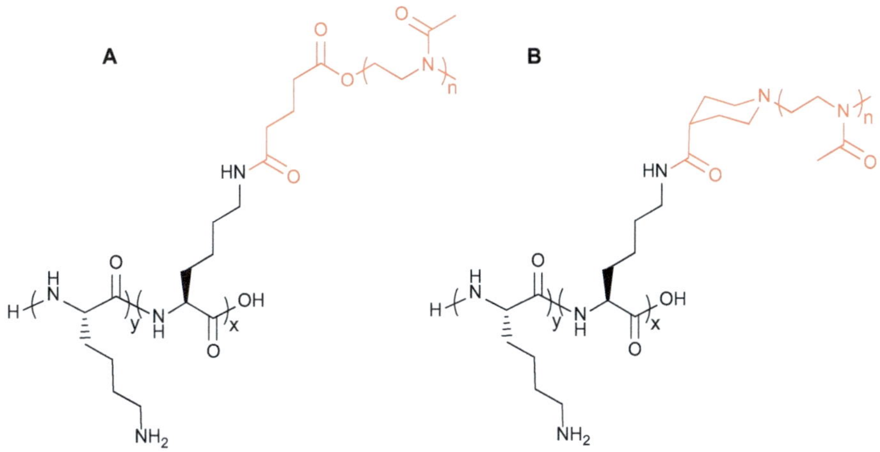

Figure 5.27 Molecular structure of two variants of PLL-*g*-PMOXA copolymers. Figure 5.27A adapted from ref. 25 with permission from the Royal Society of Chemistry.

polyoxazoline-based films were reported to be able to withstand nearly unaffected acidic and/or enzymatic hydrolysis at 37 °C, for periods of time potentially extending up to several days.[148]] On a brighter note, that work also corroborated the general observation that PLL-*g*-PEG adlayers exhibit excellent antifouling properties (against 100% FBS for 15 min at 37 °C), those being as well essentially independent of the temperature at which the coatings were deposited ($\Gamma_{37\ ^\circ\mathrm{C}} = 5.7 \pm 0.6$ ng cm^{-2} $\sim \Gamma_{80\ ^\circ\mathrm{C}} = 4.8 \pm 0.6$ ng cm^{-2}).[147] Resistance to fouling was influenced however by the enzymatic treatment, less degraded coatings assembled at higher temperature experiencing a lesser loss in antifouling performance ($\Gamma_{37\ ^\circ\mathrm{C/trypsin}} = 38.5 \pm 0.7$ ng cm^{-2} $> \Gamma_{80\ ^\circ\mathrm{C/trypsin}} = 20.4 \pm 0.7$ ng cm^{-2}).

5.2.1.4.2 Short, Non-ionic Hydroxyalkyl and MEG ATRP Monomers. Other types of short, non-ionic hydroxyalkyl monomers – related to mono-glycerol methacrylate drawn in Figure 5.20A – have also been employed to prepare 'grafted from' polymer brushes ATRP-grown from precursor thiol SAMs of ω-mercapto-undecyl α-bromoisobutyrate initiator pre-coated on gold. However, as will be seen next, these recently developed brushes were otherwise more efficient at resisting fouling from blood plasma/serum biosamples.

In 2011, Zheng in collaboration with Yu used 3-hydroxypropyl and 2-hydroxyethyl methacrylate monomers (HPMA and HEMA – Figure 5.28A and B) to prepare a series of polymer brushes for which the effect of film thickness (~10–60 nm) on antifouling properties was methodically assessed with SPR, upon 10 min exposure to both diluted and undiluted human serum and plasma.[149] Interestingly, that work revealed that optimal antifouling performance was achieved for intermediate ranges of coating thickness

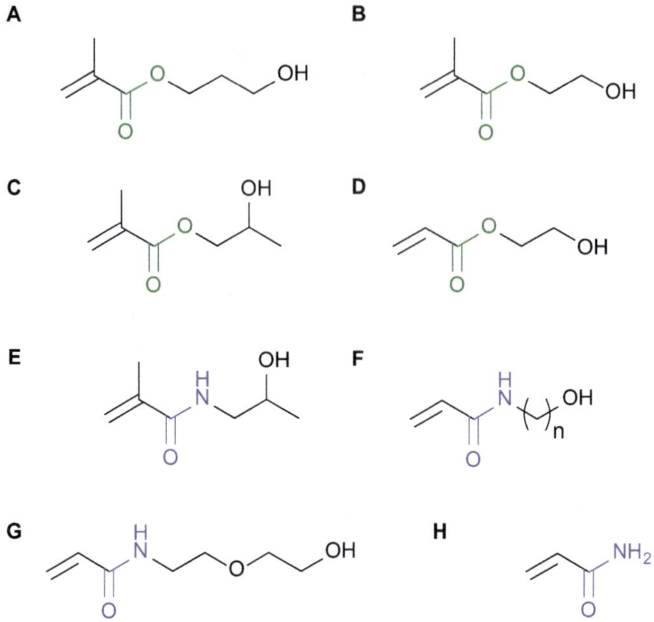

Figure 5.28 Molecular structure of short, non-ionic building block monomers for 'grafting from' brush polymerization: (A–C) 3-hydroxypropyl (HPMA), 2-hydroxyethyl (HEMA), and 2-hydroxypropyl (HPM) methacrylates, (D) 2-hydroxyethyl acrylate (HEA), (E) *N*-(2-hydroxypropyl) methacrylamide (HPMA), (F) methylene ($n=1$, HMAA), ethylene ($n=2$, HEAA), propylene ($n=3$, HPAA), and pentylidene ($n=5$, HPenAA) *N*-(n-hydroxyalkyl) acrylamides, (G) acryloylaminoethoxyethanol (AAEE), and (H) non-substituted acrylamide. Figure 5.28A–C, E & H adapted from ref. 25 with permission from the Royal Society of Chemistry.

(20–45 nm),[149,150] confirming the findings of Jiang's team in this respect for both polySBMA[105] and polyCBAA[120] systems. For both fluids diluted at 10% in PBS, polyHEMA and polyHPMA brushes offered remarkable protection against protein adsorption ($\Gamma < 0.3$ ng cm^{-2}).[149,150] With respect to undiluted serum and plasma, antifouling performances were still excellent for polyHEMA brushes with respectively $\Gamma \sim 3.0$ and 3.5 ng cm^{-2}.[149] In sharp contrast however, polyHPMA brushes adsorbed much greater amounts of proteins with $\Gamma \sim 13.5$ and 50.0 ng cm^{-2}, respectively.[149,150]

Still in 2011, Rodriguez-Emmenegger *et al.* introduced a new type of 'non-fouling' polymer brush built with *N*-(2-hydroxypropyl) methacrylamide monomer (incidentally also abbreviated 'HPMA' – Figure 5.28E).[85] These polyHPMA brushes (17–30 nm-thick) were shown to be capable of resisting fouling ($\Gamma < 0.03$ ng cm^{-2}, 15 min exposure) not only from undiluted human plasma[62,85] or serum,[91] but also FBS[62] as well as calf serum.[62] Interestingly, polyHPM brushes (30 nm-thick) resulting from the polymerization of the corresponding 2-hydroxypropyl methacrylate *ester* monomer (HPM – Figure 5.28C *vs.* HPMA – Figure 5.28E) adsorbed a considerably greater amount of plasma proteins with $\Gamma = 40.5$ ng cm^{-2}.[85] Importantly for potential biomedical/bioanalytical applications, fouling surface coverage on polyHPMA coatings was still undetectable upon long-term contact (120 *vs.* 15 min) with 100% plasma,[85] or upon exposure to single individual plasma samples obtained from different human adult donors.[89] [It was also shown in another study that ~30 nm-thick polyHPMA brushes are able to sustain exposure for 10 h to 10% human serum ($\Gamma = 7.8$ ng cm^{-2}) or 10% FBS ($\Gamma = 8.7$ ng cm^{-2}), fouling after 15 min contact being in both cases essentially undetectable ($\Gamma < 0.3$ ng cm^{-2}).[91]] Remarkably as well, multiple exposures to plasma and storage in PBS over 2 years barely affected the 'non-fouling' characteristics of the coatings, demonstrating their excellent reusability and shelf-life stability.[85] In this respect, non-ionic polyHPMA brushes easily outperformed zwitterionic polyCBAA ones ($\Gamma = 17$ *vs.* 125 ng cm^{-2} after 2 years) that were otherwise equally as antifouling.[85]

Another addition to this family of short, non-ionic hydroxyalkyl monomers that was reported by Zheng's group in late 2011 is *N*-(2-hydroxyethyl) acrylamide (HEAA – Figure 5.28F, $n = 2$).[104,151,152] All polymer brushes grown within a wide range of thicknesses (~10–40 nm) were shown by SPR to display 'non-fouling' properties as well ($\Gamma < 0.3$ ng cm^{-2}), when exposed for 10 min to either undiluted human serum or plasma. [To note, brushes ~20 nm-thick built with the 2-hydroxyethyl acrylate *ester* version of HEAA monomer (HEA – Figure 5.28D *vs.* HEAA – Figure 5.28F, $n = 2$) were not as performing but still quite effective at decreasing fouling surface coverage from 100% human plasma ($\Gamma = 4.9$ ng cm^{-2}) and 100% human serum ($\Gamma = 5.0$ ng cm^{-2})[104] – these values being well in par however with those obtained for polyHEMA brushes (*vide supra*),[149] their *meth*acrylate variant (Figure 5.28B *vs.* D).] Even more gratifying was the observation that an ultrathin polyHEAA brush only 12 nm-thick was capable of retaining such remarkable antifouling properties for a much longer period of time (60 min). Recently, in 2014, Zheng's team described a longer, OEG version of HEAA – the *N*-acryloylaminoethoxyethanol (AAEE) monomer molecule drawn in Figure 5.28G – with which were constructed polymer brushes able to reduce fouling from undiluted human plasma/serum below $\Gamma = 0.3$ ng cm^{-2} (SPR detection, 10 min exposure).[153] As was already the situation for other types of brushes – poly(L-serine),[43] polySBMA,[105] polyCBAA,[120] and polyHPMA/polyHEMA[149,150] – there also existed an optimal range of polymer thickness for antifouling polyAAEE adlayers (~10–40 nm).

In 2014 as well, Zheng in collaboration with Chen systematically investigated the influence of the alkyl chain length of a series of *N*-(*n*-hydroxyalkyl) acrylamide building block monomers (Figure 5.28F, $n = 1$–3, 5) on the antifouling performance against 100% plasma/serum of the corresponding polymer coatings.[154] All different brushes (consistently ~33 nm-thick) were shown to invariably reduce fouling below $\Gamma = 2.1$ ng cm^{-2} (SPR detection, 10 min contact), and this regardless of the nature of the undiluted testing fluid. A clear correlation between arm length and protein adsorption appeared however, brushes bearing longer side-chains being less antifouling. In this respect, polyHEAA coatings ($n = 2$) were shown once again (*vide supra*)[104,151,152] to resist fouling with $\Gamma < 0.3$ ng cm^{-2}, and so did brushes constructed with shorter *N*-(1-hydroxymethyl) acrylamide monomer (HMAA – Figure 5.28F, $n = 1$). In comparison, longer *N*-(3-hydroxypropyl) and *N*-(5-hydroxypentyl) acrylamide building block molecules (HPAA and HPenAA – Figure 5.28F, $n = 3$ and 5) yielded brushes with fouling surface coverage measurable with the SPR sensor of the study (respectively $\Gamma = 0.3–0.9$ and 0.7–2.1 ng cm^{-2}).

At last, there also is the shortest building block monomer of them all, non-substituted acrylamide (Figure 5.28H), which was recently implemented by Liu and co-workers.[155,156] Measured for both human serum and plasma (SPR, 10 min exposure), the resistance to fouling of ~39 nm-thick polyacrylamide brushes was excellent. For 10% serum and plasma, adsorption levels were indeed extremely low at $\Gamma = 0.7 \pm 0.4$ and 0.5 ± 0.4 ng cm^{-2}, respectively. Such high performance was maintained against both undiluted biosamples ($\Gamma = 2.8 \pm 0.9$ and 1.7 ± 0.7 ng cm^{-2}). In comparison, fouling on the unmodified control surface of gold SPR sensor chips was invariably greater than $\Gamma = 140$ ng cm^{-2}, irrespectively of the challenging fluid or its level of dilution.

5.2.1.4.3 EGylated Zwitterionic Hybrids

5.2.1.4.3.1 OEGylated Phosphorylcholine Polyelectrolytes. Their 2009 study on the antifouling performance of OEO-grafted poly(acrylic acid) PAA films was also the occasion for Schaaf and Mésini to assess that of various phosphorylcholine (PC)-terminated related polymer coatings.[71] Antifouling adlayers were also prepared on silicon/titanium oxide OWLS sensor chips *via* an intercalated film made of alternated poly(styrene sulfonate)/poly(allylamine hydrochloride) strata – this polyelectrolyte multilayered structure being itself anchored to the substrate through a poly(ethylene imine) adhesion coating. When subjected to 10% FBS, constructs based on PAA-*g*-(OEO_3-PC) polymer (Figure 5.29A) with a 18% grafting ratio (GR) performed rather well with $\Gamma = 15$ ng cm^{-2}. Considerably more antifouling however ($\Gamma = 1$ ng cm^{-2}) were coatings made with PAA-*g*-(OEO_3-PC) polymer at a slightly higher 25% GR.[71,157] At this GR (more precisely, 23%), films based on non-OEGylated – that is, alkylated – PAA-*g*-(*n*-hexyl-PC) polymer (Figure 5.29B) were not as effective with $\Gamma = 11$ ng cm^{-2}. Coatings constructed with PAA-*g*-(*n*-undecanyl-PC) polymer featuring longer (C_{11} *vs.* C_6) alkyl spacer arms were even more fouled with $\Gamma = 81$ ng cm^{-2} (33% GR). Finally, OEO_3-PC side-chain grafting on poly(allylamine hydrochloride) PAH – the polymer structure seen in Figure 5.29C – gave rise to a reasonably good protection against fouling ($\Gamma = 22$ ng cm^{-2}), however only when GR reached 80%. Regardless, all these coatings did display superior antifouling performance compared to those prepared with non-grafted PAA ($\Gamma = 297$ ng cm^{-2})[71,72,157] and PAH ($\Gamma = 385$ ng cm^{-2})[71] polymers.

5.2.1.4.3.2 PolyCBMA-block-*PPO*-block-*polyCBMA Poloxamers.* Using their expertise with zwitterionic materials to diversify/expand the scope of

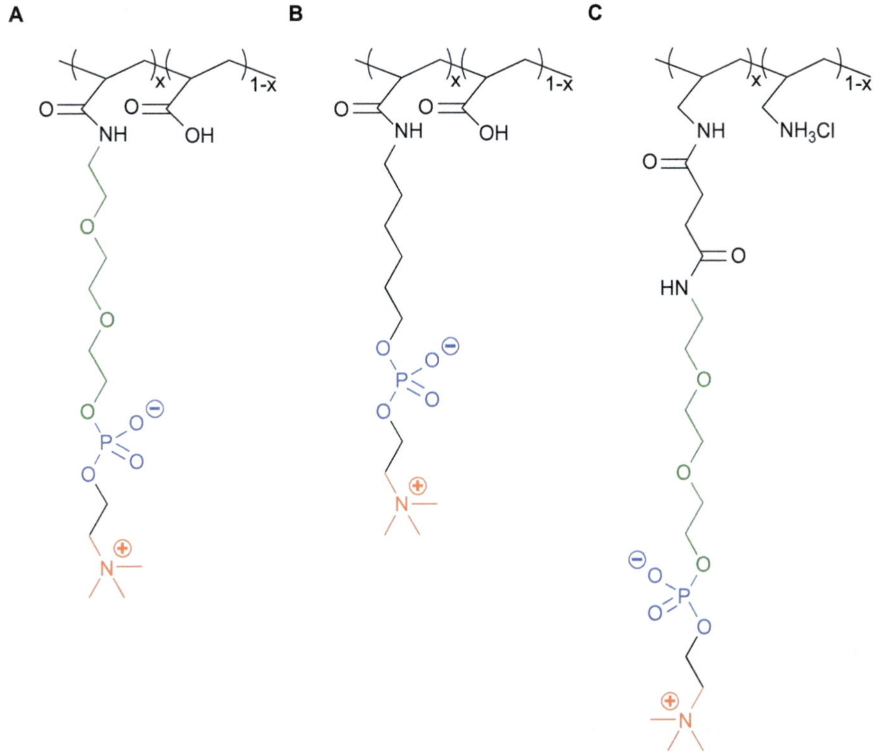

Figure 5.29 Molecular structure of (A) PAA-*g*-(OEO$_3$-PC), (B) PAA-*g*-(*n*-hexyl-PC), and (C) PAH-*g*-(OEO$_3$-PC) poly(acrylic acid)- or poly(allylamine hydrochloride)-based grafted polymers.

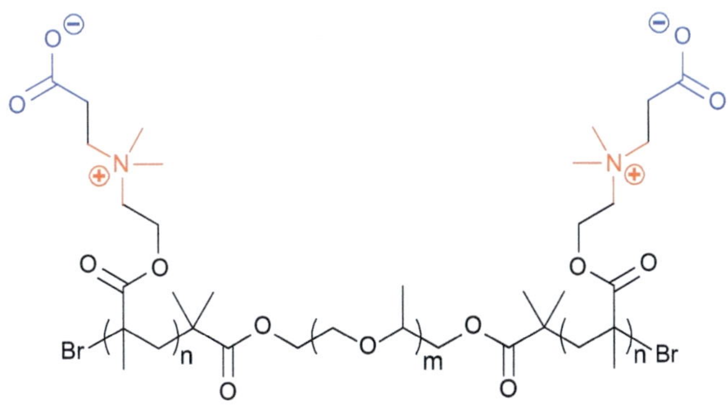

Figure 5.30 Molecular structure of the polyCBMA-*bl*-PPO-*bl*-polyCBMA zwitterionic poloxamer derivative ($n=40$, $m=48$).

available antifouling coating strategies, Jiang and co-workers recently described the synthesis of polyCBMA-*bl*-PPO-*bl*-polyCBMA triblock co-polymer, a zwitterionic poloxamer derivative (Figure 5.30).[158] Antifouling surfaces were prepared on gold SPR sensor chips upon physisorption of the middle PPO segment onto a pre-coated hydrophobic SAM of 1-undecanethiol – as such compelling the superhydrophilic, ATRP-grown polyCBMA side-chains to face outwards into the aqueous environment. Fouling from undiluted blood plasma (10 min exposure) was low at $\Gamma = 5.2 \pm 0.2$ ng cm^{-2}.

*5.2.1.4.3.3 PolyMeOEGMA-*block*-polyCBAA Copolymers.* In a recent pioneering work published in 2012, which also aimed at studying the possibility of performing living ATRP in biological milieux, Rodriguez-Emmenegger *et al.* reported a (2×15) nm-thick polyMeOEGMA-*bl*-polyCBAA diblock copolymer hybrid brush (Figure 5.31) – sequentially grown on gold SPR sensor chips *via* a precursor SAM of ω-mercapto-undecyl α-bromoisobutyrate ATRP initiator (Figure 5.20D) – capable of nearly suppressing fouling ($\Gamma < 0.03$ ng cm^{-2}, 30 min exposure at 25 °C) from various undiluted blood-based fluids whether of animal or human origin (FBS, calf serum, and human plasma).[86] In comparison, the control surface of bare gold substrates adsorbed considerably greater amounts of proteins, respectively $\Gamma = 261 \pm 9$, 280 ± 7, and 307 ± 11 ng cm^{-2}. The remarkable antifouling performance, against undiluted human plasma (for 15 min), of such hierarchically-constructed brushes (20 + 10 nm-thick) was very recently confirmed by the investigators (in 2014).[87]

5.2.1.4.4 Carboxybetaine Zwitterionic Monomer Derivatives. As evidenced throughout this chapter, research on antifouling coatings is a topic of great current interest with a high level of competitivity that is fuelled by a seemingly unlimited imagination in terms of material chemical structure and architecture. The very active field of zwitterionic polymer brushes is not spared from this trend, as illustrated by the recent work by Cheng and co-workers describing a new type of carboxybetaine methacrylamide (CBMAA) building block monomers (Figure 5.32).[159]

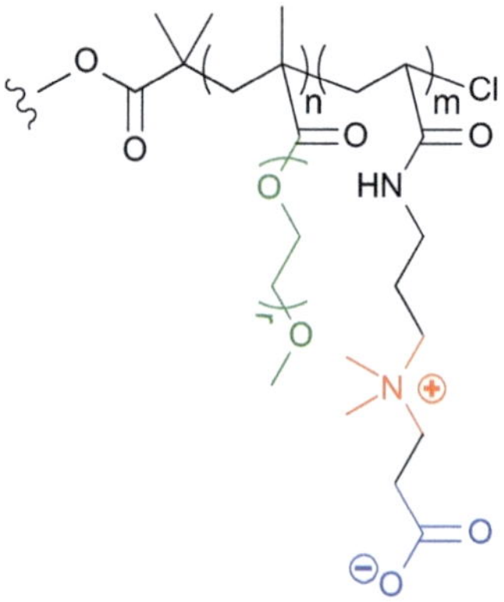

Figure 5.31 Molecular structure of the hierarchical, sequentially-grown polyMeOEGMA-*bl*-polyCBAA hybrid copolymer brush.

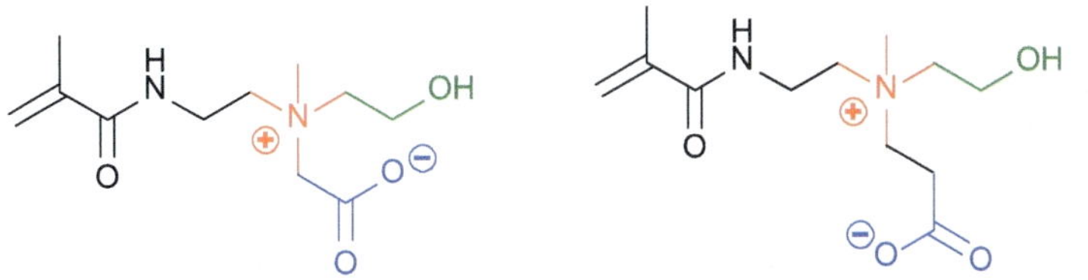

Figure 5.32 Molecular structure of two variants of CBMAA zwitterionic monomers differing with respect to the length of the spacer arm between the inner, positively-charged ammonium and terminal, negatively-charged carboxylate moieties.

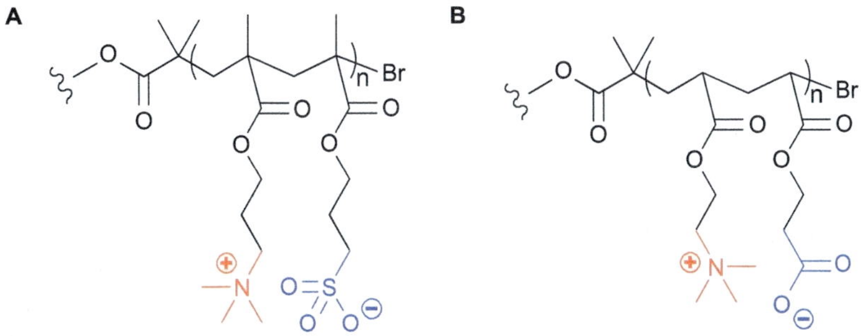

Figure 5.33 Molecular structure of the (A) ammonium/sulfonate, and (B) ammonium/carboxylate mixed-charged, pseudozwitterionic (meth)acrylate copolymer brushes.
Adapted from ref. 25 with permission from the Royal Society of Chemistry.

When challenged for 10 min either by undiluted human plasma or serum, ∼9–18 nm-thick polyCBMAA brushes (constructed photochemically on gold SPR sensor chips *via* a SI-PIMP process) exhibited 'ultralow' fouling performance with $\Gamma < 0.3$ ng cm^{-2}.

5.2.1.4.5 Pseudozwitterionic Materials.[99,160] As seen earlier, zwitterionic polymer brushes exhibit excellent protein repellence. These do not constitute, however, the only type of 'grafted from', ATRP-grown mixed-charged polymer coatings displaying antifouling properties. In fact, 'pseudozwitterionic' copolymer brushes, wherein the positively- and negatively-charged moieties are now found on two separate subunits, are also able to resist protein adsorption. This was recently demonstrated by Chang and Chen (in 2010) with 30–45 nm-thick methacrylate copolymer brushes featuring positively-charged *N*,*N*,*N*-trimethylammonium and negatively-charged sulfonate side-chains (Figure 5.33A).[58] Fouling from 20% human PPP (in PBS, pH 7.4) was measured to be very low at $\Gamma = 7.65$ ng cm^{-2} (∼15 min contact at 37 °C, SPR detection). As well, Bernards' team reported in 2012 acrylate copolymer brushes featuring positively-charged *N*,*N*,*N*-trimethylammonium and negatively-charged carboxylate subunits (Figure 5.33B).[161]

In this work, whether from 10% or undiluted FBS, fouling on ~18 nm-thick coatings was also very low (respectively $\Gamma = 2.1 \pm 2.0$ and 4.3 ± 1.7 ng cm^{-2}), as measured by SPR upon 10 min exposure.

Earlier investigations by Jiang's team with SAM coatings had already demonstrated the concept of mixed-charged material for antifouling purposes.[59] In this work, combinations of *N,N,N*-trimethylammonium (TMA), and sulfonic acid (SA) or carboxylic acid (CA) ω-functionalized alkylthiol gold surface modifiers (Figure 5.34) were investigated by means of SPR for their resistance to human serum/plasma adsorption.

At physiological pH 7.4, the best antifouling performance recorded for the TMA/SA system was achieved against 10% human serum ($\Gamma \sim 15$ ng cm^{-2}, 10 min exposure).[59] This was also the situation for the TMA/CA system ($\Gamma \sim 25$ ng cm^{-2}). Unfortunately, neither of these combinations offered antifouling protection against 100% serum ($\Gamma > 90$ ng cm^{-2}). Adsorption from human plasma was even greater, even if biosamples were diluted at 10% in PBS ($\Gamma \sim 190$–950 ng cm^{-2}).[59,83]

5.2.1.4.6 Ionic Liquids. At last, there also exists – in addition to the various (truly and pseudo) zwitterionic building block molecules discussed thus far – another type of charged material that has very recently been implemented in the preparation of antifouling coatings on gold: ionic liquids (IL). This was accomplished in 2013 by Masson's team with IL SAMs prepared with imidazolium-based thiol molecules (Figure 5.35).[162] Measured against crude serum using SPR, fouling surface coverage was relatively high at $\Gamma = 99$ ng cm^{-2}. However, although this account obviously exceeds the level of adsorption set for antifouling coatings to be presented in this chapter ($\Gamma < 30$ ng cm^{-2}), the novelty of this IL surface chemistry – and the

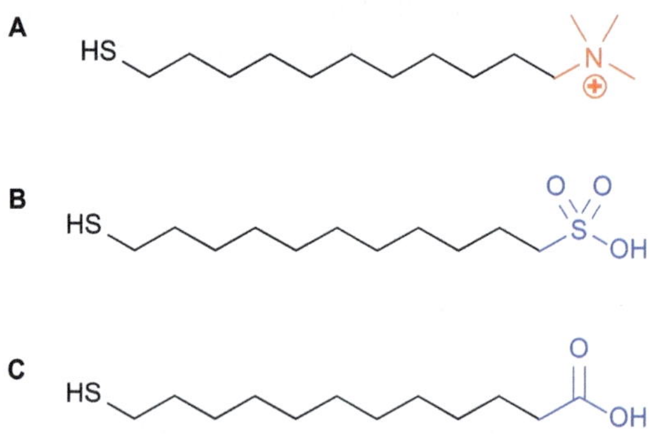

Figure 5.34 Molecular structure of the (A) *N,N,N*-trimethylammonium (TMA), (B) sulfonic acid (SA), and (C) carboxylic acid (CA) ω-functionalized alkylthiol surface modifiers for mixed, pseudozwitterionic antifouling SAM formation on gold. At physiological pH 7.4, SA and CA moieties in the mixed SAMs exist as their negatively-charged sulfonate and carboxylate forms, respectively.
Adapted from ref. 25 with permission from the Royal Society of Chemistry.

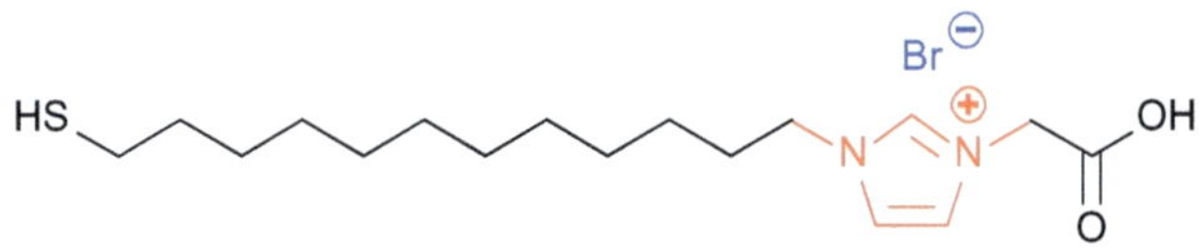

Figure 5.35 Molecular structure of the imidazolium-based IL thiol surface modifier for antifouling SAM formation on gold.

fact that the latter performed similar to benchmark PEGylated antifouling surfaces under identical conditions[162] – both make it definitely worth describing.

5.2.1.4.7 Carbohydrates: Glycocalyx Mimics. We conclude this subsection on hybrid, derivative, and other coating materials with an important class of biomimetic antifouling constructs inspired by the outer surface of biological cell membranes. This interfacial region – which consists of a non-adhesive mesh made of highly hydrated, carbohydrate-rich macromolecules (proteoglycans and glycoproteins) – is known as the 'glycocalyx' and acts as a barrier that regulates molecular and cellular interactions.[163–165]

In 1998, Marchant and co-workers set out to mimic such supramolecular assembly through the engineering of oligosaccharide surfactant copolymers for the surface modification of highly-oriented pyrolytic graphite (HOPG).[165] These macromolecules featured a poly(vinyl amine) backbone onto which hydrophobic *n*-hexanoyl and hydrophilic dextran side-chains were simultaneously grafted *via* the pendant primary amine moieties (Figure 5.36A). On HOPG, these graft-copolymers were shown to grow monomolecular epitaxial films 7–12 Å-thick. During the adsorption process, *n*-hexanoyl chains assemble on the underlying HOPG substrate through hydrophobic–hydrophobic interactions, thereby constraining the flexible poly(vinyl amine) backbone to lie parallel to the surface and the hydrophilic dextran arms to stretch outwards into the aqueous environment – in a tridimensional, comb-like overall structure. Although not quantified *per se*, protein adsorption on these glycocalyx-like assemblies was shown to be considerably lower than that recorded on the unmodified control surface of HOPG substrates (30 min exposure to 50% human PPP at 37 °C). Indeed, through monitoring of the relative absorbance of the characteristic amide I and II protein IR bands, the investigators estimated plasma protein adsorption to be reduced by at least 90%.

Continuing on, Marchant's team later conducted a more systematic investigation (in 2001), wherein a series of oligomaltose-based analogous copolymer coatings with varying saccharide side-chain length were tested for their antifouling properties – under the same, previous experimental conditions (that is, against 50% PPP for 30 min at 37 °C).[166] Still using FTIR-ATR as the detection technique (see above), the demonstration was made by the authors of this work that plasma protein adsorption steadily decreased with

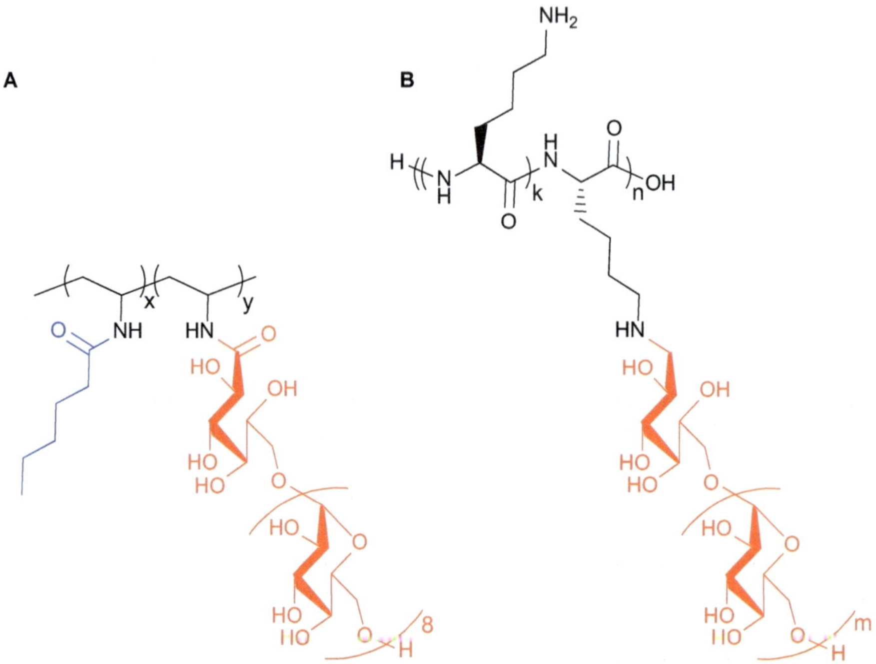

Figure 5.36 Molecular structure of the (A) poly(vinyl amine)-, and (B) poly(L-lysine)-*graft*-dextran carbohydrate-based copolymers.
Adapted from ref. 25 with permission from the Royal Society of Chemistry.

increasing film thickness – the best antifouling performance (an ~96% reduction) being the feat of the thickest films studied (42 Å).

The actual quantification of fouling surface coverage on carbohydrate-based adlayers came much later (in 2008) with the work of Spencer and co-workers, who tested the antifouling properties of a series of poly(L-lysine)-*graft*-dextran copolymer coatings (Figure 5.36B) deposited on silicon/titanium oxide OWLS sensor surfaces.[167] The best protection against fouling from human serum (30 min exposure) was rather good at $\Gamma = 13 \pm 8$ ng cm^{-2}. Of note finally is the new strategy recently devised by this research group (in 2013) to prepare dextran-based antifouling coatings, which unfortunately did not prove as effective.[64]

There is at last the approach by Kizhakkedathu and his team, who explored in 2012 the antifouling properties of a series of glycopolymer brushes ATRP-grown onto gold SPR sensor chips through an α-bromoisobutyrate-terminated thiol initiator SAM.[168] Three types of 'grafted from' coatings – constructed with mannose- (Figure 5.37A), galactose- (Figure 5.37B), or glucose-based (Figure 5.37C) pyranoside acrylamide monomers presenting varied carbohydrate stereochemistry – were examined. When challenged by undiluted blood plasma (for 6 min), it was the latter kind of brush (58.6 nm-thick) that exhibited the greatest propensity to repel plasma proteins with $\Gamma = 24.3$ ng cm^{-2}.

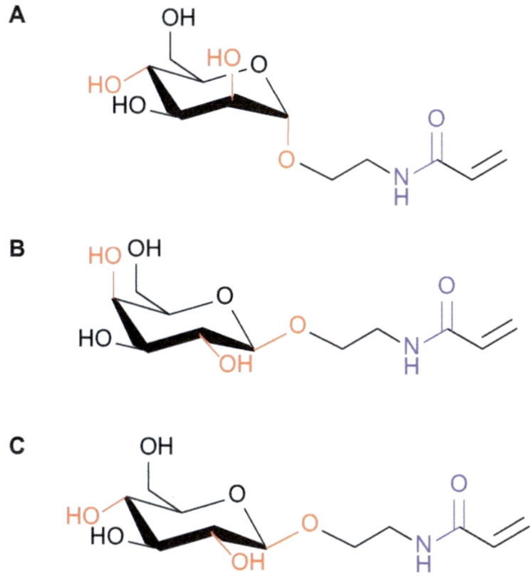

Figure 5.37 Molecular structure of the (A) mannose, (B) galactose, and (C) glucose pyranoside-based acrylamide monomers.

5.2.1.5 Antifouling and Plasma/Serum Source Variability

Prior to moving the focus of this chapter on state-of-the-art antifouling surface chemistries from blood plasma/serum to other human/animal biofluids, a parenthesis of utmost importance – especially in biosensor technology discussed later on – must be made regarding the necessity of considering the source of blood plasma/serum samples, and its variability, in the assessment of the antifouling properties of coatings. At the time this chapter was being written (in the early months of 2015), this issue had only been given little attention, but the message that had clearly emerged was that blood plasma/serum biosamples originating (1) from different species (bovine *vs.* equine sera),[94] (2) from the same species but at a different stage of maturity (adult *vs.* foetal bovine sera,[94] adult *vs.* infant human plasmas[169]), or (3) from different donors (human adults)[89,101] do/may indeed foul a given surface to various extents. In 2014, the latter situation was the focus of a pioneering study by Rodriguez-Emmenegger and co-workers, who compiled the data obtained for a selection of various antifouling polymer surface chemistries – polyHEMA (Figure 5.28B), polyHOEGMA (Figure 5.21A), polyMeOEGMA (Figure 5.21B), polyCBAA (Figure 5.22D), and polyHPMA (Figure 5.28E) brushes – challenged by human blood plasma biosamples drawn from five different individuals.[89] Despite admittedly limited sampling, unambiguous (substantial) variability in terms of fouling surface coverage was demonstrated across the different blood plasmas, most notably for polyHEMA coatings that displayed the highest dispersity between measurements with $\sim 7 < \Gamma < \sim 195$ ng cm^{-2} (followed by polyHOEGMA with $\sim 2 < \Gamma < \sim 63$ ng cm^{-2}, polyMeOEGMA with $\sim 8 < \Gamma < \sim 26$ ng cm^{-2}, and then polyCBAA with $\sim 1 < \Gamma < \sim 7\ \text{ng cm}^{-2}$) – as measured using SPR upon 15 min exposure at 25 °C. The smallest variability was observed for polyHPMA brushes for which fouling was found to be invariably below $\Gamma = 0.3\ \text{ng cm}^{-2}$. Even when the five single donations were mixed, significant variability was still observed with pooled plasma samples purchased from two other, different commercial suppliers.[89] Once more, polyHEMA coatings exhibited the highest discrepancy in terms of antifouling behavior with $\sim 7 < \Gamma < \sim 170\ \text{ng cm}^{-2}$.[89,123] However, regardless of the source of pooled plasmas (commercial or not), fouling on polyHPMA brushes was still undetectable (that is, below the SPR sensor's LOD of $0.3\ \text{ng cm}^{-2}$).

Through their methodical investigation of the antifouling performance of various polymer surface chemistries using different sources of blood plasma/serum biosamples – pooled or not – Rodriguez-Emmenegger's team provided one legitimate explanation as to why some conflicting results had appeared in the literature over the years (for the polyHEMA system).[89] From a practical point of view, the inherent variability of biological samples necessarily drawn from single donors could have serious implications in clinical diagnostics.[89]

5.2.2 Against Other Human or Animal Biofluids

Before adressing the implementation in biosensor technology of the antifouling surface chemistries described herein (and even though research activity in this domain was still modest at the time this chapter was being written), accounts also exist that report the quantified antifouling properties in terms of surface coverage of various different coatings against other human/animal biosamples than blood plasma/serum (Figure 5.4). These biofluids – which are discussed in alphabetical order, and not ranked according to their ubiquity or given importance – are presented in the following subsections, each of which will also be the object of a concise but solid description (*e.g.* composition, role) of said biological medium.

5.2.2.1 Cell Lysate

We begin with cell lysate our survey of other human or animal biological milieux for which antifouling surface chemistry has been developed or adapted. Admittedly, it can immediately be argued that cell lysates should not be considered as constituting a 'biofluid' since – as their name suggests – these are the result of the mechanical or chemical lysis of whole biological cells.[170] However, in view of the importance these biological mixtures may soon acquire in the field of biosensors applied to clinical diagnostics (see below), the description of antifouling surface chemistry against cell lysate definitely has its place in this chapter. Upon lysis, cells release cytosolic biomolecules as well as the organelle constituents of their cytoplasm (the intracellular matrix). The process is accompanied by the dismantelment of the cell bilayer membrane that endows crude, unprocessed lysates with the particularity of possessing a very high lipid content.[171] At the time this subsection was being written (early 2015), the work described next constituted, to our knowledge, the only example where the species responsible for fouling from crude cell lysate were quantified (and identified).

Based on the assumption (and previous experimental findings)[37] that crude cell lysate and serum should be expected to exhibit different fouling behavior – owing to their distinct biomolecular make-ups notably in terms of lipid *vs.* protein content – Masson and co-workers methodically revisited in 2013 the antifouling peptide SAM surface chemistry they had originally developed some years earlier (during the 2008–2012 period) to minimize fouling from 100% serum.[171] The authors' aim was to develop antifouling coatings suited for contact with crude cell lysates and capable of being implemented into SPR biosensors, for biomarker detection in lysates produced from suspected cancerous tumors sampled during solid tissue biopsy. The motivation behind this work resided in the fact that there exist cancers whose biomarker presence is difficult to detect in yet less invasive blood serum, urine, or saliva liquid biosamples.

In a first effort, a series of 3-MPA-homopentapeptide-based SAMs were prepared on gold SPR sensor chips, then exposed for 20 min to crude cell lysate (from HEK 293FT cells).[171] Lowest fouling surface coverage ($\Gamma = 194 \pm 77$ ng cm^{-2}) was recorded for the most hydrophobic SAM studied (3-MPA-Leu$_5$-OH). In comparison, fouling on such coating from undiluted serum was $\Gamma = 345 \pm 78$ ng cm^{-2}.[37] Continuing on with an investigation on the effect of peptide length, the authors next showed that – in contrast to experiments run with serum[37] – this parameter had little influence on fouling from cell lysate, the lowest adsorption being measured for 3-MPA-Leu$_2$-OH SAM ($\Gamma = 147 \pm 44$ ng cm^{-2}).[171] This relatively high level of adsorption was to be contrasted, however, with that even more pronounced recorded for serum ($\Gamma = 598 \pm 84$ ng cm^{-2}).[37] As was also the situation with serum,[38] SAMs built with 3-MPA-pentapeptides presenting binary, diblock sequences of amino acids offered greater antifouling protection, the highest performance in the present case of cell lysate being the feat of 3-MPA-(Leu)$_3$(Phe)$_2$-OH hydrophobic SAM ($\Gamma = 112 \pm 28$ ng cm^{-2}).[171]

At last, a set of experiments was also performed with *concentrated* lysate, a more challenging test for which SAMs constructed from ternary-patterned, triblock 3-MPA-(His)$_2$(Leu)$_2$(Phe)$_2$-OH hexapeptide[40] (Figure 5.38) turned out

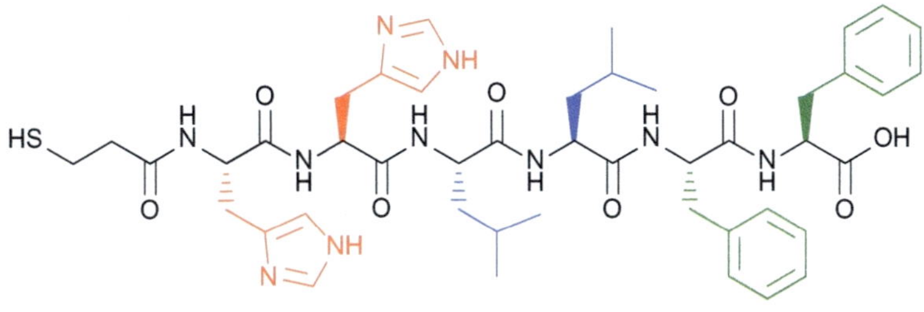

Figure 5.38 Molecular structure of the 3-MPA-(His)$_2$(Leu)$_2$(Phe)$_2$-OH ternary-patterned, triblock hexapeptide.

to be the most effective alternative ($\Gamma = 159 \pm 27$ ng cm^{-2}).[171] In comparison, the control surface of bare gold was heavily fouled ($\Gamma = 929 \pm 186$ ng cm^{-2}) – as was, to a much lesser extent however, that of a benchmark antifouling OEG SAM ($\Gamma = 294 \pm 25$ ng cm^{-2}). The antifouling performance against concentrated cell lysate of the best 3-MPA-(Leu)$_3$(Phe)$_2$-OH diblock SAM described earlier became $\Gamma = 698$ ng cm^{-2} (from $\Gamma = 112$ ng cm^{-2} with the less concentrated lysis mixture); that of the best antifouling SAM against serum – made of 3-MPA-(Leu-His-Asp)$_2$-OH hexapeptide building blocks drawn in Figure 5.6D ($\Gamma = 12 \pm 11$ ng cm^{-2})[40] – plummeted to $\Gamma = 789 \pm 107$ ng cm^{-2}.[171]

More than a still modest level of performance ($\Gamma > 100$ ng cm^{-2}), this investigation by Masson and co-workers[171] – together with previous work conducted by the same research group[36–40] – allowed for crucial pieces of information to be unravelled. Of utmost importance for biosensor technology was the unambiguous demonstration that crude cell lysate and serum do indeed display (drastically) different fouling behavior, implying the inconvenient necessity for biosensing platforms to be customized in terms of antifouling surface chemistry to a given detection biofluid. In terms of mechanism, it was concluded that a greater potential to reduce fouling from cell lysate is offered by SAMs made of hydrophobic (and/or basic) peptides, which the authors noted is in contradiction with the general trend observed in experiments performed with serum (for which hydrophilic surfaces are more effective at resisting fouling).[171]

Finally, in addition to quantifying their surface coverage using SPR, Masson's team also set out to identify the cell lysate species responsible for fouling, using a combination of enzymatic digestion followed by MALDI-TOF/TOF MS qualitative analysis.[171] The latter revealed that lipids (mostly from the phosphatidyl choline family) – and not proteins – constituted the major source of foulants. This probably was to be expected since crude cell lysate, unlike serum that is mainly composed of proteins, is essentially a concentrated mixture of highly adhesive lipids resulting from the fragmentation of biological cell membranes.[171]

5.2.2.2 *Cerebrospinal Fluid*

Cerebrospinal fluid (CSF) is mainly a product of blood plasma ultrafiltration that is secreted in specialized regions deep within the brain and flows in the space surrounding the brain and spinal cord (together known as the 'central nervous system' – CNS) back into the blood circulatory system.[172–176] In terms of composition, normal CSF – a clear, colourless liquid – is essentially an electrolyte (*e.g.* Na^+, K^+, Cl^-)-rich aqueous solution (99% of water), nearly depleted of red/white cells and proteins (0.15–1.00 mg mL^{-1} *vs.* $\sim$70 mg mL^{-1} for blood serum) as a result of ultrafiltration.[172–176] Glucose (0.40–0.70 mg mL^{-1}), neurotransmitters (*e.g.* γ-amino-butyric acid – GABA), and enzymes are also found in the metabolome of CSF.[173–177] Among its several vital functions,[25] the CSF notably creates a regulated chemical environment that supplies the CNS with essential nutrients and, conversely, disposes it of its metabolic waste products (when the CSF is re-absorbed in the bloodstream).[172–175] The CSF also serves as a vehicle for the intracerebral transport of bioactive substances.[174]

The collection of CSF specimens for clinical analysis can be performed through an uncomfortable sampling procedure known as 'lumbar puncture' (or 'spinal tap').[172,173,175,176] Lists of known and candidate CSF biomarkers of chronic neurodegenerative diseases (such as the notorious Alzheimer's disease) have recently been compiled,[176–179] analytes for which biosensor technology could be of service for diagnostic purposes.[62] A criterion, which biosensing interfaces will have to fullfil however, is the ability to resist, with appropriate antifouling surface chemistry, the signal-interfering non-specific adsorption of CSF matrix components.

To our knowledge at the time this chapter was being written, there was only one recent account that had quantified fouling surface coverage from (human) CSF. That 2012 pioneering work by Rodriguez-Emmenegger and co-workers revealed that 18 to 30 nm-thick polyHOEGMA (Figure 5.21A), polyCBAA (Figure 5.22D), polyHEMA (Figure 5.28B), and polyHPMA (Figure 5.28E) polymer brushes – ATRP-grown on gold SPR sensor chips through a precursor SAM of ω-mercapto-undecyl α-bromoisobutyrate initiator (Figure 5.20D) – all resisted fouling from undiluted human CSF ($\Gamma < 0.03$ ng cm^{-2}, 15 min exposure).[62] Even OEG_2- and OEG_6-terminated alkylthiol SAMs (both 0.9 nm-thick, Figure 5.13) – that did not offer great protection against undiluted human plasma ($\Gamma = 225 \pm 12$ and 71 ± 8 ng cm^{-2}, respectively) – exhibited pronounced antifouling properties against human CSF with, respectively, $\Gamma = 7.6 \pm 1.0$ and 4.3 ± 0.8 ng cm^{-2}.[62] This observation perhaps should not come as a surprise considering that CSF is an otherwise less proteinaceous biological medium than blood plasma (~0.35 *vs.* ~70 mg mL^{-1}), as a result of the aforementioned ultra-filtration process through which it is produced in the brain.[176]

5.2.2.3 *Egg*

Annually, over a trillion eggs – more often than not laid by chickens – are produced worldwide for food.[180] Structurally, raw eggs are essentially composed of two distinct, semi-liquid media – (1) the albumen ('egg white') that surrounds (2) the yolk – encapsulated in a hard, ~0.3 mm-thick calcareous shell.[181] The albumen has a gel-like consistency and mainly consists of water (~88%) and a variety of proteins (~10%), for instance ovalbumin (the most abundant) and avidin (well-known for its extremely high affinity for the vitamin biotin).[181] Lipids (<0.1%), carbohydrates (~1%), minerals (<1%), and vitamins are minor constituents of albumen. On the other hand, yolk is an 'oil-in-water' emulsion with a dry matter content (~51%) mainly composed of lipids (~63%) and proteins (~32%), as well as minerals (~2%), carbohydrates (~2%), and vitamins.[181] As will be discussed in the next subsection for milk, eggs are an important source of nutrition for which also exists an interest – in both fields of food quality/safety assessment and veterinary medicine – in developing biosensors for the (early) detection of pathogens/disease biomarkers.[62] A prerequisite would be to minimize, with appropriate antifouling surface chemistry, the response signal interference caused by the non-specific adsorption of egg matrix components on sensing interfaces.

As was the case for cerebrospinal fluid in the previous subsection, the abovementioned pioneering work by Rodriguez-Emmenegger's group turned out to be the only one at the time this chapter was being written to quantify fouling surface coverage from (carefully blended yolk and white) chicken egg.[62] This complex mixture proved to be otherwise more challenging however, as only the polyHOEGMA (Figure 5.21A), polyCBAA (Figure 5.22D), and polyHPMA (Figure 5.28E) polymer brushes were able this time around to resist fouling ($\Gamma < 0.03$ ng cm^{-2}, 15 min exposure). Another evidence was the extremely high level of adsorption now experienced not only by the polyHEMA brushes ($\Gamma = 51 \pm 3$ ng cm^{-2}, Figure 5.28B) but also, in particular, by both OEG_2- and OEG_6-SAMs (respectively $\Gamma = 914 \pm 76$ and 657 ± 43 ng cm^{-2}, Figure 5.13).

5.2.2.4 *Milk*

As attest the ~500 billion liters produced worldwide annually,[180] milk also is an important source of nutrition,[182] whose quality/safety it is therefore crucial to closely monitor. Compared to other biofluids (at the probable exception of blood plasma/serum), fouling by milk constituents, its quantification in terms of surface coverage, and the antifouling surface chemistries developed to combat the phenomenon have received more attention – notably from bioanalysts willing to develop biosensors for the real-time detection of pathogens/disease biomarkers with minimized response signal interference.[62] In terms of composition, milk principally consists of water (up to 87% depending on the mammal species) but also is rich in proteins (<10%) – caseins and the whey proteins β-lactoglobulin and α-lactalbumin being, in the case of cow milk, by far the most abundant (>90%).[182] Whole milk also contains sugars (~5%), mainly lactose, and diverse lipids (~4%). Also found in cow milk are organic acids, minerals and trace elements, and vitamins.

In terms of antifouling coating studies, whose objective is to quantify the minimized amount of deposited material from milk, there is again that by Rodriguez-Emmenegger's team, who demonstrated in 2012 that 30 nm-thick polyHOEGMA brushes (Figure 5.21A) were able to limit fouling from whole cow milk to $\Gamma = 3.9 \pm 0.6$ ng cm^{-2} (15 min exposure, SPR detection).[62] Much thinner (0.9 nm-thick) OEG_2- and OEG_6-terminated alkylthiol SAMs (Figure 5.13) also offered pronounced antifouling protection with $\Gamma = 15.2 \pm 1.1$ and 2.9 ± 0.5 ng cm^{-2}, respectively. More interesting was the observation that ~19 nm-thick polyCBAA (Figure 5.22D), polyHEMA (Figure 5.28B), and polyHPMA (Figure 5.28E) polymer brushes all resisted fouling from whole milk ($\Gamma < 0.03$ ng cm^{-2}). Collectively, this[62] and another

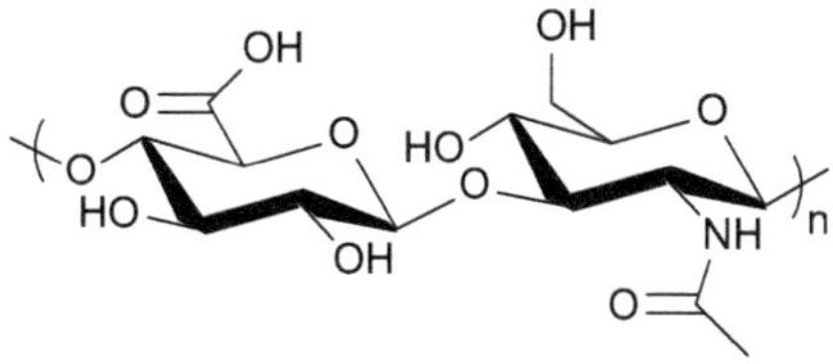

Figure 5.39 Molecular structure of hyaluronic acid.

two preceeding (in 2011)[183] or following (in 2014)[123] SPR studies by Rodriguez-Emmenegger and co-workers also revealed that the fat content (1.5 *vs.* 3.25%, *i.e.* 'semi-skimmed' *vs.* 'whole, full-fat' undiluted fresh milk) and the source of milk (fresh undiluted milk *vs.* powder milk *vs.* infant powder milk formula – the latter two reconstituted at 10% in water), if any, have little effect on the remarkable antifouling properties of OEG_6-SAMs,[183] as well as polyHOEGMA (and polyMeOEGMA),[123,183] polyHEMA,[123,183] and polyCBAA[123] polymer brushes.

Another kind of surface chemistry to combat milk fouling – based on hyaluronic acid (HA) drawn in Figure 5.39 – had also been recently developed (in 2014) at the time this chapter was being written.[184] Polysaccharidic HA films (10.5 nm-thick) were covalently grafted on gold SPR sensor chips through an intercalated, epoxide-terminated alkylthiol adhesion SAM. Fouling surface coverage upon 10 min exposure to cow milk ($\Gamma = 9.8$ ng cm^{-2})[184] was comparable to that recorded in the studies discussed above with polymer brushes and OEG_n-SAMs.[62,123,183]

5.2.2.5 *Saliva*

The secretion product of three major glands, saliva is a clear, slightly acidic (pH 6–7) biofluid mainly composed of water (99%).[185–188] Other salivary components include a variety of proteins (*e.g.* enzymes, immunoglobulins) and ions (*e.g.* Ca^{2+}, F^-, phosphates), for instance.[185–187,189] Also found in saliva, and the non-sterile environment of the oral cavity, are colonies of bacteria.[185–187,189] The adsorption of salivary material on surfaces is also a topic of great interest, not only in prosthodontics with the issue of the biocompatibility of restorative synthetic materials,[190] but also in the realm of biosensor technology as this biofluid hosts a collection of important, non-invasively collectable biomarkers.[62,185–187]

At the time this chapter was being written, studies setting out to quantify the minimized surface coverage of adsorbed salivary material were relatively scarce, and the antifouling coating strategies implemented variously effective. Successful approaches included the abovementioned one by Rodriguez-Emmenegger's and co-workers, who had recently reported (in 2012) that ~19 nm-thick polyCBAA (Figure 5.22D), polyHEMA (Figure 5.28B), and polyHPMA (Figure 5.28E) brushes on gold SPR sensor chips all resisted fouling from undiluted human saliva ($\Gamma < 0.03$ ng cm^{-2}, 15 min contact).[62]

Another polymer coating (30 nm-thick polyHOEGMA – Figure 5.21A) as well as OEG_2- and OEG_6-terminated alkylthiol SAMs (Figure 5.13), both 0.9 nm-thick, also possessed excellent antifouling properties with $\Gamma = 6.7 \pm 1.4$, 15.7 ± 1.9, and 8.2 ± 2.0 ng cm^{-2}, respectively. Others also had studied fouling from whole human saliva on methyl-[191–195] or perfluorooctyl-terminated[196,197] hydrophobic silane coatings,[198] but those surfaces did not offer as great an antifouling protection (respectively $\Gamma \sim 300$ and 110 ng cm^{-2}). In the latter case, bare substrates (silicon oxide) adsorbed a comparatively greater amount of salivary material with $\Gamma \sim 400$ ng cm^{-2}. Surprising – in view of the pronounced antifouling properties reported in this chapter and elsewhere for such type of construct – was the observation that substrates derivatized with a hydrophilic, OEG-based coating were on average more fouled than those modified with the perfluoroalkyl, hydrophobic adlayer discussed above ($\Gamma \sim 210$ *vs.* 110 ng cm^{-2}).[196,197]

5.2.2.6 *Urine*

There exists finally one last biofluid to be discussed in this chapter that has been tested for its fouling potency: urine. One important function of urine (*i.e.* the kidneys where it is produced) is to dispose the organism of water-soluble waste products. In terms of composition indeed, urine – 1.5 to 2 liters of which are collected in/eliminated from the bladder on a daily basis[199,200] – is an aqueous medium (>95% of water) containing many catabolites, most notably urea (9.3–23.3 g L^{-1}).[200,201] Other breakdown products include, for instance: creatinine (0.67–2.15 g L^{-1}), uric acid (0.04–0.67 g L^{-1}), and ammonia (0.20–0.73 g L^{-1}), as well as yellow pigments urobilin (7–90 mg L^{-1}) and bilirubin (3–30 mg L^{-1}).[201] Ions at high concentration can also be found such as Cl^- (1.87–8.40 g L^{-1}), Na^+ (1.17–4.39 g L^{-1}), and K^+ (0.75–2.61 g L^{-1}).[201] Many other (in)organic solutes, such as glucose (0.03–0.20 g L^{-1}), also make the long list of urine components.[199–201] Normal urine is a yellow/amber-coloured, nearly acellular liquid, whose analysis only reveals trace amounts[200,202] of a great variety of proteins.[199,203] Proteinuria, a condition characterized by an abnormally-high level of proteins in urine (>300 mg L^{-1}), can be readily detected through routine dipstick urinalysis.[200] Urine can also be tested for illicit substances or their metabolites, for instance performance-enhancing steroids (doping)[204] and drugs of abuse/prohibited street drugs (*e.g.* cocaine).[205] Until modern times and the advent of blood serology, urine was the primary diagnostic biofluid,[200,206] but the possibility to non-invasively collect large samples of it still makes urine a particularly convenient medium where to detect/discover biomarkers.[199,203] Biomarker detection in urine is an area where biosensor technology, again, could be of service[62] – provided that the NSA issue causing response signal interference is adressed through the application of appropriate antifouling surface chemistry.

To our knowledge, and as was already the case earlier for cerebrospinal fluid and egg, the aforementioned work by Rodriguez-Emmenegger's team pioneered the quantification of fouling by undiluted (human) urine.[62] This biofluid proved not to be an issue in this respect however ($\Gamma < 0.03$ ng cm^{-2}, 15 min contact) – for all, 18–30 nm-thick polyHOEGMA (Figure 5.21A), polyCBAA (Figure 5.22D), polyHEMA (Figure 5.28B), and polyHPMA (Figure 5.28E) tested polymer brushes. A 0.9 nm-thick OEG_6-terminated alkylthiol SAM (Figure 5.13) also displayed such performance. Even a shorter OEG_2-SAM analogue (0.9 nm-thick as well) offered great resistance against human urine with $\Gamma = 11.0 \pm 1.1$ ng cm^{-2}. This latter result is to be constrasted with that recorded for undiluted human plasma ($\Gamma = 225 \pm 12$ ng cm^{-2}),[62] but should not come as a surprise considering that blood plasma is otherwise more proteinaceous than urine (~70 *vs.* ~0.30 mg mL^{-1}).

5.2.3 At-a-glance Summary

Inventoried below in two summary tables are the best antifouling performances in terms of minimized surface coverage (ng cm^{-2}) of the various, different types of surface chemistries described in this chapter (Tables 5.1 and 5.2 respectively compile data for blood plasma/serum and for the other human/animal biofluids that are cell lysate, cerebrospinal fluid, egg, milk, saliva, and urine). Interested readers are kindly invited to consult the main text for a more detailed discussion on antifouling surface chemistries and experimental testing conditions.

5.3 Implementation in Biosensor Technology

While displaying pronounced inherent antifouling properties certainly constitutes a great achievement for a coating, the finality however – at least in the realm of biosensing – is for the imposed functionalizable surface chemistry to maintain its performance once derivatized with biomolecular ligands (to form the so-called 'biorecognition element').[87] In this respect, although perhaps an excessively low limit in some regards, it has been postulated that NSA from biological media should not exceed the value of 3 ng cm^{-2}.[10] As seen in Tables 5.1 and 5.2, such a level of performance is now easily achieved by a variety of unfunctionalized surface chemistries for a wide range of biological milieux, whether of human or animal origin. Does this still hold true upon biofunctionalization?[†] Can these surface chemistries then be implemented in biosensor devices for direct detection in crude biofluids[8,207] (that is, without any biosample/biointerface pre-analysis processing/treatment and response signal correction[208–212])? These are two questions to which an answer will next be given, insofar as possible.[123] We note that the message we wish to convey herein does not necessarily find its significance in the importance a given biosensing application may have, but rather in the potential – in terms of versatility (in the nature of capture

[†] In all fairness, there also exist situations where the pronounced antifouling properties displayed by a coating arise only upon biofunctionalization. See, for instance, refs. 72 and 162.

Table 5.1 State-of-the-art antifouling performance of various, different surface chemistries against blood plasma/serum.

Type of coating	Substrate	Adsorbed mass (ng cm^{-2})						Ref.
		Bovine serum		Human serum		Human plasma		
		10%	100%	10%	100%	10%	100%	
Single amino acid monolayer	Au	—	42	—	—	~200	—	33, 41
Peptide SAM	Au	—	12	—	—	<50	—	33, 40
Multilayered polypeptide film	Metal oxide	—	—	<1	—	—	—	42
Poly(single amino acid) brush	Au	—	—	—	<1	—	3	44–46
Polypeptoid film/SAM	Au	—	—	4	6	5	10	50
	Ti	—	—	—	~4	—	—	47
	TiO_2	—	—	—	>15	—	—	48, 49
OEG film	Au	—	—	—	—	5	24	55
	Metal oxide	1	—	—	—	—	—	71
OEG SAM	Au	—	26	~15	~85	38[a]	71	56–59, 62
PEG film	Au	>6	—	—	14	—	2	65, 79
	TiO_2	—	—	—	<5	—	—	69, 73, 74
PEG poloxamer film	Au	—	—	—	16	<2[a]	—	56, 67
OEG dendritic film	Au	—	—	—	20	—	8	81
	TiO_2	—	—	—	<2	—	—	73
Poly(OEG) brush	Au	<1	<1	<5	<1	<5	<1	59, 91–94, 97, 153
	TiO_2	<20	—	—	—	—	—	96
	Nylon	—	<1	—	—	—	<20	84
Phosphorylcholine-based zwitterionic polymer brush	Au	>13	—	—	—	>345[b]	—	65, 100, 101

Sulfobetaine-based zwitterionic polymer brush	Au	—	—	~1	<1	2[a]	<1	57, 104, 106
	SiO_2	—	26	—	—	—	—	108
Carboxybetaine-based zwitterionic polymer brush	Au	—	<1	<1	<1	<1	<1	59, 62, 83–85, 100, 101, 113, 118–120, 123, 159
	SiO_2	—	<5	—	12	—	6	115, 117
PEG-polypeptide hybrid (and derivatives) film	Metal oxides	—	5	—	<2	—	—	127–130, 132–134, 136, 137, 139, 140, 143–147
	PDMS	—	—	—	<1	—	—	142
Poly(short hydroxyalkyl) brush	Au	<1	<1	<1	<1	<1	<1	62, 85, 89, 91, 149–152, 154–156
EGylated zwitterionic hybrid polymer film/brush	Au	—	<1	—	—	—	<1	86
	Metal oxide	1	—	—	—	—	—	71, 157
Pseudozwitterionic copolymer brush	Au	2	4	—	—	8[a]	—	58, 161
Pseudozwitterionic mixed SAM	Au	—	—	~15	>90	>190	>250	59, 83
Ionic liquid SAM	Au	—	—	—	99	—	—	162
Polysaccharide film	Metal oxide	—	—	—	13	—	—	167
Poly(carbohydrate) brush	Au	—	—	—	—	—	24	168

[a]20% platelet-poor plasma.
[b]33% in PBS.

Table 5.2 State-of-the-art antifouling performance of various, different surface chemistries against other human or animal biofluids including cell lysate, cerebrospinal fluid, egg, milk, saliva, and urine.

		Adsorbed mass (ng cm^{-2})						
Type of coating	Substrate	Cell lysate	CSF[a]	Egg	Milk	Saliva	Urine	Ref.
Peptide SAM	Au	159	—	—	—	—	—	171
OEG SAM	Au	294	4	657	3	8	<1	62, 171
Poly(OEG) brush	Au	—	<1	<1	4	7	<1	62
Zwitterionic brush (carboxybetaine)	Au	—	<1	<1	<1	<1	<1	62
Poly(short hydroxyalkyl) brush	Au	—	<1	<1	<1	<1	<1	62
Polysaccharide film	Au	—	—	—	10	—	—	184

[a]CSF stands for 'cerebrospinal fluid'.

bioreceptors/detectable analytes) and transferability (to other sensor substrates/detection techniques) – of the methodology that allowed this application to see the light of day in the first place.

5.3.1 Biofunctionalization of Peptide SAMs

The implementation of antifouling peptide SAM surface chemistry (see subsection 5.2.1.1) for biosensor detection in biological milieux has been reported by Masson and co-workers. In this 2010 proof-of-concept study, binary-patterned 3-MPA-His$_3$-Asp$_2$-OH pentapeptide SAMs grafted on gold SPR sensor chips were biofunctionalized with antibodies against human matrix metalloproteinase-3 (MMP-3),[38] through standard *N*-ethyl-*N*′-(3-dimethylamino-propyl)-carbodiimide/*N*-hydroxy-succinimide (EDC/NHS) coupling chemistry. With this SPR immunosensor prototype, the investigators could detect MMP-3, an enzyme involved in important pathologies, at low-nanomolar concentration in undiluted bovine serum (Figure 5.40). As shown as well in this figure, the control verification was also made with unspiked serum that fouling was (still) negligible upon biofunctionalization ($\Gamma = 32 \pm 5$ ng cm^{-2} before antibody immobilization).

A year earlier (in 2009), Masson's team had already employed this strategy to detect in crude bacterial cell lysate the presence at submicromolar concentration of β-lactamase, an enzyme linked to antibiotic resistance.[37] In that work, biofunctionalization with anti-β-lactamase antibodies was done on 3-MPA-Ser$_5$-OH homopentapeptide SAMs (Figure 5.6B).

Finally, to be pointed out as well is the 2014 proof-of-concept work by Mendes and co-workers, who prepared biotinylated mixed SAMs on gold SPR sensor chips (using the thiol building block molecules drawn in Figure 5.41) to study their specific interactivity with neutravidin target in the presence of 10% FBS and under switching conditions of applied electrical potential.[63] NSA of unspiked serum samples on these biotinylated surfaces was measured to be lower than $\Gamma = 23$ ng cm^{-2}.

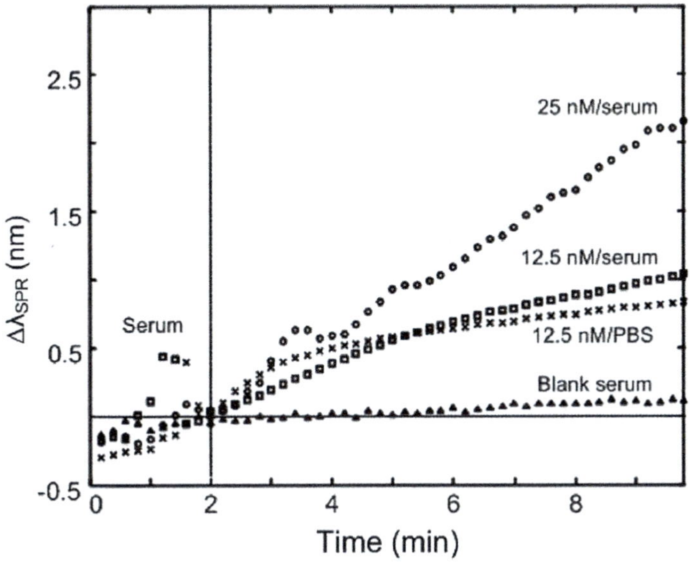

Figure 5.40 Low-nanomolar serological detection of human matrix metalloproteinase-3 with a peptide SAM-based SPR immunosensor.
Reprinted with permission from ref. 38. Copyright 2010 American Chemical Society.

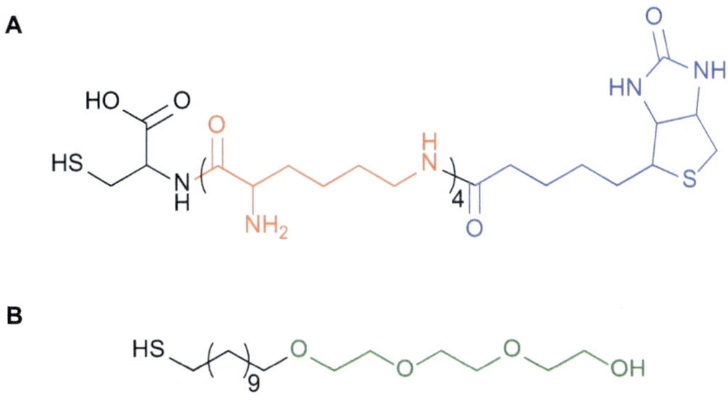

Figure 5.41 Molecular structure of (A) the biotinyl-terminated lysine peptide-based thiol linker for specific neutravidin binding, and (B) the OEG_3-terminated alkylthiol spacing diluent for antifouling background formation.

5.3.2 Biofunctionalization of Polyelectrolyte Films

In a 2013 study dedicated to developing a nanoarchitectonic bioaffinity sensor interface, biotinylated polyelectrolyte coatings on OWLS chips served to anchor a layer of multivalent streptavidin proteins on top of which model biotinylated antibodies (against ovalbumin) were finally captured, as illustrated in Figure 5.42A.[72] On the accompanying 'adsorbed mass *vs.* time' graph (Figure 5.42B) that shows the specific binding of ovalbumin for the

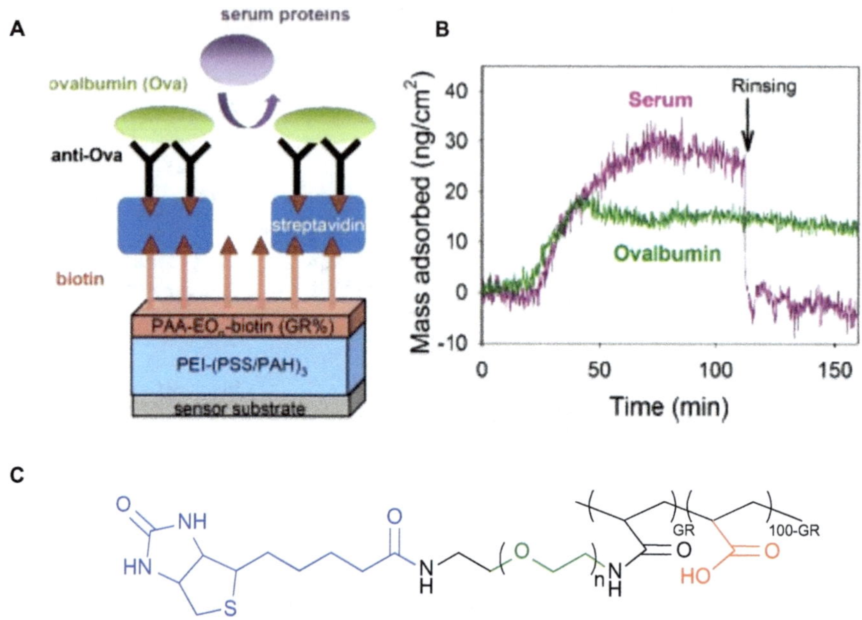

Figure 5.42 (A, B) Detection of ovalbumin with a polyelectrolyte-based sandwich immunosensor exhibiting negligible non-specific adsorption of foetal bovine serum (diluted at 10%). PAA, EO, GR, PEI, PSS, and PAH respectively stand for 'poly(acrylic acid)', 'ethylene oxide', 'grafting ratio', 'poly(ethylene imine)', 'poly(styrene sulfonate)', and 'poly(allylamine hydrochloride)'. Reprinted with permission from ref. 72. Copyright 2013 American Chemical Society. (C) Molecular structure of the biotinylated polyelectrolyte adsorbate featuring an OEG spacer arm (optimal biosensor performance was reached for $n = 9$ and $GR = 25\%$).

optimized structure (Figure 5.42C), it can also be seen that serum adsorption (10% FBS) on such sandwich biosensing platform was not an issue. In contrast, serum fouling on the non-biotinylated polyelectrolyte surface was substantial ($\Gamma = 217\ \text{ng cm}^{-2}$).[72]

5.3.3 Biofunctionalization of PLL-*g*-PEG Films

Antifouling PLL-*g*-PEG surface chemistry (Figure 5.26, $\Gamma < 2\ \text{ng cm}^{-2}$) has also been employed in the context of biosensing. In these studies – whether functionalized with biotinyl residues for the development of streptavidin sandwich immunoassays (Figure 5.43A),[141] or with nitrilotriacetic acid (NTA) ligands for that of a sensing interface relying on Ni^{II}/His-tag docking chemistry (Figure 5.44)[140] – the demonstration was made that the resulting derivatized PLL-*g*-PEG coatings still offered excellent protection against undiluted human serum, fouling being in each case within the LOD of the OWLS detection technique used ($\sim 1\ \text{ng cm}^{-2}$).[136,140,141]

In the former work, it was also verified that completing the construction of the multilayered immunosensing platform (through deposition of the

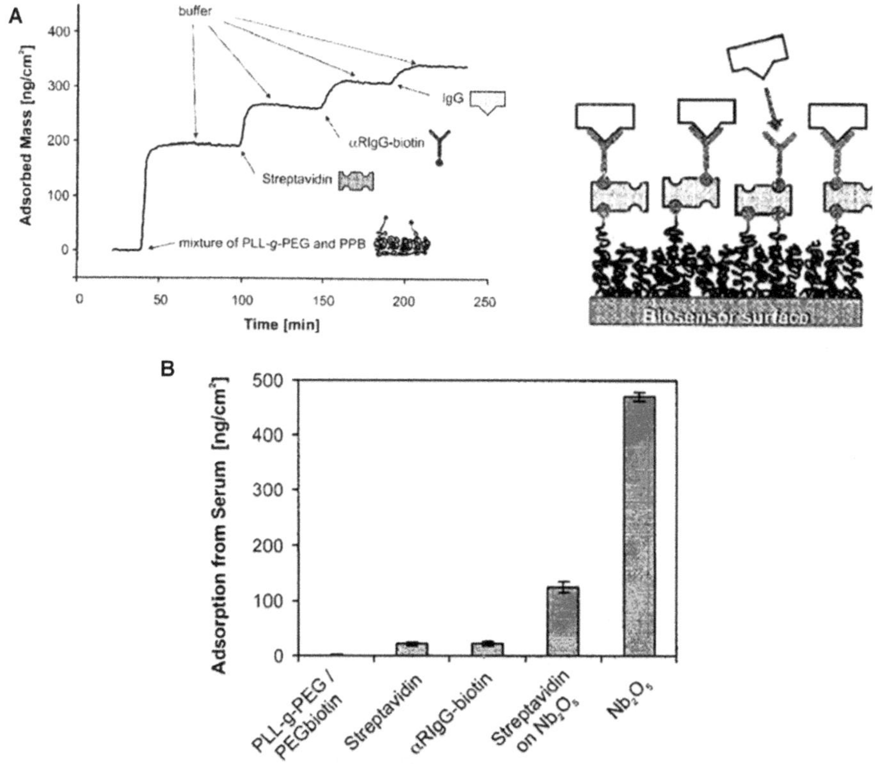

Figure 5.43 (A) *In situ*, three-step construction of a sandwich immunoassay based on PLL-*g*-PEG surface chemistry. 'PPB' is biotinylated PLL-*g*-PEG. (B) Antifouling performance against human serum of the three successively built PLL-*g*-PEG-based coatings (left) *vs.* that of streptavidin-coated and bare Nb_2O_5 substrates (right).

intercalated streptadivin layer, then the final capture of anti-IgG biotinylated antibodies) had little effect on the antifouling performance of the coatings against serum (Figure 5.43B).

Finally, of note as well is the ability to maintain their pronounced antifouling properties against full plasma/serum upon biofunctionalization, which was repeatedly displayed by PLL-*g*-PEG coatings derivatized with cell-adhesive oligopeptides containing the integrin receptor-binding arginine/glycine/aspartate (RGD) amino acid sequence (for cell attachment purposes).[130,132,134,139,213]

5.3.4 Biofunctionalization of Ionic Liquid SAMs

In a recent 2013 account, Masson and Schmitzer developed a proof-of-concept SPR immunosensor relying on a new class of SAM built from the carboxylic acid-terminated ionic liquid (IL) thiol molecule shown earlier in

Figure 5.35.[162] The construction of the biosensing platform was completed upon biofunctionalization with anti-human IgG antibodies using the classic EDC/NHS coupling chemistry (the protocol also including the final deactivation of residual NHS ester moieties with ethanolamine). NSA from crude serum ($\sim$70 mg proteins mL^{-1}) on such an interface was measured to be $\Gamma = 45$ ng cm^{-2} – a value that was, incidentally, significantly lower than that recorded before functionalization ($\Gamma = 99$ ng cm^{-2}).[162] Under these conditions, it was possible for the investigators to detect human IgG spiked at 100 nM in bovine serum.

To note, this IL SAM surface chemistry was also implemented by the authors to improve the performance in terms of activity of an enzymatic biosensor based on human dihydrofolate reductase (hDHFR) for the detection of methotrexate, a commonly prescribed chemotherapy inhibitor drug, which this enzyme involved in the mechanism of cell proliferation strongly interacts with.[162] In this scenario, biofunctionalization of the IL SAM coating with hDHFR was performed using the aforementioned NTA/His-tag immobilization chemistry (Figure 5.44).

5.3.5 Biofunctionalization of Polysaccharidic Coatings

In the world of biosensing, the commercial Biacore system featuring carbohydrate-based carboxymethyl dextran surface chemistry has encountered great success and popularity. Such surface functionalization strategy is known however for not always being able to offer a satisfactory degree of protection against NSA from complex biological media.[184,212,214] This situation naturally led some researchers, such as Huang and Su, to try and develop alternative polysaccharidic coatings.[184] In this recently published proof-of-concept work (in 2014), the functionalizable precursor film prepared on SPR sensor chips was based on hyaluronic acid (Figure 5.39) and used to immobilize antibodies against bovine serum albumin (BSA) using the standard EDC/NHS coupling chemistry (the procedure being completed with the final deactivation of unreacted NHS ester moieties with ethanolamine). The levels of NSA on this immunosensing platform were rather low for both cow milk and 10% serum at, respectively, $\Gamma = 16$ and 17 ng cm^{-2} (from $\Gamma = 10$ ng cm^{-2} for cow milk pre-antibody attachment); this, however, was not the case for undiluted serum[214] for which NSA attained $\Gamma = 60$ ng cm^{-2}.[‡] In terms of analytical performance, this SPR immunosensor prototype was shown to exhibit very high selectivity for BSA and a rather wide 15–700 nM dynamic range (Figure 5.45A), as well as the capability to be regenerated for repeated measurements (Figure 5.45B).

[‡]Hyaluronic acid coatings were also shown to display great antifouling performance against two food extract media: diluted soybean milk ($\Gamma = 0.6$ ng cm^{-2}) and undiluted orange juice ($\Gamma = 16.1$ ng cm^{-2}).

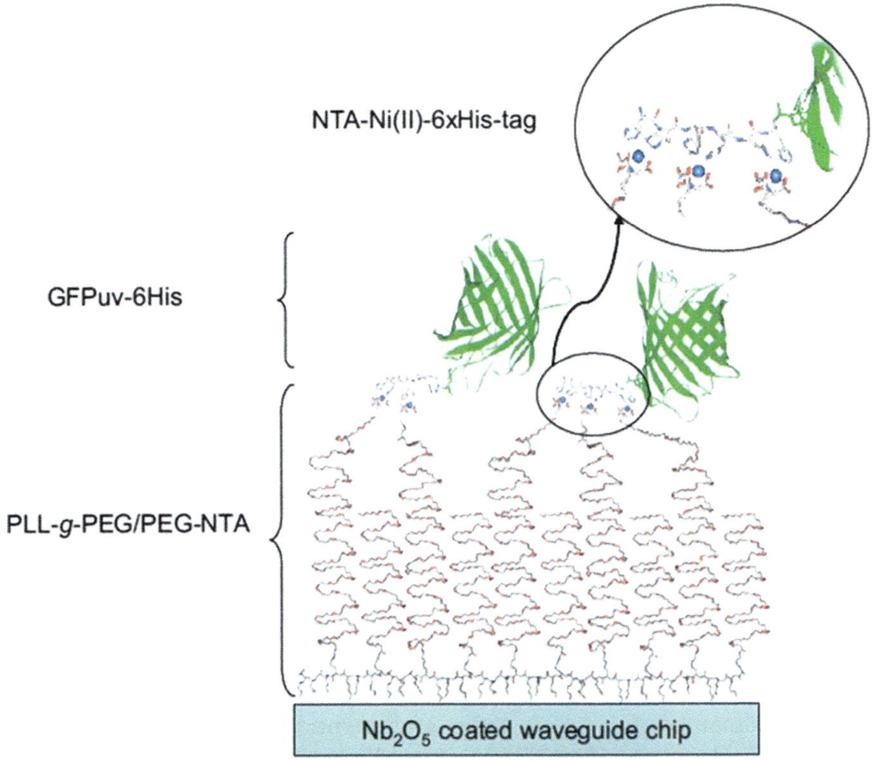

Figure 5.44 The PLL-*g*-PEG/NTA/Ni^{II} docking surface chemistry on an OWLS chip for detection of His-tagged proteins (in this example, His-tagged green fluorescent protein GFPuv was captured).
Reproduced from ref. 140 with permission of John Wiley & Sons. Copyright © 2006 Wiley-VCH Verlag GmbH & Co. KGaA, Weinheim.

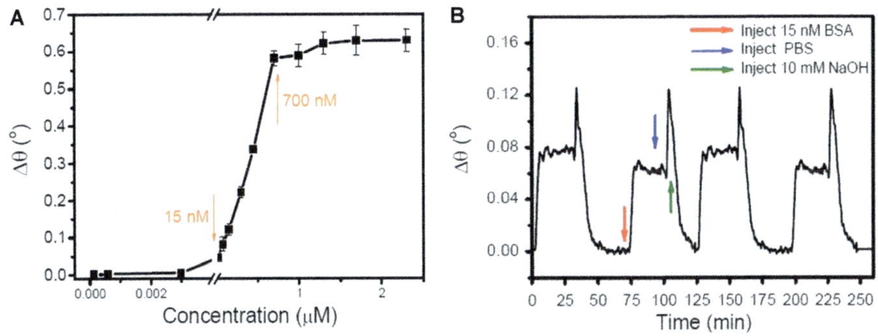

Figure 5.45 Nanomolar detection of BSA (in buffer) with a regenerable SPR immunosensor based on hyaluronic acid polysaccharide surface chemistry.
Reprinted with permission from ref. 184. Copyright 2014 American Chemical Society.

5.3.6 Biofunctionalization of Polymer Brushes

5.3.6.1 Built from OEGylated Monomers:[215] HOEGMA and MeOEGMA

The biofunctionalization of antifouling polyOEGMA brushes has also been the object of several recent studies. Rodriguez-Emmenegger and co-workers for instance used such surface chemistry in 2011 to immobilize, *via* a covalently pre-coated layer of streptavidin, model antibodies and oligonucleotides conjugated to biotin.[84] In that work with polyHOEGMA brushes, functionalization occurred through the hydroxyl terminal group of OEG side-chains.[123] In a more recent 2014 account with polyMeOEGMA surface chemistry,[88] the investigators adopted a different biofunctionalization strategy in view of the impossibility – in this case – for immobilization to take place through pendant OEG arms, which are methoxy-capped hence unreactive (Figure 5.21B *vs.* A). The attachment procedure – which now necessarily had to involve the only functionalizable, distal end of the polymer brush backbones – hinged on an ingenious 'click' reaction based on Diels–Alder [4 + 2] cycloaddition (Figure 5.46). In this latter proof-of-concept work, demonstration was made using SPR that antifouling properties against undiluted human plasma were maintained upon functionalization ($\Gamma < 21$ ng cm^{-2}, 15 min exposure).

With respect to actual biosensing in biological media, the investigators later implemented (in 2014) their aforementioned sandwich platform strategy based on polyHOEGMA/streptavidin surface chemistry[84] – using the oligonucleotide variant to immobilize antigen bioreceptors through complementary hybridization (Figure 5.47A) – to the pioneering serological detection of Epstein–Barr virus (EBV) antibody biomarkers directly in clinical samples drawn from patients with suspected infection (and from healthy donors as negative controls).[90] Using different antigen probes to recognize different antibodies, the authors could rapidly diagnose with SPR whether a collected serum sample presented signs of EBV infection and, if so, at what stage of development. For example, the SPR measurements shown in Figure 5.47B revealed that this particular serum sample belonged to an individual in the primary phase of the infection. It is fair to note however – in regards to the guidelines set in the introductory paragraph of this subsection on the implementation of antifouling surface chemistry in biosensing technology – that (optimal) SPR detection was performed in 1% serum, with PBS containing 1% BSA as both the diluting and carrier solution. As well, the response signal was corrected to account for residual fouling ($\Gamma \sim 20$ ng cm^{-2} pre-biofunctionalization, against undiluted serum) using reference platforms lacking antigen ligands. At last, biosensor regenerability for repeated measurements was also demonstrated.

Finally, it is worth also mentioning the approach by Chilkoti's team to employ antifouling polyHOEGMA surface chemistry in an antibody microarray sandwich assay for the quantification of a variety of protein

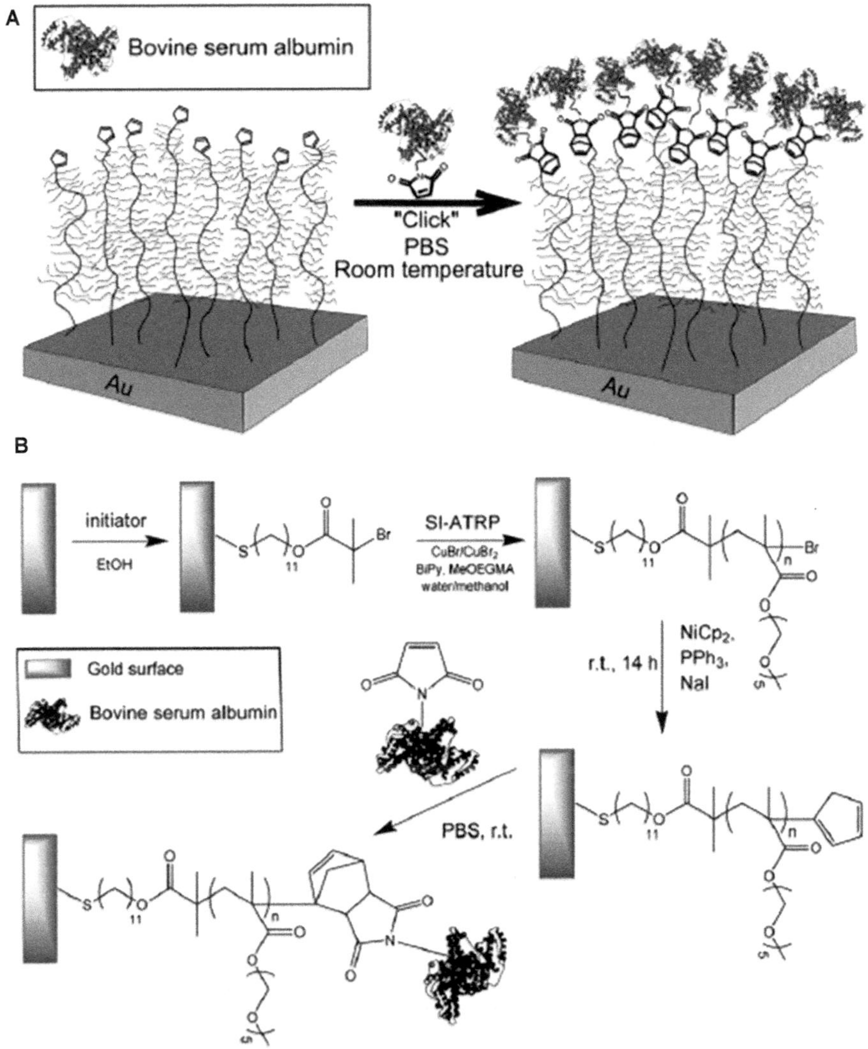

Figure 5.46 (A) PolyMeOEGMA brush biofunctionalization (with BSA) through Diels–Alder $[4+2]$ cycloaddition. (B) In order for this 'click' reaction to take place, the ATRP-grown brush was derivatized post-polymerization with cyclopentadienyl 'diene' residues (through substitution of the reactive bromine backbone terminii), and BSA conjugated to maleimide 'dienophile'.
Reproduced from ref. 88 with permission of the Royal Society of Chemistry.

biomarkers (such as human interleukin-6, tumor necrosis factor, and osteoprotegerin) in undiluted serum.[216] The detection principle relied on the interception by surface-captured analytes of biotinylated secondary antibodies, an interaction that would be revealed through final incubation

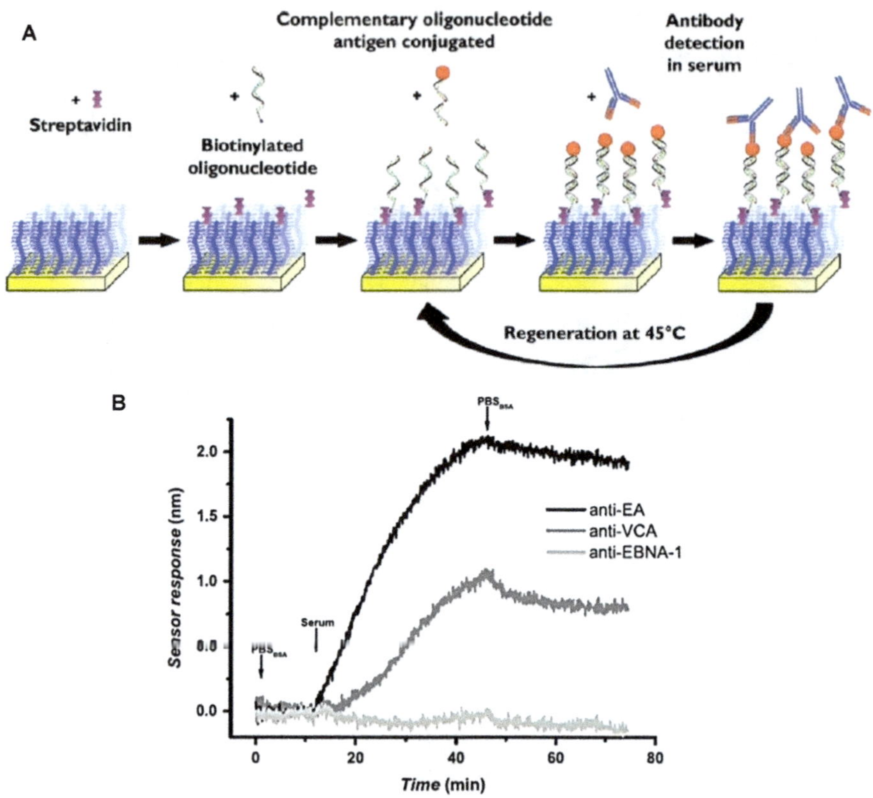

Figure 5.47 (A) Multi-step construction of a regenerable biosensing interface based on polyHOEGMA/streptavidin/complementary oligonucleotide sandwich surface chemistry for the serological detection of Epstein–Barr virus antibody biomarkers. (B) Diagnostically, this serum sample, for which anti-EA and anti-VCA but no anti-EBNA antibodies are detected using SPR, belongs to a donor afflicted by EBV infection in its early stage of development. 'EA', 'VCA', and 'EBNA' respectively stand for 'early antigen', 'viral capside antigen', and 'Epstein–Barr nuclear antigen'.
Reprinted from T. Riedel, C. Rodriguez-Emmenegger, A. de los Santos Pereira, A. Bedajankova, P. linoch, P. M. Boltovets, E. Brynda, Diagnosis of Epstein–Barr virus infection in clinical serum samples by an SPR biosensor Assay, *Biosens. Bioelectron.*, **55**, 278–284, Copyright 2013 with permission from Elsevier.

with fluorescently-labelled streptavidin (Figure 5.48A). This immunoassay performed extraordinarily well as femtomolar LODs and dynamic ranges spanning several orders of magnitude in concentration were reported, even when measurements were done directly in whole blood (Figure 5.48B).[216] Another strength of this ultrasensitive assay resided in the simplicity with which the microarray platform was fabricated, through mere antibody spotting then *absorption* into the nanogel structure of the polymer

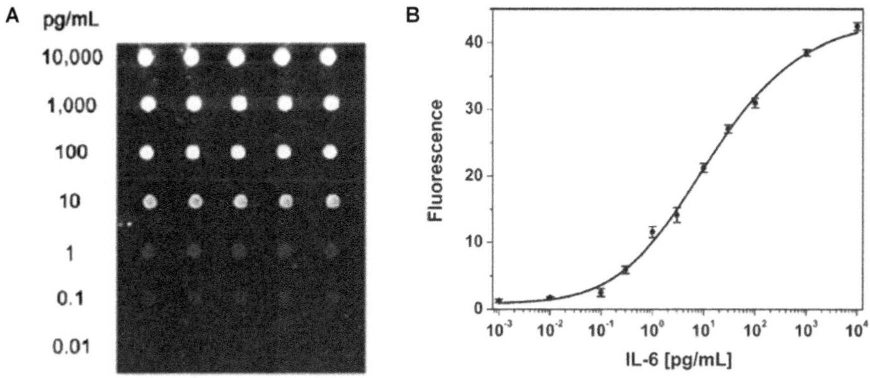

Figure 5.48 (A) Fluorescent development of an antibody microarray sandwich assay dedicated to the serological detection of protein biomarkers. In this example with interleukin-6 (IL-6) analyte, the LOD in undiluted FBS is approximately 100 fg mL^{-1} (or ~5 fM). (B) In whole blood (rabbit, heparinized), the resulting dose–response curve also reveals a femtomolar LOD for IL-6, together with a dynamic range spanning several orders of magnitude in concentration.
Reproduced from ref. 216 with permission of John Wiley & Sons. Copyright © 2009 Wiley-VCH Verlag GmbH & Co. KGaA, Weinheim.

brush matrix. Even though the non-covalent nature of antibody immobilization was deliberate – so as to eliminate the required, extra steps of coating chemical activation/deactivation – the demonstration was made that this physical entrapment protocol yielded robust interfaces capable of withstanding sonication/surfactant release conditions. Another source of satisfaction was their uncompromised analytical performance despite lengthy storage.

5.3.6.2 Built from Short Hydroxyalkyl Monomers: HEMA and HPMA

The monoethylene glycol version of polyHOEGMA brushes – polyHEMA (Figure 5.28B *vs.* 5.21A) – has also been adopted in the construction of biosensing platforms for analysis in biological media. Indeed, this surface chemistry was implemented in 2011 by Rodriguez-Emmenegger *et al.* in an SPR immunosensor dedicated to the detection of *Cronobacter sakazakii* bacteria in milk.[183] As can be seen in Figure 5.49A–C, this opportunistic pathogen, which is notoriously known for being particularly lethal to neonates following consumption of contaminated milk, could be successfully detected at concentrations down to ~10^6 cells mL^{-1} in samples prepared from three different commercial sources of milk. Figure 5.49A–C also encompasses the verification made by the investigators that fouling from all three types of consumer milk samples was still invariably negligible

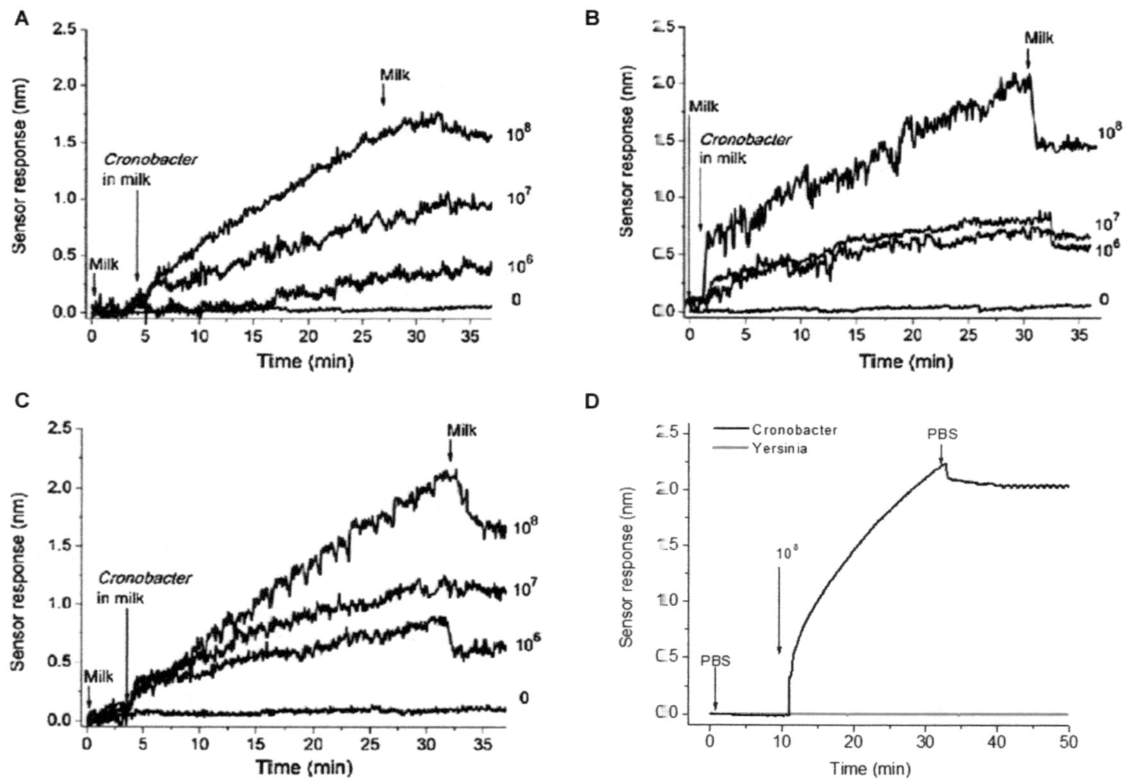

Figure 5.49 (A–C) Detection of *Cronobacter sakazakii* bacterium at concentrations down to 10^6 cells mL^{-1} in three different types of dairy product samples – A: fresh undiluted full-fat (3.25%) milk, B: whole-fat powder milk (10% in water), and C: infant powder milk formula (reconstituted at 10% in water as well) – using an SPR immunosensor based on polyHEMA surface chemistry. (D) Resistance to non-specific bacterial adhesion: an SPR experiment demonstrating the great selectivity of this immunosensor towards *Cronobacter sakazakii*, through the lack of cross-reactivity with another bacterium (*Yersinia enterocolitica*). Reprinted from C. Rodriguez-Emmenegger, O. A. Avramenko, E. Brynda, J. Skvor, A. B. Alles, Poly(HEMA) brushes emerging as a new platform for direct detection of food pathogen in milk samples, *Biosens. Bioelectron.*, **26**, 4545–4551, Copyright 2011 with permission from Elsevier.

(*i.e.* below the LOD of their SPR instrument) following the biofunctionalization with antibodies of the primer polyHEMA brush – thereby eliminating the need to subtract from the response signal the NSA contribution of the milk matrix components that would be measured with an independent reference channel.[183] [As shown in Figure 5.49D, the immunosensor's great selectivity towards *Cronobacter sakazakii* was also demonstrated upon exposure and lack of cross-reactivity with another bacterium (*Yersinia enterocolitica*).]

It is worth noting that such a feat was not a quality shared by another polyHEMA-based immunosensing interface featuring anti-*Escherichia coli O157:H7* antibodies, a shortcoming that did not prevent however the investigators from detecting this bacterium with a comparable LOD (8×10^5 cells mL^{-1}) – in fresh undiluted half-fat (1.5%) milk.[123] To note, an otherwise more antifouling biofunctionalized SPR platform based on polyCBAA zwitterionic surface chemistry even allowed for this latter detection limit to be improved by an order of magnitude (to 6×10^4 cells mL^{-1}).[123]

Other ultralow fouling polymer brushes built with short, non-ionic hydroxyalkyl monomers have also been implemented in the construction of SPR immunosensors. This notably was the case in 2011 for polyHPMA surface chemistry (Figure 5.28E) with another pioneering piece of work by Rodriguez-Emmenegger and colleagues.[85] Detection was that of a peptidoglycan-polysaccharide antigen unique to *G Streptococcus* bacterium, the protocol also involving a step of SPR signal amplification through interception of secondary antibodies (Figure 5.50A). Even though analysis was carried out in buffer as a proof of concept, the demonstration was made by the investigators that the non-specific adsorption of adventitious species on immunosensing interfaces from such a highly fouling medium as undiluted human plasma would not be an issue (Figure 5.50B – from $\Gamma < 0.03$ ng cm^{-2} pre-biofunctionalization), should measurements be performed in such a real-life, biological environment.

5.3.6.3 Built from Zwitterionic Monomers

Arguably, if one type of antifouling surface chemistry had to be singled out for biosensing purposes in complex biological media, the consensus most likely would be to opt for zwitterionic polymer brushes. As will be seen next, research in this domain constitutes, as a matter of fact, one among the most actively pursued and diversified.

5.3.6.3.1 Carboxybetaine Methacrylate (CBMA). With respect to brushes polymerized with CBMA monomer (Figure 5.22B), the 2010 work by Jiang's team is particularly worth noting. In this study, polyCBMA brushes prepared on gold SPR sensor chips were decorated with antibodies against 'activated leukocyte cell adhesion molecule' (ALCAM) using standard EDC/NHS bioconjugation chemistry (Figure 5.51A).[114] Using such immunosensing platform, it was possible for the authors to detect this potential

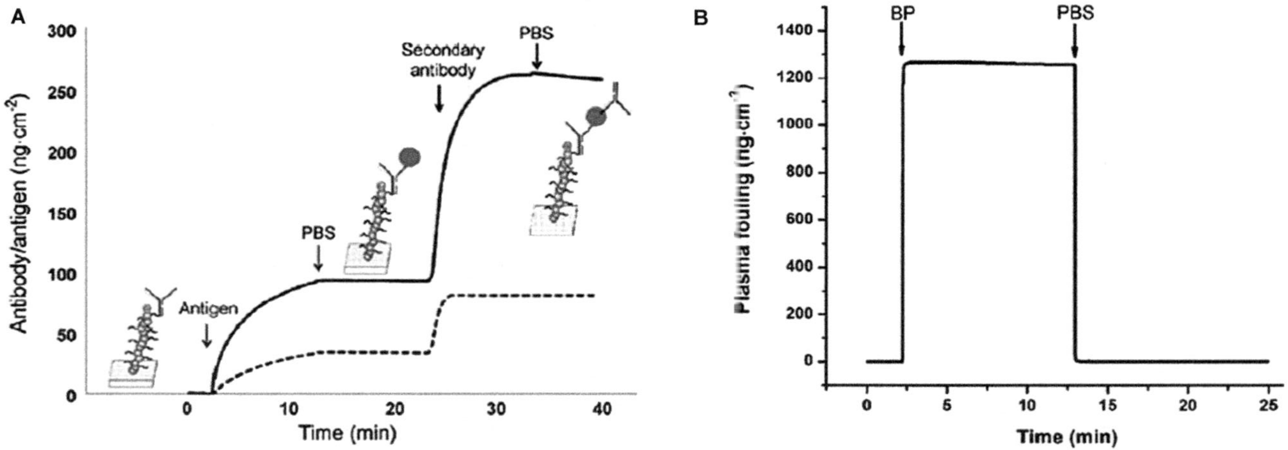

Figure 5.50 (A) Detection (in buffer) of an antigen unique to *G Streptococcus* bacterium at concentrations down to 600 ng mL^{-1} (dashed line) using an SPR immunosensor based on polyHPMA surface chemistry (the solid line corresponds to an antigen concentration of 6000 ng mL^{-1}). The SPR response signal was also amplified through the binding of secondary antibodies by surface-captured antigen analytes. (B) An SPR experiment with undiluted human plasma proving that fouling from this biofluid on antibody-biofunctionalized polyHPMA-based sensing platforms is negligible (*i.e.* below the reported SPR apparatus' LOD of 0.03 ng cm^{-2}), as was already the situation pre-biofunctionalization.
Reproduced from ref. 85 with permission of John Wiley & Sons. Copyright © 2011 Wiley-VCH Verlag GmbH & Co. KGaA, Weinheim.

carcinoma biomarker directly in 100% human plasma, with a reported LOD of $\sim$30 ng mL^{-1} (Figure 5.51B). We note at last – to comply with the recurrent theme of this subsection as to whether biofunctionalization impairs antifouling performance – that evidence was provided by the investigators that no loss in such property was observed upon antibody immobilization on the primer polyCBMA brush ($\Gamma_{\text{before}} = 8.9 \pm 3.4$ ng cm^{-2} *vs.* $\Gamma_{\text{after}} = 9.5 \pm 4.1$ ng cm^{-2} – Figure 5.51C *vs.* D).

A final comment deserves to be made regarding the successful transfer of this versatile methodology to the surface modification of other substrate materials than gold – notably silicon oxide – in immunosensors devoted to the label-free detection, with reported LODs of similar order of magnitude, once again of ALCAM[115,117] but also other antigen[116] analytes present in undiluted plasma or serum.

5.3.6.3.2 Carboxybetaine Acrylamide (CBAA). An anti-ALCAM SPR immunosensor – based on a similar construction protocol and working principle, but on a different polymer brush variant built from CBAA monomer (Figure 5.22D) – had already been devised in 2008 by Jiang and colleagues to detect this biomarker in undiluted human plasma[118] (or undiluted human serum in 2009),[120] with a reported LOD in the low nanomolar range as well.[119] This methodology was also employed later on for the detection of other analytes in undiluted human plasma such as streptavidin (in 2009),[121] thyroid stimulating hormone (in 2011),[119,124] or human chorionic gonadotropin (in 2013),[119] or *Escherichia coli O157:H7* in undiluted milk (in 2014).[123] In all instances, the verification was made that biofunctionalization with antibodies did not (significantly) alter antifouling performance.[119,121,123,124] This feat notably allowed for thyroid stimulating hormone (TSH) to be monitored at concentrations as low as $\sim$1 ng mL^{-1} (Figure 5.52A). Another attractive feature of this SPR immunosensor resided in the environmentally-friendly photochemical process through which the polyCBAA matrix for antibody immobilization was grown (Figure 5.52C), a polymerization mechanism to be contrasted with that of the traditional ATRP growth technique that relies on catalysis by potentially toxic transition metals.[124] Finally, as mentioned above, ultralow fouling properties were maintained following antibody immobilization (Figure 5.52B).

5.3.6.3.3 Hierarchical Architectures. At last, there also exists a particular class of multilayered zwitterionic polymer brushes that have been implemented for biosensing purposes and for which can immediately be distinguished two types of architectures, depending on whether or not the monomers used to sequentially build such hierarchical structures are of the same chemical nature.

On the one hand, 'two-layer' homopolymer brushes solely based on CBAA monomer have been engineered by Jiang and co-workers for the SPR immunosensing detection of thyroid stimulating hormone (TSH) model analyte.[125,126] The main focus of this proof-of-concept investigation was not so much the detection aspect, but rather the controlled construction of a biosensing platform that would be characterized by a thin and highly packed lower polyCBAA brush on top of which would be subsequently polymerized another, thicker but looser, such brush (Figure 5.53A) – the motivation and aim being to build novel 'dual-functional' immunosensing interfaces combining both ultralow fouling background yet high antibody loading capability, as opposed to the current 'one-layer' technology for which a trade-off is to be conceded in terms of these properties

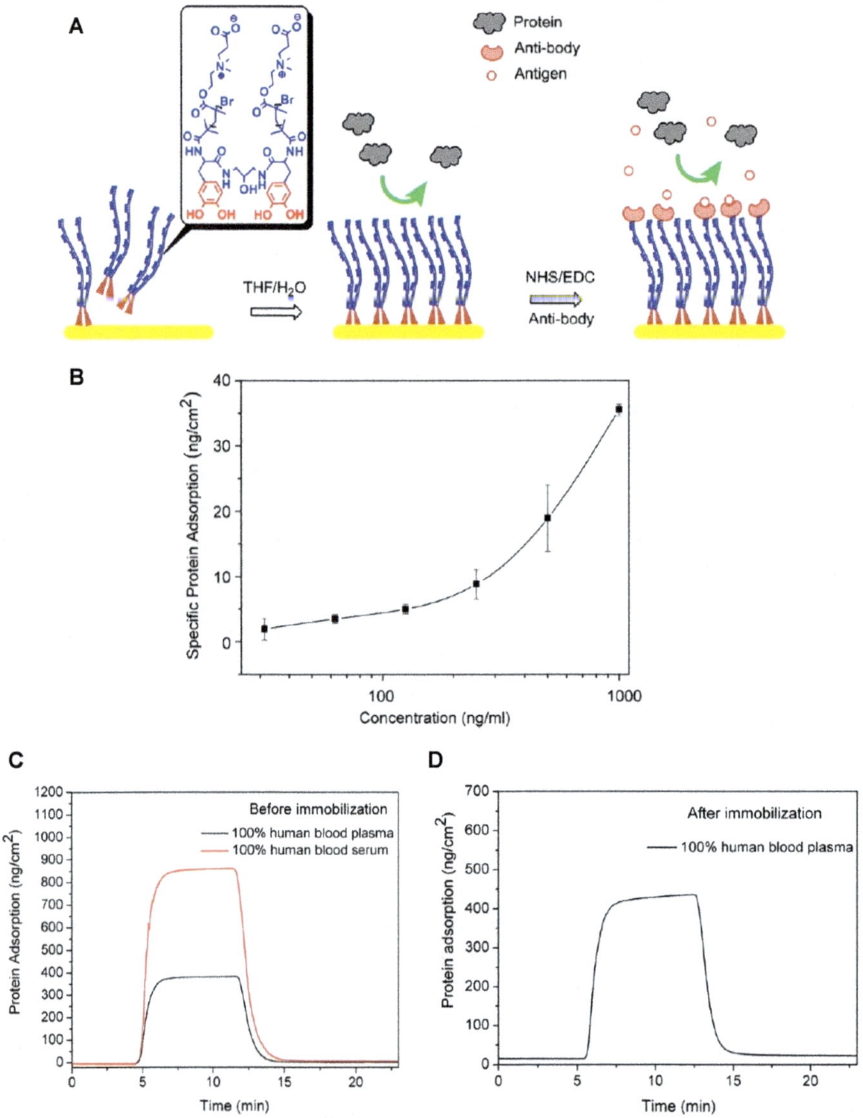

(Figure 5.53A *vs.* B).[125,126] In this study, the authors would embark on a systematic investigation and comparison (Figure 5.53C) of antibody immobilization and subsequent antigen binding capacities, as well as of fouling from undiluted human plasma both pre- and post-biofunctionalization – for both regular 'one-layer' and new 'two-layer' polyCBAA matrix, either polymerized through the traditional, transition metal-catalyzed SI-ATRP process or that, catalyst-free, promoted by UV-photoirradiation (known as SI-PIMP).[125] The unambiguous conclusion of this work was that 'two-layer' immunosensing platforms grown photochemically displayed highest antibody loading ($\sim$800 ng cm^{-2}, Figure 5.53C – left) and highest antigen binding ($\sim$130 ng cm^{-2}, Figure 5.53C – right) while displaying lower fouling from undiluted human plasma ($\Gamma < 5$ ng cm^{-2}, Figure 5.53C – middle), a level of adsorption essentially identical to that observed pre-biofunctionalization.[125,126]

To finish, there also exists a second kind of hierarchical platforms that rely on the sequential polymerization of a combination of different monomer building blocks. One example of such 'mixed' polymer brush coating was recently reported (in 2014) by Rodriguez-Emmenegger's group in a proof-of-concept investigation, wherein polyCBAA atop polyMeOEGMA brushes were grown in an overall diblock configuration (Figure 5.54A).[87] The upper polyCBAA film was then derivatized with a layer of neutravidin, using classic EDC/NHS coupling chemistry, for the subsequent capture of goat model antibodies through biotinyl residues. In terms of analytical performance, this SPR immunosensor prototype allowed for mouse IgG analyte to be detected in undiluted human plasma in the μg mL^{-1} range (Figure 5.54B). As always, the focus is last given to determine whether biofunctionalization impaired antifouling performance – a situation that, as it turned out, was not encountered in that work since fouling from undiluted plasma only rose to $\Gamma = 4.9 \pm 0.6$ ng cm^{-2}, from an undetectable level prior to antibody attachment.

Figure 5.51 (A) Construction and working principle of an immunosensor – based on polyCBMA surface chemistry featuring bioinspired catechol residues for substrate anchorage – dedicated to the SPR detection of ALCAM antigen, a potential cancer biomarker. (B) 'Reference-compensated' dose–response curve for ALCAM in 100% human plasma. It is indeed important to note that the contribution to each response signal of ALCAM naturally present in blood (at $\sim$84 ng mL^{-1})[114,115,117–120] was eliminated by subtracting the reference signal obtained with undiluted plasma that was not spiked with exogenous ALCAM (see also ref. 115). (C, D) Representative SPR sensograms, recorded before and after biofunctionalization of polyCBMA surface chemistry with anti-ALCAM antibodies, showing that such a transformation had no effect on the level of adsorption of undesirable species from 100% human plasma. Reprinted from C. Gao, G. Li, H. Xue, W. Yang, F. Zhang, S. liang, Functionalizable and ultra-low fouling zwitterionic surfaces via adhesive mussel mimetic linkages, *Biomaterials*, 7, 1486–1492, Copyright 2010 with permission from Elsevier.

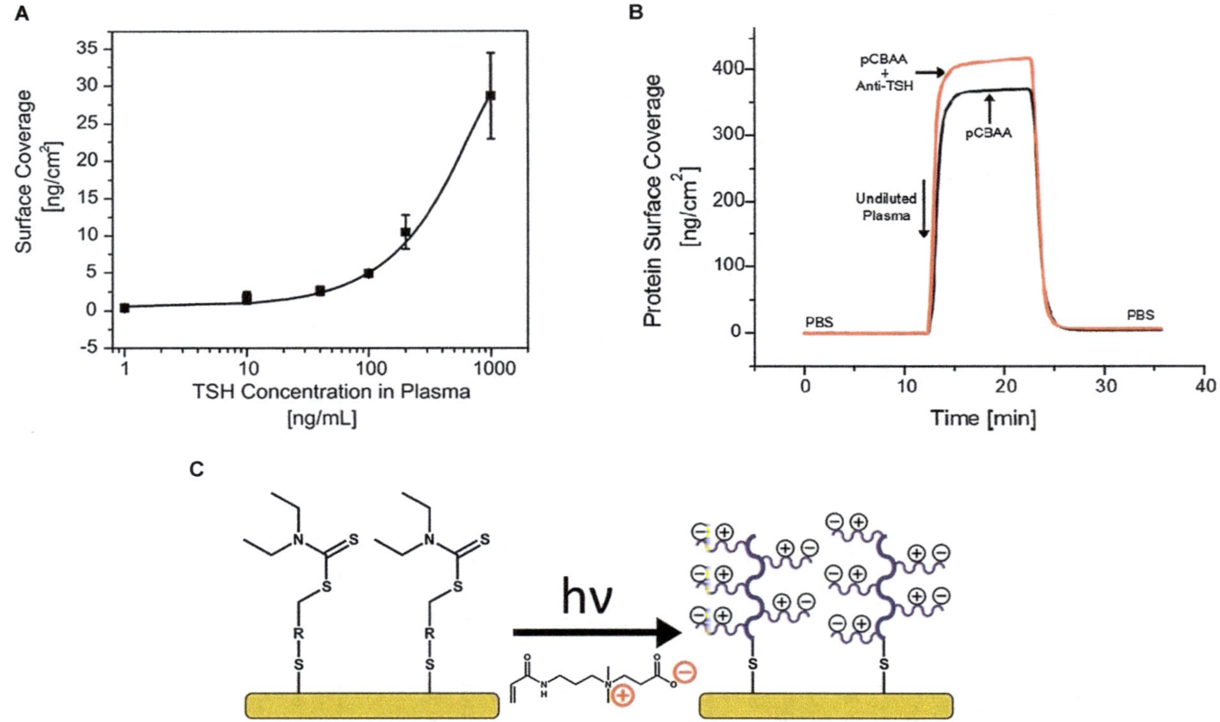

Figure 5.52 (A) Detection of thyroid stimulating hormone (TSH) in undiluted human plasma using an SPR immunosensor relying on polyCBAA surface chemistry. This dose–response curve was normalized by subtracting for each point the signal obtained from a reference platform (also biofunctionalized with antibodies) dedicated to measuring NSA from TSH-unspiked plasma flowed in an independent channel. (B) SPR experiments demonstrating the great resistance to fouling from 100% human plasma of polyCBAA surface chemistry both pre- and post-biofunctionalization with anti-TSH antibodies. (C) Schematic illustration of the photochemical process used to grow the primer polyCBAA zwitterionic brush onto an initiator SAM for subsequent antibody immobilization.
Reprinted with permission from ref. 124. Copyright 2011 American Chemical Society.

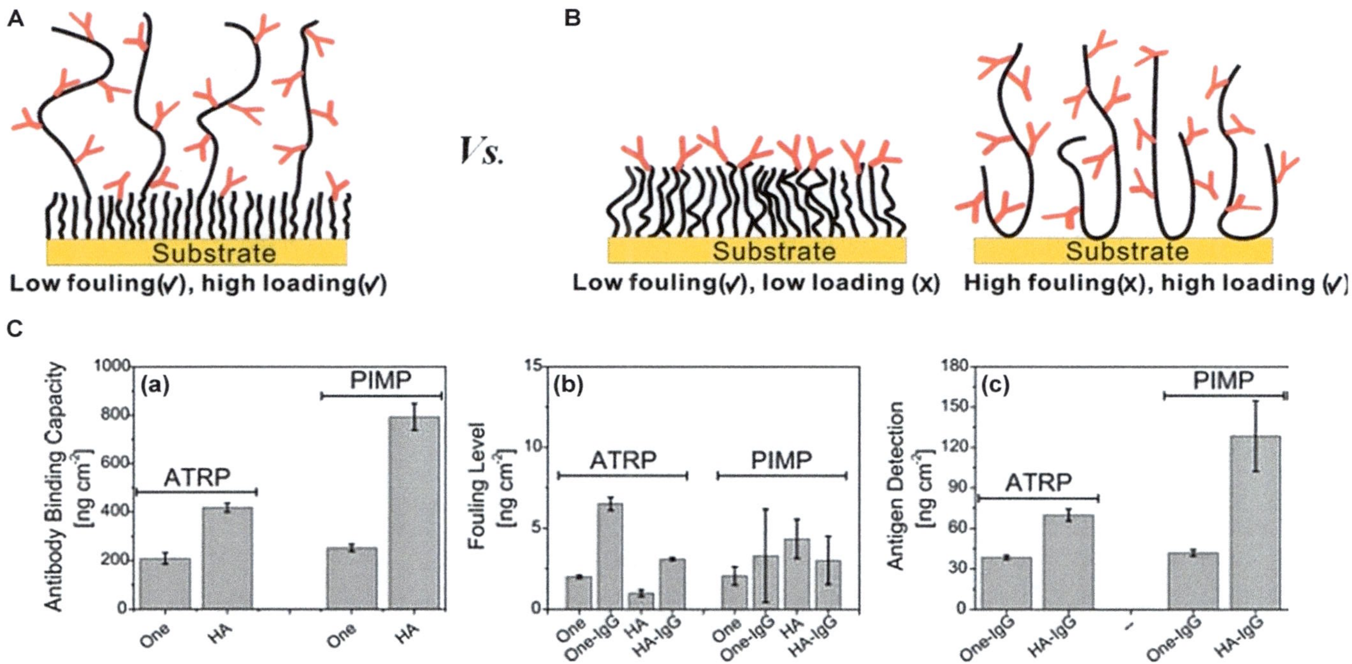

Figure 5.53 (A, B) The 'two-layer' approach – which features a thick, low-density polyCBAA brush polymerized atop a thinner but high-density such surface chemistry – provides biosensing platforms with 'dual functionality'; that is, both high resistance to fouling and high biorecognition ligand loading capacity. This situation is to be contrasted with that of the 'one-layer' strategy, wherein a compromise is to be made in terms of these properties. Reproduced from ref. 126 with permission of the American Chemical Society. (C) Anti-TSH antibody immobilization capacity (left), both pre- and post-biofunctionalization fouling levels from undiluted human plasma (middle), and TSH antigen binding capability in buffer (right) – of/on 'one-layer' and 'two-layer', ATRP- or PIMP-grown polyCBAA matrix-based biosensing platforms. 'ATRP', 'PIMP', 'one', 'HA', and 'IgG' respectively stand for/correspond to 'atom transfer radical polymerization', 'photoiniferter-mediated polymerization', 'one-layer (platform)', 'hierarchical (two-layer platform)', and 'antibody-biofunctionalized'. Reproduced from ref. 125 with permission of John Wiley & Sons. Copyright © 2012 Wiley-VCH Verlag GmbH & Co. KGaA, Weinheim.

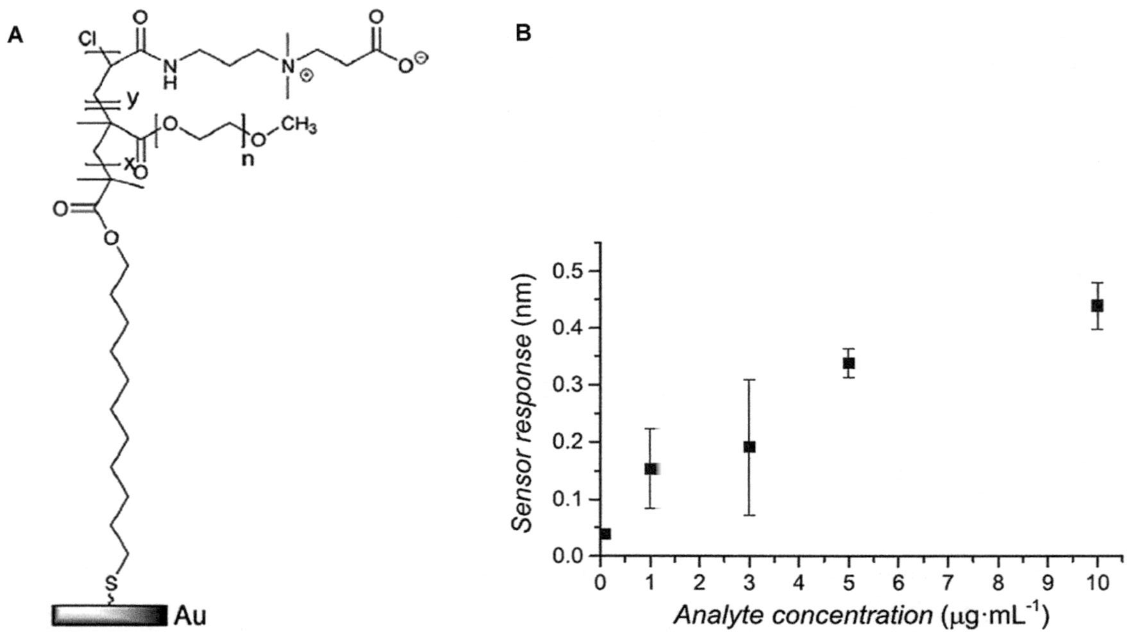

Figure 5.54 (A) Schematic representation of the 'mixed', diblock poly(MeOEGMA-*bl*-CBAA) brush coating (20 + 10 nm-thick) for antibody immobilization and immunosensing on gold SPR chips. (B) SPR detection of mouse IgG in undiluted human plasma. This dose–response curve was corrected by subtracting for each data point the signal obtained from a simultaneous measurement performed with unspiked plasma in an independent reference channel.
Reprinted from Hierarchical antifouling brushes for biosensing applications, *Sens. Actuators, B*, **202**, 1313–1321, Copyright 2014 with permission from Elsevier.

5.4 Conclusion and Outlook

The functionality of bioanalytical devices (and life expectancy in the case of implant application for continuous *in vivo* monitoring) critically depends on the interactivity of their exogenous detection platform with the biofluid multi-components they are exposed to, the outcome largely hinging on the physicochemical makeup of the artificial material contacting said biological environment. The recurrent, technical issue for biosensors – this genuine plague the technology deeply suffers from – more specifically is that of the non-specific adsorption and consequential signal interference of the biofluid matrix that hampers the detection, not to mention the quantitation, of target analytes present at considerably lower concentration. It is in this context that emerged the idea to first comprehensively identify the state-of-the-art surface chemistries developed to date (January 2015) to minimize fouling surface coverage from both human and animal biosamples, and to then move on to describing these stealth artificial coatings as they are implemented into biosensing applications or the like to combat NSA. Pronounced antifouling properties (*i.e.* a few ng cm^{-2} or less) were imparted to a variety of both inorganic and organic underlying substrates through deposition/growth of (sub)nanometer-thin SAMs or thicker polymer films/brushes, a rather large repertoire of which is now available as demonstrated in this chapter. Attention was first focused on covering antifouling surface chemistry against the most investigated biofluids – blood plasma and serum – whose provenance and variability (in terms of species, age, and single *vs.* pooled donations) were also shown to impact their fouling potency. The quantification of minimized fouling surface coverage by appropriate surface chemistry was also a topic of interest for another six, much less explored human or animal biological milieux: cell lysate, cerebrospinal fluid, egg, milk, saliva, and urine. Transferring such technology into the realm of biosensing was made possible thanks to the pivotal successful maintenance of pronounced repellent properties upon primer coating biofunctionalization, which allowed for ng mL^{-1} LODs to be reached (in undiluted human plasma). Despite such a tremendous achievement, the truth of the matter is that applications in this nascent field remain scarce, most being as well prototypical in nature and limited to analysis in blood plasma/serum. It is therefore our hope that the present chapter will not only encourage continued both fundamental and applied research on antifouling surface chemistries for these biological media (as well as the aforementioned cell lysate, cerebrospinal fluid, egg, milk, saliva, and urine), but also trigger an equivalent interest for some others – a non-exhaustive list of 25 potential candidates with bioanalytical relevance is proposed hereinafter: amniotic fluid,[217] bile,[218] breath,[219,220] cerumen,[221] chyle,[222] endolymph,[223] feces,[220,224] flatus/intestinal gas,[225] gastric juice,[226] honey,[227] ink,[228] interstitial fluid,[229] karyoplasm,[230] lymph,[231] ocular fluids (aqueous and vitreous humours),[232] pericardial fluid,[233] perilymph,[234] peritoneal fluid,[235] pleural fluid,[236] sebum,[220,237] semen,[238] sputum,[239] synovial fluid,[240,241] sweat,[220] and tears.[242] As antifouling surface chemistry against complex biological milieux continues to mature and diversify, the day inches closer when NSA signal interference will no longer be an issue and bioanalytical applications will know no boundary in terms of detection biofluids and detectable concentrations of target analytes. Maybe, the world of commercial biosensors will then open up and no longer be restricted to niche applications. Only time will tell.

References and Notes

1. S. J. Setford and J. D. Newman, Enzyme Biosensors, in *Methods in Biotechnology*, ed. J. L. Barredo, Microbial Enzymes and Biotransformations, Humana Press Inc., USA, 2005, vol. 17, pp. 29–60.
2. J. Wang, *Electroanalysis*, 2001, **13**, 983.
3. E.-H. Yoo and S.-Y. Lee, *Sensors*, 2010, **10**, 4558.
4. A. Turner, *Trends Biotechnol.*, 2013, **31**, 119.
5. A. P. F. Turner, *Chem. Soc. Rev.*, 2013, **42**, 3184.
6. J. H. T. Luong, K. B. Male and J. D. Glennon, *Biotechnol. Adv.*, 2008, **26**, 492.
7. M. Thompson, S. Sheikh, C. Blaszykowski and A. Romaschin, Biosensor Technology and the Clinical Biochemistry Laboratory – Issue of Signal Interference from the Biological Matrix, in *Detection Challenges in Clinical Diagnostics*, ed. P. Vadgama and S. Peteu, RSC Detection Science Series, Royal Society of Chemistry Publishing, UK, 2013, ch. 1, pp. 1–34.
8. J.-F. Masson and S. S. Zhao, Plasmonic Sensors for Analysis of Proteins and an Oncologic Drug in Human Serum, in *Nanobiosensors and Nanobioanalyses*, ed. M. C. Vestergaard, K. Kerman, I.-M. Hsing and E. Tamiya, Springer, Japan, 2015, ch. 15, pp. 305–333.
9. C. I. L. Justino, T. A. Rocha-Santos and A. C. Duarte, *TrAC, Trends Anal. Chem.*, 2010, **29**, 1172.
10. P. Gong and D. W. Grainger, Nonfouling Surfaces: A Review of Principles and Applications for Microarray Capture Assay Designs, in *Microarrays Volume 1 Synthesis Methods*, ed. J. B. Rampal, Methods in Molecular Biology, Humana Press Inc., USA, 2nd edn, 2007, ch. 3, vol. 381, pp. 59–92.
11. M. R. Malone, J.-F. Masson, M. Barhnart, S. Beaudoin and K. S. Booksh, *Proc. SPIE*, 2005, **6007**, 60070V.
12. S. Choi and J. Chae, *J. Micromech. Microeng.*, 2010, **20**, 075015.
13. B. Kasemo and J. Gold, *Adv. Dent. Res.*, 1999, **13**, 8.
14. S. P. Nichols, A. Koh, W. L. Storm, J. H. Shin and M. H. Schoenfisch, *Chem. Rev.*, 2013, **113**, 2528.
15. P. H. Kvist and H. E. Jensen, *J. Diabetes Sci. Technol.*, 2007, **1**, 746.
16. C. Blaszykowski, S. Sheikh and M. Thompson, *Trends Biotechnol.*, 2014, **32**, 61.
17. P. Thevenot, W. Hu and L. Tang, *Curr. Top. Med. Chem.*, 2008, **8**, 270.

18. J. M. Anderson, A. Rodriguez and D. T. Chang, *Semin. Immunol.*, 2008, **20**, 86.
19. B. D. Ratner and S. J. Bryant, *Annu. Rev. Biomed. Eng.*, 2004, **6**, 41.
20. A. Barfidokht and J. J. Gooding, *Electroanalysis*, 2014, **26**, 1182.
21. Fouling/adsorption surface coverage, expressed in terms of adsorbed mass per unit area, can sometimes be found 'quantified' using sensor devices based on acoustic wave physics – a yet non-gravimetric technique, when implemented in the liquid phase.
22. J. Homola, *Chem. Rev.*, 2008, **108**, 462.
23. M. Couture, S. S. Zhao and J.-F. Masson, *Phys. Chem. Chem. Phys.*, 2013, **15**, 11190.
24. M. Swierczewska, G. Liu, S. Lee and X. Chen, *Chem. Soc. Rev.*, 2012, **41**, 2641.
25. C. Blaszykowski, S. Sheikh and M. Thompson, *Chem. Soc. Rev.*, 2012, **41**, 5599.
26. W. F. Ganong, Circulating Body Fluids, *Review of Medical Physiology*, Lange Medical Books/McGraw-Hill (Medical Publishing Division), USA, 22nd edn, 2005, pp. 515–546.
27. A. C. Guyton and J. E. Hall, *Textbook of Medical Physiology*, Elsevier Saunders, USA, 11th edn, 2006, pp. 419–468.
28. J. F. Dailey, *Dailey's Notes on Blood*, Medical Consulting Group, USA, 3rd edn, 1996.
29. D. U. Silverthorn, Blood, in *Human Physiology – An Integrated Approach*, Pearson-Benjamin Cummings Publisher, USA, 3rd edn, 2004, ch. 16, pp. 523–545.
30. Y. Shen, J. Kim, E. F. Strittmatter, J. M. Jacobs, D. G. Camp II, R. Fang, N. Tolié, R. J. Moore and R. D. Smith, *Proteomics*, 2005, **5**, 4034.
31. J. D. Andrade and V. Hlady, *Ann. NY Acad. Sci.*, 1987, **516**, 158.
32. J. N. Adkins, S. M. Varnum, K. J. Auberry, R. J. Moore, N. H. Angell, R. D. Smith, D. L. Springer and J. G. Pounds, *Mol. Cell. Proteomics*, 2002, **1**, 947.
33. M. Lestelius, P. Tengvall and I. Lundström, *J. Colloid Interface Sci.*, 1995, **171**, 3.
34. M. Lestelius, B. Liedberg, I. Lundström and P. Tengvall, *J. Biomed. Mater. Res.*, 1994, **28**, 871.
35. P. Tengvall, M. Lestelius, B. Liedberg and I. Lundström, *Langmuir*, 1992, **8**, 1236.
36. O. R. Bolduc and J.-F. Masson, *Langmuir*, 2008, **24**, 12085.
37. O. R. Bolduc, C. M. Clouthier, J. N. Pelletier and J.-F. Masson, *Anal. Chem.*, 2009, **81**, 6779.
38. O. R. Bolduc, J. N. Pelletier and J.-F. Masson, *Anal. Chem.*, 2010, **82**, 3699.
39. O. R. Bolduc, D. Correia-Ledo and J.-F. Masson, *Langmuir*, 2012, **28**, 22.
40. O. R. Bolduc, P. Lambert-Lanteigne, D. Y. Colin, S. S. Zhao, C. Proulx, D. Boeglin, W. D. Lubell, J. N. Pelletier, J. Féthière, H. Ong and J.-F. Masson, *Analyst*, 2011, **136**, 3142.

41. Q. Zou, L. L. Kegel and K. S. Booksh, *Anal. Chem.*, 2015, **87**, 2488.
42. L. Richert, P. Lavalle, D. Vautier, B. Senger, J.-F. Stoltz, P. Schaaf, J.-C. Voegel and C. Picart, *Biomacromolecules*, 2002, **3**, 1170.
43. Q. Liu, A. Singh and L. Liu, *Biomacromolecules*, 2013, **14**, 226.
44. Q. Liu, W. Li, A. Singh, G. Cheng and L. Liu, *Acta Biomater.*, 2014, **10**, 2956.
45. W. Li, Q. Liu and L. Liu, *J. Biomater. Sci. Polym. Edn.*, 2014, **25**, 1730.
46. W. Li, Q. Liu and L. Liu, *Langmuir*, 2014, **30**, 12619.
47. A. R. Statz, R. J. Meagher, A. E. Barron and P. B. Messersmith, *J. Am. Chem. Soc.*, 2005, **127**, 7972.
48. A. R. Statz, A. E. Barron and P. B. Messersmith, *Soft Matter*, 2008, **4**, 131.
49. A. R. Statz, J. Kuang, C. Ren, A. E. Barron, I. Szleifer and P. B. Messersmith, *Biointerphases*, 2009, **4**, FA22.
50. S. Lin, B. Zhang, M. J. Skoumal, B. Ramunno, X. Li, C. Wesdemiotis, L. Liu and L. Jia, *Biomacromolecules*, 2011, **12**, 2573.
51. I. Banerjee, R. C. Pangule and R. S. Kane, *Adv. Mater.*, 2011, **23**, 690.
52. P. Vermette and L. Meagher, *Colloids Surf., B*, 2003, **28**, 153.
53. E. Ostuni, R. G. Chapman, R. E. Holmlin, S. Takayama and G. M. Whitesides, *Langmuir*, 2001, **17**, 5605.
54. K. Knop, R. Hoogenboom, D. Fischer and U. S. Schubert, *Angew. Chem., Int. Ed.*, 2010, **49**, 6288.
55. L. Cao, M. Chang, C.-Y. Lee, D. G. Castner, S. Sukavaneshvar, B. D. Ratner and T. A. Horbett, *J. Biomed. Mater. Res., Part A*, 2007, **81A**, 827.
56. Y. Chang, W.-L. Chu, W.-Y. Chen, J. Zheng, L. Liu, R.-C. Ruaan and A. Higuchi, *J. Biomed. Mater. Res., Part A*, 2010, **93A**, 400.
57. Y. Chang, S.-C. Liao, A. Higuchi, R.-C. Ruaan, C.-W. Chu and W.-Y. Chen, *Langmuir*, 2008, **24**, 5453.
58. Y. Chang, S.-H. Shu, Y.-J. Shih, C.-W. Chu, R.-C. Ruaan and W.-Y. Chen, *Langmuir*, 2010, **26**, 3522.
59. J. Ladd, Z. Zhang, S. Chen, J. C. Hower and S. Jiang, *Biomacromolecules*, 2008, **9**, 1357.
60. See also: B. Zhu, T. Eurell, R. Gunawan and D. Leckband, *J. Biomed. Mater. Res.*, 2001, **56**, 406.
61. See also: C. Subramani, A. Bajaj, O. R. Miranda and V. M. Rotello, *Adv. Mater.*, 2010, **22**, 5420.
62. C. Rodriguez-Emmenegger, M. Houska, A. B. Alles and E. Brynda, *Macromol. Biosci.*, 2012, **12**, 1413.
63. For the antifouling properties against 10% FBS (30 min exposure) of the corresponding, 1.7 nm-thick OEG_3-SAM on gold, see: M. Lashkor, F. J. Rawson, J. A. Preece and P. M. Mendes, *Analyst*, 2014, **139**, 5400.
64. For the antifouling properties against human serum of 20 kDa PEG (~3.9 nm-thick), 2000 kDa dextran (~7.1 nm-thick), and 27 kDa poly(vinyl alcohol) (~0.9 nm-thick) polymeric films spin-coated onto polyallylamine-*g*-perfluorophenylazide adhesion coatings (~1.8 nm-thick) themselves anchored to Si/SiO_2 wafers through multiple

electrostatic interactions, see: Â. Serrano, O. Sterner, S. Mieszkin, S. Zürcher, S. Tosatti, M. E. Callow, J. A. Callow and N. D. Spencer, *Adv. Funct. Mater.*, 2013, **23**, 5706.
65. K. Yoshimoto, T. Hirase, J. Madsen, S. P. Armes and Y. Nagasaki, *Macromol. Rapid Commun.*, 2009, **30**, 2136.
66. O. Pop-Georgievski, D. Verreault, M.-O. Diesner, V. Proks, S. Heissler, F. Rypáček and P. Koelsch, *Langmuir*, 2012, **28**, 14273.
67. L. M. Feller, S. Cerritelli, M. Textor, J. A. Hubbell and S. G. P. Tosatti, *Macromolecules*, 2005, **38**, 10503.
68. J. P. Bearinger, S. Terrettaz, R. Michel, N. Tirelli, H. Vogel, M. Textor and J. A. Hubbell, *Nat. Mater.*, 2003, **2**, 259.
69. V. Zoulalian, S. Zürcher, S. Tosatti, M. Textor, S. Monge and J.-J. Robin, *Langmuir*, 2010, **26**, 74.
70. V. Zoulalian, S. Monge, S. Zürcher, M. Textor, J.-J. Robin and S. Tosatti, *J. Phys. Chem. B*, 2006, **110**, 25603.
71. A. Reisch, J.-C. Voegel, E. Gonthier, G. Decher, B. Senger, P. Schaaf and P. J. Mésini, *Langmuir*, 2009, **25**, 3610.
72. J. Davila, D. Toulemon, T. Garnier, A. Garnier, B. Senger, J.-C. Voegel, P. J. Mésini, P. Schaaf, F. Boulmedais and L. Jierry, *Langmuir*, 2013, **29**, 7488.
73. T. Gillich, E. M. Benetti, E. Rakhmatullina, R. Konradi, W. Li, A. Zhang, A. D. Schlüter and M. Textor, *J. Am. Chem. Soc.*, 2011, **133**, 10940.
74. J. L. Dalsin, L. Lin, S. Tosatti, J. Vörös, M. Textor and P. B. Messersmith, *Langmuir*, 2005, **21**, 640.
75. S. Zürcher, D. Wäckerlin, Y. Bethuel, B. Malisova, M. Textor, S. Tosatti and K. Gademann, *J. Am. Chem. Soc.*, 2006, **128**, 1064.
76. J.-Y. Wach, B. Malisova, S. Bonazzi, S. Tosatti, M. Textor, S. Zürcher and K. Gademann, *Chem. - Eur. J.*, 2008, **14**, 10579.
77. B. Malisova, S. Tosatti, M. Textor, K. Gademann and S. Zürcher, *Langmuir*, 2010, **26**, 4018.
78. For the antifouling properties against 50% FBS (2 h incubation) of ~9 kDa PEG multisorbate films (~4 nm-thick) tethered to a variety of metal substrates (Au, Si, Ti) through catechol graft residues, see also: H. Lee, K. D. Lee, K. B. Pyo, S. Y. Park and H. Lee, *Langmuir*, 2010, **26**, 3790.
79. For the remarkable ($\Gamma < 30$ ng cm^{-2}), long-term (10 days) antifouling properties against undiluted human blood plasma/serum of PEO coatings grafted at high temperatures (≥90 °C melts) onto dopamine–melanin anchoring polymer films (12 nm-thick) pre-coated on gold surfaces, see also: O. Pop-Georgievski, Š. Popelka, M. Houska, D. Chvostová, V. Proks and F. Rypáček, *Biomacromolecules*, 2011, **12**, 3232; for a more recent (2012), related study minimizing fouling from 100% human blood plasma to $\Gamma = 7 \pm 2$ ng cm^{-2} with 14.1 nm-thick PEO_{20kDa} brushes (as measured by SPR upon 20 min exposure), see reference 66.

80. For a more recent study on the antifouling properties against full human serum (20 min incubation) of ~2 kDa PEG-dopamine brushes (~62 Å-thick) grafted onto titanium oxide through the pioneering electro-reduction of *o*-benzoquinone anchoring residues, see also: S. Liu, L. Chen, L. Tan, F. Cao, L. Bai and Y. Wang, *J. Mater. Chem. B*, 2014, **2**, 6758.
81. G. Gunkel, M. Weinhart, T. Becherer, R. Haag and W. T. S. Huck, *Biomacromolecules*, 2011, **12**, 4169.
82. For the antifouling properties against 100% human PPP of a closely related poly-*N*-[(2,3-dihydroxy-propyl)acrylamide] polymer brush on silicon, see: Y. Zou, P.-Y. J. Yeh, N. A. A. Rossi, D. E. Brooks and J. N. Kizhakkedathu, *Biomacromolecules*, 2010, **11**, 284.
83. Z. Zhang, M. Zhang, S. Chen, T. A. Horbett, B. D. Ratner and S. Jiang, *Biomaterials*, 2008, **29**, 4285.
84. C. Rodriguez-Emmenegger, O. Kylián, M. Houska, E. Brynda, A. Artemenko, J. Kousal, A. B. Alles and H. Biederman, *Biomacromolecules*, 2011, **12**, 1058.
85. C. Rodriguez-Emmenegger, E. Brynda, T. Riedel, M. Houska, V. Šubr, A. B. Alles, E. Hasan, J. E. Gautrot and W. T. S. Huck, *Macromol. Rapid Commun.*, 2011, **32**, 952.
86. C. Rodriguez-Emmenegger, E. Hasan, O. Pop-Georgievski, M. Houska, E. Brynda and A. B. Alles, *Macromol. Biosci.*, 2012, **12**, 525.
87. A. de los Santos Pereira, T. Riedel, E. Brynda and C. Rodriguez-Emmenegger, *Sens. Actuators, B*, 2014, **202**, 1313.
88. A. R. Kuzmyn, A. de los Santos Pereira, O. Pop-Georgievski, M. Bruns, E. Brynda and C. Rodriguez-Emmenegger, *Polym. Chem.*, 2014, **5**, 4124.
89. A. de los Santos Pereira, C. Rodriguez-Emmenegger, F. Surman, T. Riedel, A. B. Alles and E. Brynda, *RSC Adv.*, 2014, **4**, 2318.
90. T. Riedel, C. Rodriguez-Emmenegger, A. de los Santos Pereira, A. Bědajánková, P. Jinoch, P. M. Boltovets and E. Brynda, *Biosens. Bioelectron.*, 2014, **55**, 278.
91. C. Rodriguez-Emmenegger, A. Decker, F. Surman, C. M. Preuss, Z. Sedláková, N. Zydziak, C. Barner-Kowollik, T. Schwartz and L. Barner, *RSC Adv.*, 2014, **4**, 64781.
92. H. Ma, J. Hyun, P. Stiller and A. Chilkoti, *Adv. Mater.*, 2004, **16**, 338.
93. C. Rodriguez-Emmenegger, A. Jäger, E. Jäger, P. Stepanek, A. B. Alles, S. S. Guterres, A. R. Pohlmann and E. Brynda, *Colloids Surf. B*, 2011, **83**, 376.
94. J. E. Gautrot, B. Trappmann, F. Oceguera-Yanez, J. Connelly, X. He, F. M. Watt and W. T. S. Huck, *Biomaterials*, 2010, **31**, 5030.
95. H. Ma, D. Li, X. Sheng, B. Zhao and A. Chilkoti, *Langmuir*, 2006, **22**, 3751.
96. J. E. Raynor, T. A. Petrie, A. J. García and D. M. Collard, *Adv. Mater.*, 2007, **19**, 1724.
97. O. Pop-Georgievski, C. Rodriguez-Emmenegger, A. de los Santos Pereira, V. Proks, E. Brynda and F. Rypáček, *J. Mater. Chem. B*, 2013, **1**, 2859.

98. For the antifouling properties against 33% foetal calf serum (15 min incubation) of polyCBAA brushes grafted on cellulose (filter paper), see also: T. Tischer, C. Rodriguez-Emmenegger, V. Trouillet, A. Welle, V. Schueler, J. O. Mueller, A. S. Goldmann, E. Brynda and C. Barner-Kowollik, *Adv. Mater.*, 2014, **26**, 4087.
99. S. Jiang and Z. Cao, *Adv. Mater.*, 2010, **22**, 920.
100. N. Y. Kostina, C. Rodriguez-Emmenegger, M. Houska, E. Brynda and J. Michálek, *Biomacromolecules*, 2012, **13**, 4164.
101. C. Rodriguez-Emmenegger, E. Brynda, T. Riedel, Z. Sedlakova, M. Houska and A. B. Alles, *Langmuir*, 2009, **25**, 6328.
102. For the antifouling properties ($\Gamma > 100$ ng cm^{-2}) against 10% FBS (3 h exposure at 37 °C) of photoreactive PCMA-based block terpolymer coatings grafted to stainless steel, see: Y. Iwasaki, A. Matsumoto and S.-I. Yusa, *ACS Appl. Mater. Interfaces*, 2012, **4**, 3254.
103. For the antifouling properties against 10% FBS (30 min exposure) of $[PCMA/Zr^{4+}]_n$ multilayered films self-assembled through ionic complexation on silicon wafers, see: S. Pang, C. Zhu, F. Xu, C. Chen and J. Ji, *Colloids Surf. B*, 2012, **94**, 22.
104. C. Zhao, J. Zhao, X. Li, J. Wu, S. Chen, Q. Chen, Q. Wang, X. Gong, L. Li and J. Zheng, *Biomaterials*, 2013, **34**, 4714.
105. W. Yang, S. Chen, G. Cheng, H. Vaisocherová, H. Xue, W. Li, J. Zhang and S. Jiang, *Langmuir*, 2008, **24**, 9211.
106. G. Li, H. Xue, G. Cheng, S. Chen, F. Zhang and S. Jiang, *J. Phys. Chem. B*, 2008, **112**, 15269.
107. G. Li, G. Cheng, H. Xue, S. Chen, F. Zhang and S. Jiang, *Biomaterials*, 2008, **29**, 4592.
108. F. T. Limpoco and R. C. Bailey, *J. Am. Chem. Soc.*, 2011, **133**, 14864.
109. Y. Chang, W.-Y. Chen, W. Yandi, Y.-J. Shih, W.-L. Chu, Y.-L. Liu, C.-W. Chu, R.-C. Ruaan and A. Higuchi, *Biomacromolecules*, 2009, **10**, 2092.
110. S. Choi, B.-C. Choi, C. Xue and D. Leckband, *Biomacromolecules*, 2013, **14**, 92.
111. Y. Chang, Y.-J. Shih, C.-J. Lai, H.-H. Kung and S. Jiang, *Adv. Funct. Mater.*, 2013, **23**, 1100.
112. Q. Liu, W. Li, H. Wang and L. Liu, *J. Appl. Polym. Sci.*, 2014, **131**, 40789.
113. G. Cheng, G. Li, H. Xue, S. Chen, J. D. Bryers and S. Jiang, *Biomaterials*, 2009, **30**, 5234.
114. C. Gao, G. Li, H. Xue, W. Yang, F. Zhang and S. Jiang, *Biomaterials*, 2010, **31**, 1486.
115. N. D. Brault, C. Gao, H. Xue, M. Piliarik, J. Homola, S. Jiang and Q. Yu, *Biosens. Bioelectron.*, 2010, **25**, 2276.
116. See also: J. T. Kirk, N. D. Brault, T. Baehr-Jones, M. Hochberg, S. Jiang and D. M. Ratner, *Biosens. Bioelectron.*, 2013, **42**, 100.
117. See also: M. G. von Muhlen, N. D. Brault, S. M. Knudsen, S. Jiang and S. R. Manalis, *Anal. Chem.*, 2010, **82**, 1905.

118. H. Vaisocherová, W. Yang, Z. Zhang, Z. Cao, G. Cheng, M. Piliarik, J. Homola and S. Jiang, *Anal. Chem.*, 2008, **80**, 7894.
119. N. D. Brault, A. D. White, A. D. Taylor, Q. Yu and S. Jiang, *Anal. Chem.*, 2013, **85**, 1447.
120. W. Yang, H. Xue, W. Li, J. Zhang and S. Jiang, *Langmuir*, 2009, **25**, 11911.
121. H. Vaisocherová, Z. Zhang, W. Yang, Z. Cao, G. Cheng, A. D. Taylor, M. Piliarik, J. Homola and S. Jiang, *Biosens. Bioelectron.*, 2009, **24**, 1924.
122. N. D. Brault, H. S. Sundaram, Y. Li, C.-J. Huang, Q. Yu and S. Jiang, *Biomacromolecules*, 2012, **13**, 589.
123. H. Vaisocherová, V. Ševců, P. Adam, B. Špačková, K. Hegnerová, A. de los Santos Pereira, C. Rodriguez-Emmenegger, T. Riedel, M. Houska, E. Brynda and J. Homola, *Biosens. Bioelectron.*, 2014, **51**, 150.
124. J. E. Krause, N. D. Brault, Y. Li, H. Xue, Y. Zhou and S. Jiang, *Macromolecules*, 2011, **44**, 9213.
125. C.-J. Huang, N. D. Brault, Y. Li, Q. Yu and S. Jiang, *Adv. Mater.*, 2012, **24**, 1834.
126. C.-J. Huang, Y. Li and S. Jiang, *Anal. Chem.*, 2012, **84**, 3440.
127. G. L. Kenausis, J. Vörös, D. L. Elbert, N.-P. Huang, R. Hofer, L. Ruiz-Taylor, M. Textor, J. A. Hubbell and N. D. Spencer, *J. Phys. Chem. B*, 2000, **104**, 3298.
128. N.-P. Huang, R. Michel, J. Vörös, M. Textor, R. Hofer, A. Rossi, D. L. Elbert, J. A. Hubbell and N. D. Spencer, *Langmuir*, 2001, **17**, 489.
129. S. Pasche, S. M. De Paul, J. Vörös, N. D. Spencer and M. Textor, *Langmuir*, 2003, **19**, 9216.
130. S. Tosatti, Z. Schwartz, C. Campbell, D. L. Cochran, S. VandeVondele, J. A. Hubbell, A. Denzer, J. Simpson, M. Wieland, C. H. Lohmann, M. Textor and B. D. Boyan, *J. Biomed. Mater. Res., Part A*, 2004, **68A**, 458.
131. K. N. Hansson, S. Tosatti, J. Isaksson, J. Wetterö, M. Textor, T. L. Lindahl and P. Tengvall, *Biomaterials*, 2005, **26**, 861.
132. M. Schuler, G. Rh. Owen, D. W. Hamilton, M. de Wild, M. Textor, D. M. Brunette and S. G. P. Tosatti, *Biomaterials*, 2006, **27**, 4003.
133. R. R. Maddikeri, S. Tosatti, M. Schuler, S. Chessari, M. Textor, R. G. Richards and L. G. Harris, *J. Biomed. Mater. Res., Part A*, 2008, **84A**, 425.
134. D. Vonwil, M. Schuler, A. Barbero, S. Ströbel, D. Wendt, M. Textor, U. Aebi and I. Martin, *Eur. Cells Mater.*, 2010, **20**, 316.
135. J. Pei, H. Hall and N. D. Spencer, *Biomaterials*, 2011, **32**, 8968.
136. N.-P. Huang, S. M. De Paul and M. Textor, *Biomacromolecules*, 2011, **12**, 4213.
137. B. Pidhatika, M. Rodenstein, Y. Chen, E. Rakhmatullina, A. Mühlebach, C. Acikgöz, M. Textor and R. Konradi, *Biointerphases*, 2012, **7**, 1.
138. R. Ogaki, O. Z. Andersen, G. V. Jensen, K. Kolind, D. C. E. Kraft, J. S. Pedersen and M. Foss, *Biomacromolecules*, 2012, **13**, 3668.

139. S. VandeVondele, J. Vörös and J. A. Hubbell, *Biotechnol. Bioeng.*, 2003, **82**, 784.
140. G. Zhen, D. Falconnet, E. Kuennemann, J. Vörös, N. D. Spencer, M. Textor and S. Zürcher, *Adv. Funct. Mater.*, 2006, **16**, 243.
141. N.-P. Huang, J. Vörös, S. M. De Paul, M. Textor and N. D. Spencer, *Langmuir*, 2002, **18**, 220.
142. S. Lee and J. Vörös, *Langmuir*, 2005, **21**, 11957.
143. R. Konradi, C. Acikgoz and M. Textor, *Macromol. Rapid Commun.*, 2012, **33**, 1663.
144. R. Konradi, B. Pidhatika, A. Mühlebach and M. Textor, *Langmuir*, 2008, **24**, 613.
145. B. Pidhatika, J. Möller, V. Vogel and R. Konradi, *Chimia*, 2008, **62**, 264.
146. B. Pidhatika, J. Möller, E. M. Benetti, R. Konradi, E. Rakhmatullina, A. Mühlebach, R. Zimmermann, C. Werner, V. Vogel and M. Textor, *Biomaterials*, 2010, **31**, 9462.
147. O. Guillaume-Gentil, T. Zambelli and J. A. Vorholt, *Lab Chip*, 2014, **14**, 402.
148. H. P. C. Van Kuringen, J. Lenoir, E. Adriaens, J. Bender, B. G. De Geest and R. Hoogenboom, *Macromol. Biosci.*, 2012, **12**, 1114.
149. C. Zhao, L. Li, Q. Wang, Q. Yu and J. Zheng, *Langmuir*, 2011, **27**, 4906.
150. C. Zhao, L. Li and J. Zheng, *Langmuir*, 2010, **26**, 17375.
151. C. Zhao and J. Zheng, *Biomacromolecules*, 2011, **12**, 4071.
152. C. Zhao, Q. Chen, K. Patel, L. Li, X. Li, Q. Wang, G. Zhang and J. Zheng, *Soft Matter*, 2012, **8**, 7848.
153. H. Chen, M. Zhang, J. Yang, C. Zhao, R. Hu, Q. Chen, Y. Chang and J. Zheng, *Langmuir*, 2014, **30**, 10398.
154. J. Yang, M. Zhang, H. Chen, Y. Chang, Z. Chen and J. Zheng, *Biomacromolecules*, 2014, **15**, 2982.
155. Q. Liu, A. Singh, R. Lalani and L. Liu, *Biomacromolecules*, 2012, **13**, 1086.
156. L. Liu, Q. Liu and A. Singh, Polyacrylamide: Evaluation of Ultralow Fouling Properties of a Traditional Material, in *Proteins at Interfaces III State of the Art*, ed. T. Horbett, J. L. Brash and W. Norde, ACS Symposium Series, American Chemical Society, USA, 2012, ch. 30, vol. 1120, pp. 661–676.
157. See also: A. Reisch, J. Hemmerlé, J.-C. Voegel, E. Gonthier, G. Decher, N. Benkirane-Jessel, A. Chassepot, D. Mertz, P. Lavalle, P. Mésini and P. Schaaf, *J. Mater. Chem.*, 2008, **18**, 4242.
158. Y. Li, A. J. Keefe, M. Giarmarco, N. D. Brault and S. Jiang, *Langmuir*, 2012, **28**, 9707.
159. B. Cao, L. Li, Q. Tang and G. Cheng, *Biomaterials*, 2013, **34**, 7592.
160. S. Chen, L. Li, C. Zhao and J. Zheng, *Polymer*, 2010, **51**, 5283.
161. T. Tah and M. T. Bernards, *Colloids Surf. B*, 2012, **93**, 195.
162. M. Ratel, A. Provencher-Girard, S. S. Zhao, J. Breault-Turcot, J. Labrecque-Carbonneau, M. Branca, J. N. Pelletier, A. R. Schmitzer and J.-F. Masson, *Anal. Chem.*, 2013, **85**, 5770.

163. S. Reitsma, D. W. Slaaf, H. Vink, M. A. M. J. van Zandvoort and M. G. A. oude Egbrink, *Pflügers Archiv. – Eur. J. Physiol.*, 2007, **454**, 345.
164. S. Weinbaum, J. M. Tarbell and E. R. Damiano, *Annu. Rev. Biomed. Eng.*, 2007, **9**, 121.
165. N. B. Holland, Y. Qiu, M. Ruegsegger and R. E. Marchant, *Nature*, 1998, **392**, 799.
166. M. A. Ruegsegger and R. E. Marchant, *J. Biomed. Mater. Res.*, 2001, **56**, 159.
167. C. Perrino, S. Lee, S. W. Choi, A. Maruyama and N. D. Spencer, *Langmuir*, 2008, **24**, 8850.
168. K. Yu, B. F. L. Lai and J. N. Kizhakkedathu, *Adv. Healthcare Mater.*, 2012, **1**, 199.
169. R. M. Cornelius, J. G. Archambault, L. Berry, A. K. C. Chan and J. L. Brash, *J. Biomed. Mater. Res.*, 2002, **60**, 622.
170. In *Cell Lysis Technical Handbook (Featuring Cell Lysis Reagents and Detergents)*, Thermo Scientific Pierce, 2009.
171. A. Aubé, J. Breault-Turcot, P. Chaurand, J. N. Pelletier and J.-F. Masson, *Langmuir*, 2013, **29**, 10141.
172. D. U. Silverthorn, The Central Nervous System, *Human Physiology – An Integrated Approach*, Pearson-Benjamin Cummings Publisher, USA, 3rd edn, 2004, ch. 9, pp. 284–320.
173. F. Fischbach and M. B. Dunning III, Cerebrospinal Fluid Studies, *A Manual of Laboratory and Diagnostic Tests*, Wolters Kluwer Health/Lippincott Williams and Wilkins Publishers, USA, 8th edn, 2008, ch. 5, pp. 310–337.
174. R. D. Terlizzi and S. Platt, *Vet. J.*, 2006, **172**, 422.
175. J. Valenta and P. Fiala, Ventricular System and Cerebrospinal Fluid, *Central Nervous System – Overview of Anatomy*, Karolinum Press, Czech Republic, 1st edn, 2012, ch. 9, pp. 93–98.
176. M. H. Maurer, *Mass Spectrom. Rev.*, 2010, **29**, 17.
177. R. Mandal, A. C. Guo, K. K. Chaudhary, P. Liu, F. S. Yallou, E. Dong, F. Aziat and D. S. Wishart, *Genome Med.*, 2012, **4**, 38.
178. H. Tumani, C. Teunissen, S. Süssmuth, M. Otto, A. C. Ludolph and J. Brettschneider, *Expert Rev. Mol. Diagn.*, 2008, **8**, 479.
179. A. M. Fagan and R. J. Perrin, *Biomarkers Med.*, 2012, **6**, 455.
180. *Food and Agriculture Organization of the United Nations (UNFAO) – Statistics Division*, http://faostat.fao.org.
181. H.-D. Belitz, W. Grosch and P. Schieberle, *Food Chemistry*, Springer-Verlag, Germany, 3rd edn, 2004, pp. 551–565.
182. H.-D. Belitz, W. Grosch and P. Schieberle, *Food Chemistry*, Springer-Verlag, Germany, 3rd edn, 2004, pp. 505–550.
183. C. Rodriguez-Emmenegger, O. A. Avramenko, E. Brynda, J. Skvor and A. B. Alles, *Biosens. Bioelectron*, 2011, **26**, 4545.
184. X. Liu, R. Huang, R. Su, W. Qi, L. Wang and Z. He, *ACS Appl. Mater. Interfaces*, 2014, **6**, 13034.

185. J. M. Yoshizawa, C. A. Schafer, J. J. Schafer, J. J. Farrell, B. J. Paster and D. T. W. Wong, *Clin. Microbiol. Rev.*, 2013, **26**, 781.
186. P. D. V. de Almeida, A. M. T. Grégio, M. A. N. Machado, A. A. S. de Lima and L. R. Azevedo, *J. Contemp. Dent. Pract.*, 2008, **9**, 72.
187. S. P. Humphrey and R. T. Williamson, *J. Prosthet. Dent.*, 2001, **85**, 162.
188. M. A. Catalán, T. Nakamoto and J. E. Melvin, *J. Med. Invest.*, 2009, **56**, 192.
189. T. Arnebrant, Protein Adsorption in the Oral Environment, in *Biopolymers at Interfaces*, ed. M. Malmsten, Marcel Dekker Inc., USA, 2nd edn, 2003, ch. 29, pp. 811–856.
190. M. Edgerton and M. J. Levine, *J. Prosthet. Dent.*, 1993, **69**, 406.
191. N. Vassilakos, T. Arnebrant and P.-O. Glantz, *Scand. J. Dent. Res.*, 1992, **100**, 346.
192. N. Vassilakos, T. Arnebrant and P.-O. Glantz, *Scand. J. Dent. Res.*, 1993, **101**, 133.
193. See also: N. Vassilakos, P.-O. Glantz and T. Arnebrant, *Scand. J. Dent. Res.*, 1993, **101**, 339.
194. See also: L. Lindh, T. Arnebrant, P.-E. Isberg and P.-O. Glantz, *Biofouling*, 1999, **14**, 189.
195. See also: L. Lindh, P.-O. Glantz, P.-E. Isberg and T. Arnebrant, *Biofouling*, 2001, **17**, 227.
196. R. Müller, G. Gröger, K.-A. Hiller, G. Schmalz and S. Ruhl, *Appl. Environ. Microbiol.*, 2007, **73**, 2653.
197. See also: R. Müller, K.-A. Hiller, G. Schmalz and S. Ruhl, *Anal. Biochem.*, 2006, **359**, 194.
198. For oral film formation on polydimethylsiloxane (PDMS) coatings, see: R. E. Baier and P.-O. Glantz, *Acta Odontol. Scand.*, 1978, **36**, 289.
199. S. Decramer, A. Gonzalez de Peredo, B. Breuil, H. Mischak, B. Monsarrat, J.-L. Bascands and J. P. Schanstra, *Mol. Cell. Proteomics*, 2008, **7**, 1850.
200. S. Bouatra, F. Aziat, R. Mandal, A. C. Guo, M. R. Wilson, C. Knox, T. C. Bjorndahl, R. Krishnamurthy, F. Saleem, P. Liu, Z. T. Dame, J. Poelzer, J. Huynh, F. S. Yallou, N. Psychogios, E. Dong, R. Bogumil, C. Roehring and D. S. Wishart, *PLoS One*, 2013, **8**, e73076.
201. D. F. Putnam, Composition and Concentrative Properties of Human Urine, *NASA Contractor Report NASA CR-1802*, National Aeronautics and Space Administration Publisher, USA, 1971.
202. M. Houser, *J. Pediatr.*, 1984, **104**, 845.
203. T. Pisitkun, R. Johnstone and M. A. Knepper, *Mol. Cell. Proteomics*, 2006, **5**, 1760.
204. M. K. Parr and W. Schänzer, *J. Steroid Biochem. Mol. Biol.*, 2010, **121**, 528.
205. K. Wolff, M. Farrell, J. Marsden, M. G. Monteiro, R. Ali, S. Welch and J. Strang, *Addiction*, 1999, **94**, 1279.
206. J. A. Armstrong, *Kidney Int.*, 2007, **71**, 384.
207. C. Zhao, L.-Y. Li, M.-M. Guo and J. Zheng, *Chem. Pap.*, 2012, **66**, 323.

208. SPR immunosensing of staphylococcal enterotoxin B in milk using mixed carboxyl-/hydroxyl-terminated alkyl SAM surface chemistry: J. Homola, J. Dostálek, S. Chen, A. Rasooly, S. Jiang and S. S. Yee, *Int. J. Food Microbiol.*, 2002, **75**, 61.
209. SPR immunosensing of mycocardial infarction biomarkers in bovine serum using carboxyl-terminated alkyl SAM surface chemistry: J.-F. Masson, T. M. Battaglia, P. Khairallah, S. Beaudoin and K. S. Booksh, *Anal. Chem.*, 2007, **79**, 612.
210. SPRi biomarker immunosensing in mouse embryonic stem cell lysate using PEG surface chemistry: D. Tyagi, J. B. Perez, A. Nand, C. Zhiqiang, P. Wang, J. Na and J. Zhu, *Anal. Biochem.*, 2015, **471**, 29.
211. SPRi immunosensing in crude mouse brain cell lysate using PEG surface chemistry: M. Kyo, K. Usui-Aoki and H. Koga, *Anal. Chem.*, 2005, **77**, 7115.
212. SPR immunosensing of interleukin-8 protein in human saliva using carboxymethyl dextran carbohydrate surface chemistry: C.-Y. Yang, E. Brooks, Y. Li, P. Denny, C.-M. Ho, F. Qi, W. Shi, L. Wolinsky, B. Wu, D. T. W. Wong and C. D. Montemagno, *Lab Chip*, 2005, **5**, 1017.
213. S. Tosatti, S. M. De Paul, A. Askendal, S. VandeVondele, J. A. Hubbell, P. Tengvall and M. Textor, *Biomaterials*, 2003, **21**, 4949.
214. J.-F. Masson, T. M. Battaglia, M. J. Davidson, Y.-C. Kim, A. M. C. Prakash, S. Beaudoin and K. S. Booksh, *Talanta*, 2005, **67**, 918.
215. A. Hucknall, S. Rangarajan and A. Chilkoti, *Adv. Mater.*, 2009, **21**, 2441.
216. A. Hucknall, D.-H. Kim, S. Rangarajan, R. T. Hill, W. M. Reichert and A. Chilkoti, *Adv. Mater.*, 2009, **21**, 1968.
217. I. A. Buhimschi, R. Christner and C. S. Buhimschi, *BJOG*, 2005, **112**, 173.
218. A. Farina, J.-M. Dumonceau, P. Antinori, I. Annessi-Ramseyer, J.-L. Frossard, D. F. Hochstrasser, M. Delhaye and P. Lescuyer, *Biochim. Biophys. Acta*, 2014, **1844**, 1018.
219. S. A. Kharitonov and P. J. Barnes, *Am. J. Respir. Crit. Care Med.*, 2001, **163**, 1693.
220. M. Shirasu and K. Touhara, *J. Biochem.*, 2011, **150**, 257.
221. M. A. Feig, E. Hammer, U. Völker and N. Jehmlich, *J. Proteomics*, 2013, **83**, 119.
222. X. Cai, G. Pacheco-Rodriguez, Q.-Y. Fan, M. Haughey, L. Samsel, S. El-Chemaly, H.-P. Wu, J. P. McCoy, W. K. Steagall, J.-P. Lin, T. N. Darling and J. Moss, *Am. J. Respir. Crit. Care Med.*, 2010, **182**, 1410.
223. S. H. Kim, J. Y. Kim, H. J. Lee, M. Gi, B. G. Kim and J. Y. Choi, *PLoS One*, 2014, **9**, e111039.
224. A. D. Sutherland, R. B. Gearry and F. A. Frizelle, *Dis. Colon Rectum*, 2008, **51**, 1283.
225. C. Malagù, B. Fabbri, S. Gherardi, A. Giberti, V. Guidi, N. Landini and G. Zonta, *Sensors*, 2014, **14**, 18982.
226. L. Cui, X. Zhang, G. Ye, T. Zheng, H. Song, H. Deng, B. Xiao, T. Xia, X. Yu, Y. Le and J. Guo, *Cancer*, 2013, **119**, 1618.

227. S.-R. Won, D.-C. Lee, S. H. Ko, J.-W. Kim and H.-I. Rhee, *Food Res. Int.*, 2008, **41**, 952.
228. C. D. Derby, *Mar. Drugs*, 2014, **12**, 2700.
229. P. Gromov, I. Gromova, C. J. Olsen, V. Timmermans-Wielenga, M.-L. Talman, R. R. Serizawa and J. M. A. Moreira, *Biochim. Biophys. Acta*, 2013, **1834**, 2259.
230. A. Urruticoechea, I. E. Smith and M. Dowsett, *J. Clin. Oncol.*, 2005, **23**, 7212.
231. Z. Meng and T. D. Veenstra, *Proteomics Clin. Appl.*, 2007, **1**, 747.
232. F. H. Grus, S. C. Joachim and N. Pfeiffer, *Proteomics Clin. Appl.*, 2007, **1**, 876.
233. K. Karatolios, S. Pankuweit and B. Maisch, *Heart Failure Rev.*, 2013, **18**, 337.
234. A. C. Lysaght, S.-Y. Kao, J. A. Paulo, S. N. Merchant, H. Steen and K. M. Stankovic, *J. Proteome Res.*, 2011, **10**, 3845.
235. R. A. Siciliano, M. F. Mazzeo, V. Spada, A. Facchiano, A. d'Acierno, M. Stocchero, P. De Franciscis, N. Colacurci, N. Sannolo and N. Miraglia, *Gynecol. Endocrinol.*, 2014, **30**, 872.
236. J. M. Porcel, *Clin. Chest Med.*, 2013, **34**, 27.
237. F. González-Illán, G. Ojeda-Torres, L. M. Díaz-Vázquez and O. Rosario, *J. Anal. Toxicol.*, 2011, **35**, 232.
238. B. Behrouzi, S. Kenigsberg, N. Alladin, S. Swanson, J. Zicherman, S.-H. Hong, S. I. Moskovtsev and C. L. Librach, *Syst. Biol. Reprod. Med.*, 2013, **59**, 153.
239. S. D. Sagel, B. D. Wagner, M. M. Anthony, P. Emmett and E. T. Zemanick, *Am. J. Respir. Crit. Care Med.*, 2012, **186**, 857.
240. J. B. Catterall, T. V. Stabler, C. R. Flannery and V. B. Kraus, *Arthritis Res. Ther.*, 2010, **12**, R229.
241. C. Deirmengian, N. Hallab, A. Tarabishy, C. Della Valle, J. J. Jacobs, J. Lonner and R. E. Booth Jr., *Clin. Orthop. Relat. Res*, 2010, **468**, 2017.
242. N. von Thun und Hohenstein-Blaul, S. Funke and F. H. Grus, *Exp. Eye Res.*, 2013, **117**, 126.

CHAPTER 6

Prevention of Deleterious Biofluid–Surface Interactions in Detection and Medical Devices: A Look into the Future

6.1 On the Mechanism of Surface Resistance to Protein Adsorption: Schools of Thought and Rationalization

As seen in previous chapters, the spontaneous adsorption of proteins onto foreign surfaces and the resulting conformational changes potentially leading to the exposure of cryptic binding sites for cell surface receptors is a ubiquitous, problematic phenomenon – a genuine biotechnological plague – that affects the performance and fate of both bioanalytical and biomedical equipment, implants or devices in contact with blood or other biofluids. It is thus of no surprise that considerable efforts have been devoted over the last few decades to engineering so-called 'protein-resistant' surfaces, most traditionally through the application of organic coatings onto bulk substrates (as thoroughly discussed in Chapter 5). Despite a surge of interest in zwitterionic coatings in recent years, poly- and oligoethylene oxide/glycol (PEO or PEG, and OEO or OEG) non-charged constructs still, to this day, remain the most widely studied for implementation into real-world applications.[1–7] Years of research have not allowed, however, to fully elucidate the mechanism behind the antifouling properties of such surface chemistry;

RSC Detection Science Series No. 8
Biological Fluid–Surface Interactions in Detection and Medical Devices
By Michael Thompson, Christophe Blaszykowski, Sonia Sheikh, Cesar Rodriguez-Emmenegger and Andres de los Santos Pereira

Published by the Royal Society of Chemistry, www.rsc.org

a mechanism that still remains somewhat obscure but the object of great interest for both fundamental and applied biophysical chemistry research.[7–9] The origin of the 'PEO/OEO effect' has long been debated, and two main schools of thought – that consider protein repellence from a 'physical' or a 'chemical' point of view – have emerged.[10] There have been several contributions to both views of protein resistance; however, only the key ones will be discussed hereinafter.

6.1.1 Physical View

This view heavily hinges on notions developed in the 1970–1980s embodied in the Alexander–de Gennes theory of polymer interfaces.[10–14] The theory, which focuses on steric effects, was later used by Andrade and Jeon to construct a theoretical model, wherein the effect of long and flexible PEO chains on protein adsorption was studied.[15,16] This early 1990s work essentially showed that, as a protein approaches a surface, PEO chains would experience compression – a situation entropically unfavorable resulting in the generation of repulsive forces. This 'steric repulsion' (or 'excluded volume') effect is widely considered to be the main mechanism responsible for protein repellence. Although this highly influential work continues to be heavily referred to, it has been pointed out that such a simplistic model suffers from some limitations.[10] For instance, it would only be valid for very long, heavy PEO chains with a repeat unit length $n > 1000$ and molecular weight MW > 44000 g mol^{-1}, as well as for high PEO chain surface densities.[10] Another shortcoming would be that proteins are meaninglessly modelled as mere spheres or as having infinite size.[15,16] At the turn of the millennium, Szleifer and colleagues worked at developing – using the single chain mean field theory – an improved model taking into account all interactions PEO chains may experience.[10,17–21] This work – which assumed protein-PEO, protein–solvent, and PEO–solvent interactions to be identical – concluded that, unlike surface grafting density, the length of PEO chains had little effect on protein resistance. Another milestone in the 'physical' view of protein repellence was reached by Halperin in 1999, who followed Andrade–Jeon's and Szleifer's efforts.[22] In Halperin's model, the effective interaction experienced by a protein approaching a PEO brush-coated surface is described as being the sum of: (i) a purely repulsive force between the protein and the PEO brush, and (ii) a purely attractive force between the protein and the bare surface. Protein adsorption is discussed as a function of the PEO chain grafting density, brush thickness, and protein size.

6.1.2 Chemical View: the Pivotal Cloaking Role of Water

The 'physical' models discussed thus far relied on steric repulsion and entropic effects to explain the resistance of PEO coatings to protein adsorption. Although these models were built in a hydrated state (rather than in vacuum), the effect of water was neglected. In fact, molecules of water were represented as non-interacting spheres, a situation which clearly does not accurately describe the reality of hydrated systems, in which hydrogen bonding interactions with water are an important inherent aspect.[10] This recognition naturally led to the 'chemical' view of protein resistance in which the general consensus – according to which water through surface hydration would play a key role in antifouling – has now been reached. The hypothesis is relevant not only for PEG/OEG coatings (through hydration *via* hydrogen bonding) but also zwitterionic films (through hydration *via* ionic solvation).[3,4,9,23–33] Even though the effect of surface hydration would affect coatings made of either long PEG or shorter OEG chains, its relative contribution to antifouling would be expected to be much greater for the latter type of construct as the role of entropy due to chain flexibility/compression would be less significant in this case.[3] Regardless, the precise underlying mechanism still is a matter of some debate, owing to the sheer number of influential parameters on both surface and protein ends.[9,23,24,28–33] There have been as well several milestones in the evolution of the chemical view of protein repellence. One example is the theory proposed in the 1990s by Besseling *et al.* that takes into account hydration forces between surfaces, and the dependence of water molecule orientation on the interactions.[34–36] The investigators suggested that the surface, water molecules, and incoming proteins communicate through hydrogen bonding interactions, whose magnitude would depend on the number of electron donor (*i.e.* oxygen atoms) and electron acceptor (*i.e.* hydrogen atoms) sites involved. The model proposed that incoming proteins would be repelled from highly hydrated PEG or OEG surfaces as a result of the enthalpic cost associated with the removal of water molecules hydrogen-bonded to the grafted PEG/OEG chains. More recently, Grunze and co-workers also contributed to developing the chemical view of protein resistance by investigating the effect of chain conformation through both experimental and theoretical means.[37–39] The authors prepared, on both gold and silver substrates, OEG self-assembled monolayers (SAMs) containing three ethylene glycol units, and used FT-IR spectroscopy to look at chain conformation while assessing fibrinogen adsorption through force measurements.[37,39] It was found that OEG-SAMs prepared on gold substrates displayed amorphous and helical chain conformations, and were resistant to fibrinogen adsorption. Conversely, OEG-SAMs prepared on silver substrates formed tightly-packed assemblies with high surface density, preferred an '*all-trans*' chain conformation, and adsorbed fibrinogen. *All-trans* OEG residues were driven into such conformation as a result of their perpendicular packing orientation with respect to the silver substrate, as opposed to the 30° chain tilt (relative to the normal) observed in the case of gold. Another consequence was a smaller cross-sectional area per molecule on silver (19.1 $Å^2$) than on gold (21.4 $Å^2$).[37] To understand the underlying raison d'être behind their experimental observations, Grunze and co-workers conducted a complementary study using computer simulations.[38,40] The investigators concluded that, unlike their *trans* counterparts, OEG chains adopting a helical conformation offer a conducive environment for interchain water absorption – the related OEG-SAMs on gold displaying a tighter interfacial layer of water. In 2003, Grunze's team expanded on this model by demonstrating that, along with OEG chain conformation, the hydrophilicity of both the terminal group and inner chain units, as well as the lateral packing density of the chain, influenced the ability of OEG-SAMs to coordinate water and thus resist protein adsorption.[39] The authors found that protein adsorption on OEG-SAMs increased as internal or interfacial water coordination decreased. This led to the argument of a 'water barrier', wherein embedded and interfacial water molecules are tightly bound and organized into permeated structures that have an energy cost in terms of disturbance.[3] Another argument rather favors the notion of 'interfacial energy matching', according to which there is no net energy gain for solutes – which are fully solvated in the bulk aqueous medium where they reside – to adsorb on hydrated surfaces.[2,24,28] In both cases, water in contact with surfaces may form a phase physically distinct from ordinary bulk water,[24,25,28,33] as supported by empirical evidence.[41–43] The nature and extent of this special zone of hydration have also been extensively debated.[24,28,43–45] For instance, while some have described physically distinct water interphases projecting up to several hundred microns into the contiguous aqueous medium,[43] others, in stark contrast, have reported the phenomenon to be much more limited, to a few layers of water or less.[24,28,41,42] Understandably, the contentious question of surface hydration – and its connection to antifouling – has been, and continues to be, relentlessly researched by many.[3,24]

6.1.3 Surface Hydration and Antifouling: A Rationalization of their Connection through both Empirical and Computational Evidence

Water is essential to Life. This understatement stems from the vital role this ubiquitous fluid plays in a wide variety of biological and other processes due to its many unique dissolving, dissociating/lysing, and solvating properties. As discussed above, it is thus not surprising that water is hypothesized to also play a key role in the antifouling of surfaces, although the precise underlying mechanism is, as mentioned, still a matter of debate. Recently, we also have set out to try and solve this puzzle through the aid of a series of both experimental and computational studies.[46–48] Initial experiments implemented the use of ultra-high frequency bulk acoustic wave sensor technology to investigate the antifouling properties against full serum of new organosiloxane surface chemistry on quartz. A key result was the observation that, among a series of ultrathin coatings with systematically-varied chemical structure (Figure 6.1), only the monoethylene glycol (MEG) variety incorporating a single, internal ether atom of oxygen in the chains was able to dramatically alter serum adsorption, the synergy being strongest for the MEG-OH system possessing distal hydroxyl moieties.[46]

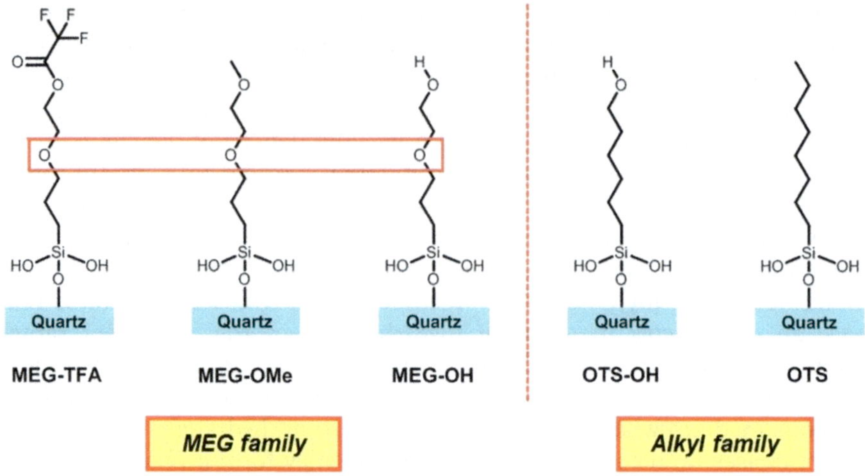

Figure 6.1 Systematic surface modification of quartz (SiO_2) with ultrathin organosiloxane adlayers.

Conversely, the antifouling behavior was less pronounced for alkylated adlayers lacking the internal ether oxygen atom in the chains. Further comparative analysis suggested that antifouling properties would also be deeply rooted in (the state of) surface hydration, wherein the internal ether atom of oxygen in the MEG backbones – a feature alkylated systems lack – is postulated to play a key role through the instigation of a special intrafilm zone of hydration.[46] Ensuing neutron reflectometry experiments conducted with the antifouling MEG-OH system and its less effective OTS-OH alkylated analogue corroborated this mechanistic hypothesis by revealing the existence of distinct hydration patterns schematically depicted in Figure 6.2.[47] For the OTS-OH system, water appeared to be unable to penetrate within the adlayer, but formed a short-ranged interphase physically distinct from bulk water (Figure 6.2, right). This region also seemed to be characterized by a highly-structured molecular organization with a sharp transition towards ordinary bulk water. This is to be contrasted with the continuous, longer-range and diffusive transition zone of water stemming from within the highly-absorbent MEG-OH adlayer (Figure 6.2, left). Collectively, acoustic wave sensing and neutron reflectometry experiments provided convincing empirical evidence that the internal ether atom of oxygen in the MEG chains would play a determining role in (the state of) surface hydration, hence antifouling.

We next complemented this experimental probing of surface hydration with a series of molecular dynamics simulations that allowed for a better understanding of antifouling with respect to surface hydration to emerge; and a molecular-level mechanism – well in line with that generally invoked in the literature – to be rationalized in terms of a set of basic requirements, those presented in Figure 6.3. In this study, Sheikh *et al.* showed that in

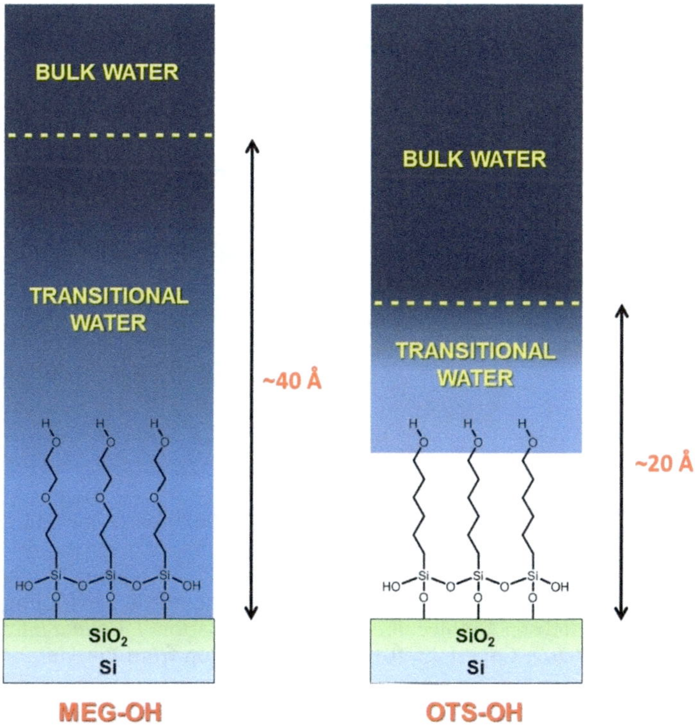

Figure 6.2 Distinct hydration patterns for MEG-OH and OTS-OH siloxane coatings on Si/SiO_2 proposed from the interpretation of neutron reflectometry data. Note: this schematic representation (not to scale) merely depicts MEG-OH and OTS-OH surface chemistry, not the actual surface coverage/patchiness of these adlayers, nor the anchorage nature and degree of order/packing of the surface-modifying residues within.

order to be pronouncedly antifouling, a coating must display both internal and interfacial hydrophilicity, and kosmotropicity (the ability to structure water at the molecular level). Furthermore, hydration water must be tightly coordinated with limited dynamicity (*i.e.* lability and mobility).[48]

The satisfaction of all these criteria would give rise to a stable permeant network of water, whose disturbance *via* dehydration upon protein adsorption would constitute an energetic penalty and generate repulsive forces. A 'substrate effect' to surface hydration, which would take place at defected/damaged sites and participate in endowing antifouling adlayers with hydrogel-like characteristics, was also proposed.[48] In contrast, no significant role in antifouling was assigned to the entropy linked to adlayer flexibility/compressibility for such ultrathin structures made of short MEG building blocks.[48] Even though one may disagree on the specifics of this paradigm, it is safe to say that the concept of surface hydration will (continue to) hold a key position in the educated design of antifouling surfaces.

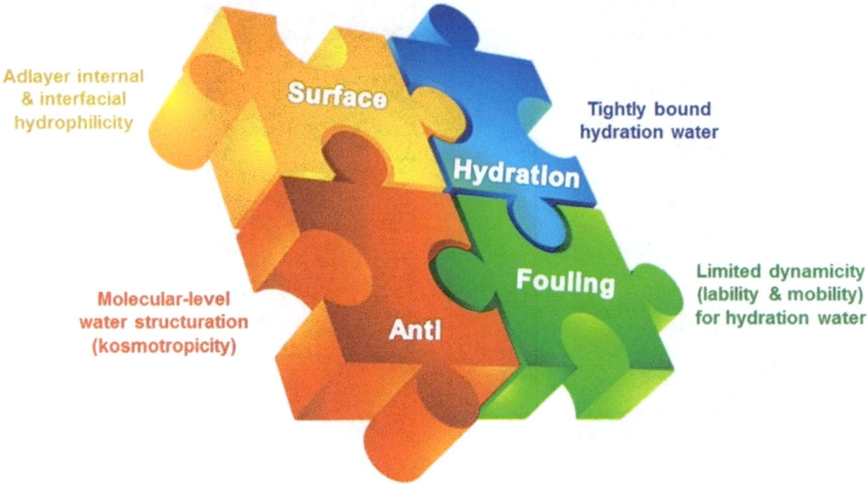

Figure 6.3 The 'surface hydration/antifouling connection' puzzle solved in terms of a set of basic requirements.
Reprinted from, S. Sheikh, C. Blaszykowski, R. Nolan, D. Thompson, M. Thompson, On the hydration of subnanometric antifouling organosilane adlayers: A molecular dynamics simulation, *J. Colloid Interface Sci.*, **437**, 197-207, Copyright 2014 with permission from Elsevier.

6.2 Ultrathin Surface Chemistry for Biocompatibility Enhancement

The pronounced antifouling behavior against a complex biological medium such as blood serum is only one remarkable feature of the MEG-OH nanogel surface chemistry developed by Sheikh *et al.* in 2012.[46] Demonstrated soon after was the outstanding antithrombogenicity exhibited by ultrathin MEG-OH coatings prepared on polycarbonate, a plastic polymer widely used in the biomedical industry.[49] Antithrombogenic properties were assessed after 5 min of continuous exposure to fluorescently-labelled whole human blood dispensed through a perfusion chamber at a shear rate of 1000 s^{-1}, a condition of haemodynamics conducive to platelet accumulation/thrombus growth. Remarkably, platelet adhesion, aggregation and thrombus formation on MEG-OH surface chemistry was greatly inhibited (>97% decrease in surface coverage) compared to the non-derivatized substrate material. Most importantly, thrombogenesis was nearly non-existent, as can be seen in Figure 6.4. Such an achievement was recently repeated with poly(vinyl chloride) as the substrate, a material that constitutes another highly praised plastic polymer in the biomedical industry. This remarkable result was obtained regardless of the applied shear rate (300, 900, and 1500 s^{-1}) or blood exposure time (now lasting up to 60 min).[50] Moreover, a closer look at the few observable micro-aggregates on MEG-OH surface chemistry also revealed that these entities, which ordinarily spanned less than 5 μm across,[49,50]

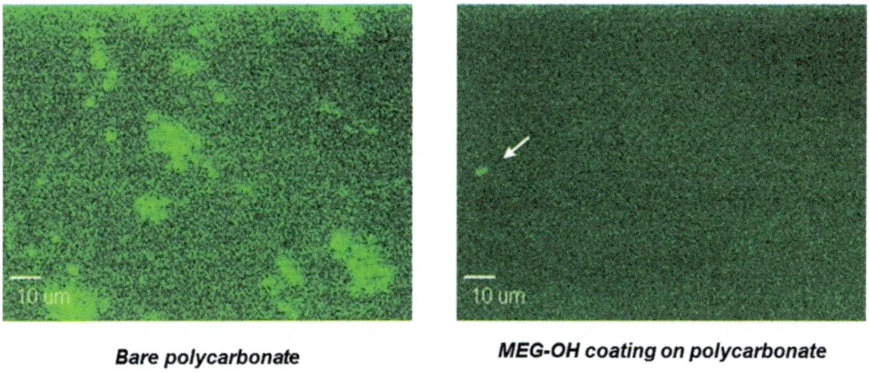

Figure 6.4 Platelet adhesion, aggregation and thrombus formation, or lack of, on bare (left) and MEG-OH-derivatized (right) polycarbonate substrates after five minutes exposure to fluorescently-labelled whole human blood at a shear rate of 1000 s^{-1}.
Reprinted with permission from ref. 49. Copyright 2014 American Chemical Society.

appeared to only be comprised of a limited number of platelets (individual size of 2–4 μm), making them unlikely to cause any harm associated with blood vessel occlusion should they detach as circulating thromboemboli.

Despite their preliminary nature, these results by Thompson and collaborators with MEG-OH surface chemistry were quite remarkable considering that blood was not given, nor apparently required, any anticoagulant treatment to prevent clotting (besides its standard collection and storage in heparinized Vacutainers) in spite of being exposed *in vitro* to a foreign surface.[49,50] There is immense potential packed in this nascent ultrathin surface chemistry awaiting to be tapped to impart virtually any substrate material with two highly praised and sought-after features in the realm of biomaterials: antifouling behavior against blood serum, and antithrombogenicity when exposed to whole blood. Far ahead still is, however, the day when MEG-OH coatings will be implemented in real-world bioanalytical/biomedical applications, but the potential to cause a biotechnological revolution in the field is already quite real.

State-of-the-art polymer brushes capable of preventing fouling from blood plasma were also found to resist the attachment of leukocytes and platelets, which makes such chemical surface modification strategy another promising avenue for the design of haemocompatible biomaterials. However, regardless of the type of surface-modifying chemistry implemented, such an achievement only represents the tip of the iceberg as long-term interaction (days, weeks, months, and even years), the desirable non-utilization of anticoagulant agents, and the combined effect of blood cellular and humoral response systems constitute many areas that often still remain to be investigated. Moreover, rheological conditions may also have an effect, either negative, as fluid dynamics can alter mass transport characteristics over a surface, or positive, by preventing the accumulation of activated complement and coagulation mediators. The promise of chemical surface modification ideally displaying no thrombogenicity, immunogenicity, complement system activation, and platelet/leukocyte attachment would have profound implications both for short- and long-term applications. The success of its translation into the fabrication of actual devices would still require to be validated through an extra testing step, in compliance with strict governmental regulations/international standards. For use in extracorporeal circulation apparatuses (*e.g.* in cardiopulmonary bypass surgery and haemodialysis), requirements may not be as stringent as the duration of such procedures only typically lasts a few hours. In contrast, permanent implants intended to stay inserted for long periods of time, potentially reaching decades, demand a much more delicate level of control of their interaction with surrounding biological tissue so as to effectively direct the healing response towards the desired outcome. Whichever the application may be, the functionality and life expectancy of biomedical equipment, implants or devices will critically depend on a thorough understanding of the interactivity and interplay of their exogenous synthetic surface with the multi-components of the biofluid matrix they are exposed to or the tissue they are contacting.

6.3 Rational Design of Biocompatible Materials

We live in a golden age, wherein anything like rarely before seems technologically achievable. Yet, predictably endowing synthetic materials with biocompatibility – for use in biomedical equipment, for example – is still a challenge at the beginning of the 21st century. Failure can partially be attributed to a certain oblivion of the fact that bioinertness is (1) the contextual property of an interacting system (*i.e.* a given material/biological environment system) and not solely an inherent property of said material's surface;[51] and 2) determined at the most fundamental, (sub)molecular level.[52] Attempts have been made in the literature to try and raise awareness that it is indeed at this nanometer scale that thrombogenic/immune biological processes/responses[53] with potentially deleterious repercussion are triggered upon the surface-mediated activation of adsorbed/adsorbing proteins, through conformational restructuration with exposure of cryptic bioactive sites.[52–55] One can read, nonetheless, that antifouling – in this context, the ability of an artificial material's surface to merely prevent the natural adsorption and accumulation of proteins – would constitute a prerequisite to the biocompatibility of a material. However, as recently argued and demonstrated elsewhere, the often misidentified key to achieving such property would not be to try and permanently minimize the (total) amount of adsorbed proteins, but rather to subtly control the degree of unfolding (*i.e.* control the conformational state) that relevant proteins – whichever they may be[56] and no matter how long the unfolding process may otherwise take[57] – experience upon co-adsorption with other, bystanding proteins (as schematically depicted in

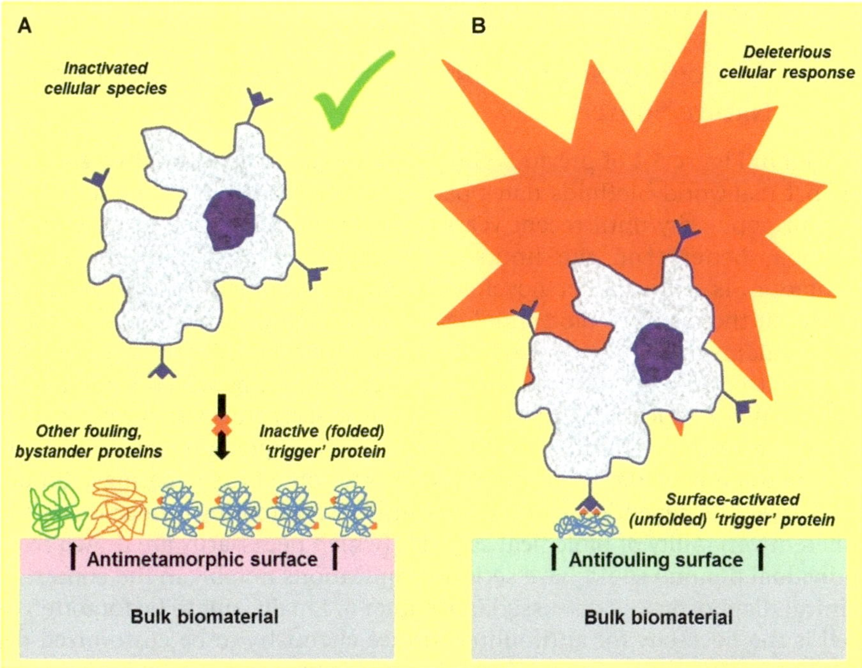

Figure 6.5 Material biocompatibility and the 'dynamics *vs.* amount of protein adsorption' paradigm. (A) A biocompatible 'antimetamorphic' surface prevents proteins capable of initiating a biological process/response – and co-adsorbed among other fouling, bystander matrix proteins – from unfolding and exposing hitherto concealed bioactive sites. (B) Conversely, the 'antifouling' surface – although much more efficient at reducing the (total) amount of protein adsorption – is not intrinsically engineered to inhibit the conformational restructuration of residual adsorbed proteins resulting in bioactive site exposure.
Reprinted from, C. Blaszykowski, S. Sheikh, M. Thompson, Biocompatibility and antifouling: is there really a link, *Trends Biotechnol.*, **32**, 61–62, Copyright 2014 with permission from Elsevier.

Figure 6.5).[53] In this 'dynamics *vs.* amount of protein adsorption' paradigm,[58] the adjectives 'biocompatible' and 'antifouling' are not synonyms (Figure 6.5).[53] In practice, there exist two promising routes to advantageously alter the nanoscale physicochemical features of the surface of materials so as to inhibit the undesirable conformational rearrangement and activation of adsorbed proteins: a physical method that could consist in the direct nanopatterning of biomaterial substrates with commensurate protein dimensions,[59–61] and a chemical one that would hinge on substrate modification with hydrogel surface chemistry (in order to present proteins with an energetic continuum in terms of the solvation environment).[53,62,63] Admittedly, compared to the common trial-and-error 'random' approach, lengthier can be expected to be this road leading to the rational design of bioinert materials, but so is sometimes the path to genuine success.

6.4 The Future in Practice

6.4.1 Antifouling Surface Chemistry: Coating *vs.* Biosample Nature/Source

As seen in Figure 5.4 of previous Chapter 5, research on antifouling surfaces against real-world biofluids dates back several decades (1970s) but did not gain in popularity until recent years.[7] Short-term resistance to protein adsorption, better than the limits of detection of state-of-the-art instrumentation, is still a rather novel achievement, so much that the potential impact in the realm of Biotechnology of such highly effective repellent surfaces remains to be fully realized. Much biophysical chemistry research also remains to be conducted regarding the mechanisms by which some apparently similar surfaces display such dramatic discrepancies in terms of their resistance to the adsorption of blood plasma components, such as is the case between polySBMA and polyCBAA zwitterionic brush coatings. On another level, the dependence of a given coating's antifouling properties on the inherent variability of biological samples (which necessarily are drawn from individual donors) could have serious implications notably in the context of clinical diagnostics, as discussed in Chapter 5. Finally, not to be forgotten as well is the necessity for antifouling surface chemistry to be customized depending on the nature of the biofluid in which bioanalysis is intended to be performed.

The design of new antifouling surfaces with improved performance will require a deeper understanding of the molecular-level organization (*e.g.* distribution, conformation) of molecular/polymer building blocks within such supramolecular assemblies, their interaction with water molecules and ions, and the resulting combined effect in terms of repulsive forces/barriers to proteins and other adsorbing species. As was the case in the study of internal and interfacial water arrangement within/atop MEG-OH adlayers discussed earlier in Subsection 6.1.3, neutron reflectometry experiments and molecular dynamics computer simulations can be expected to deliver valuable insight in this matter. Clearly, the multiplicity of influential parameters on both surface and interacting system ends entails future investigations to be highly interdisciplinary in their approach to probe the various interactions at play. In this regard, novel developments in characterization methodologies and techniques could provide additional avenues to try and fulfil such an arduous task. Leaving aside their validity in the perspective of real-world applications with complex biological media, it is a fact for example that single protein adsorption experiments traditionally focus on the quantitative and kinetic aspect of the phenomenon, paying much less attention to the equally important information about the strength or type of the interaction. New developments in AFM technology – which have enabled the measurement of individual binding events, notably those involving bacteria – could possibly be of service here.

6.4.2 Implants: Tissue Healing and Bacterial Infection

Preventing protein adsorption, opsonization, and cell attachment – as a means to inhibit thrombogenicity and the foreign body reaction – could represent the most straightforward strategy towards the functional integration of implants. Another opposite approach would be to actually promote and mediate the specific attachment and proliferation of the cells necessary to normal tissue function. In the case of implanted objects expected to intimately contact blood for extended periods of time (*e.g.* artificial vascular grafts and stents), this latter strategy refers to the supported growth of new endothelial lining tissue. While tissue engineering advances may eventually make it possible to tailor ready-to-use fully biological implants *in vitro*, the tolerance of an artificial material by its biological surrounding can also be envisioned to be provided *in vivo* through surface modification with a dual-functional coating exhibiting both antifouling/biocompatible and biorecognition capability. Derivatized in this manner, implanted scaffolds will signal specific cells to attach, differentiate, and regenerate tissue without being hampered by the detrimental interference of other matrix species. In practice, the suppression of adverse reactions combined with the enhancement of tissue healing could bring the integration of bioimplants and associated technology onto a whole new level. For instance, particularly welcome would be the controlled avoidance of the aforementioned 'foreign body reaction', notably for indwelling biosensors (*e.g.* for continuous glucose monitoring), as such an accomplishment would allow to considerably prolong the operational life expectancy of implanted detection devices.

Always accompanying any implantation/insertion procedure (*e.g.* of a stent or catheter) is the risk for bacterial infection. Indeed, as explained in previous chapters, proteins spontaneously adsorbed onto indwelling material may form a conditioning film conducive to bacteria attachment and colonization. As a consequence, there is currently great interest in engineering antibacterial surfaces capable of preventing such proteinaceous layers and resultant biofilms to materialize. Considering the great concern the phenomenon causes in the biomedical world, this field of research is not expected to fade any time soon.

6.4.3 Bioanalysis and the Universalization of Biosensor Technology

As a consequence of the customizable combination of effective antifouling layers against complex biological media with improved functionalization strategies, the applicability of label-free affinity biosensor technology for the detection in real-world biosamples of virtually any type of target analyte is ever so slightly coming within reach, as seen in Chapter 5. Technical improvements of transducer components have also resulted in enhanced sensitivities and limits of detection, as well as new sensor configurations allowing high-throughput parallel detection (*e.g.* SPR imaging).

The combination of such advanced devices with effective dual-functional sensing interfaces (*i.e.* presenting both antifouling and biorecognition capabilities) can be expected to soon revolutionize the diagnostic toolbox available to physicians. Indeed, one such sensing device could in principle be capable of detecting hundreds of different biomarkers within minutes without prior sample preparation. Also in sight is another goal of biosensor technology, that of being capable of performing point-of-care testing with droplet size samples in order to yield rapid and early information about a disease, and efficiently track its progress, so as to anticipate the onset of symptoms and improve the prognosis by implementing appropriate personalized therapy. Label-free affinity biosensors based on surface plasmon resonance and quartz crystal microbalance detection technologies (and their variants) have already made a substantial contribution in bioanalysis. In the ever-increasing technological world we live in, the research on portable biosensor devices for healthcare purposes is promised a bright future.

References

1. B. D. Ratner, D. G. Castner, T. A. Horbett, T. J. Lenk, K. B. Lewis and R. J. Rapoza, *J. Vac. Sci. Technol., A*, 1990, **8**, 2306.
2. P. Gong and D. W. Grainger, Nonfouling Surfaces: A Review of Principles and Applications for Microarray Capture Assay Designs, in *Microarrays Volume 1 Synthesis Methods*, ed. J. B. Rampal, Methods in Molecular Biology, Humana Press Inc., USA, 2nd edn, 2007, vol. 381, pp. 59–92.
3. S. Chen, L. Li, C. Zhao and J. Zheng, *Polymer*, 2010, **51**, 5283.
4. C. Zhao, L.-Y. Li, M.-M. Guo and J. Zheng, *Chem. Pap.*, 2012, **66**, 323.
5. P. Vermette and L. Meagher, *Colloids Surf. B*, 2003, **28**, 153.
6. E. Ostuni, R. G. Chapman, R. E. Holmlin, S. Takayama and G. M. Whitesides, *Langmuir*, 2001, **17**, 5605.
7. C. Blaszykowski, S. Sheikh and M. Thompson, *Chem. Soc. Rev.*, 2012, **41**, 5599.
8. M. Thompson, S. Sheikh and C. Blaszykowski, *Austin J. Nanomed. Nanotechnol.*, 2014, **2**, 2.
9. E. A. Vogler, *Biomaterials*, 2012, **33**, 1201.
10. M. Morra, *J. Biomater. Sci., Polym. Ed.*, 2000, **11**, 547.
11. S. Alexander, *J. Phys. France*, 1977, **38**, 977.
12. P.-G. de Gennes, *Macromolecules*, 1981, **14**, 1637.
13. P.-G. de Gennes, *Macromolecules*, 1982, **15**, 492.
14. P.-G. de Gennes, *Ann. Chim.*, 1987, **77**, 389.
15. S. I. Jeon, J. H. Lee, J. D. Andrade and P.-G. de Gennes, *J. Colloid Interface Sci.*, 1991, **142**, 149.
16. S. I. Jeon and J. D. Andrade, *J. Colloid Interface Sci.*, 1991, **142**, 159.
17. T. McPherson, A. Kidane, I. Szleifer and K. Park, *Langmuir*, 1998, **14**, 176.
18. I. Szleifer, *Curr. Opin. Colloid Interface Sci.*, 1996, **1**, 416.
19. I. Szleifer, *Biophys. J.*, 1997, **72**, 595.
20. I. Szleifer, *Phys. A*, 1997, **244**, 370.

21. I. Szleifer, *Curr. Opin. Solid State Mater. Sci.*, 1997, **2**, 337.
22. A. Halperin, *Langmuir*, 1999, **15**, 2525.
23. R. A. Latour, *J. Biomed. Mater. Res., A*, 2006, **78**, 843.
24. M. J. Stevens and G. S. Grest, *BioInterphases*, 2008, **3**, FC13.
25. W. Feng, M.-P. Nieh, S. Zhu, T. A. Harroun, J. Katsaras and J. L. Brash, *BioInterphases*, 2007, **2**, 34.
26. Y. He, Y. Chang, J. C. Hower, J. Zheng, S. Chen and S. Jiang, *Phys. Chem. Chem. Phys.*, 2008, **10**, 5539.
27. L. Li, S. Chen, J. Zheng, B. D. Ratner and S. Jiang, *J. Phys. Chem. B*, 2005, **109**, 2934.
28. A. E. Ismail, G. S. Grest and M. J. Stevens, *Langmuir*, 2007, **23**, 8508.
29. J. Zheng, L. Li, S. Chen and S. Jiang, *Langmuir*, 2004, **20**, 8931.
30. J. C. Hower, M. T. Bernards, S. Chen, H.-K. Tsao, Y.-J. Sheng and S. Jiang, *J. Phys. Chem. B*, 2009, **113**, 197.
31. T. Hayashi, Y. Tanaka, Y. Koide, M. Tanaka and M. Hara, *Phys. Chem. Chem. Phys.*, 2012, **14**, 10196.
32. S. N. Jamadagni, R. Godawat and S. Garde, *Langmuir*, 2009, **25**, 13092.
33. M. W. A. Skoda, R. M. J. Jacobs, J. Willis and F. Schreiber, *Langmuir*, 2007, **23**, 970.
34. N. A. M. Besseling and J. M. H. M. Scheutjens, *J. Phys. Chem.*, 1994, **98**, 11597.
35. N. A. M. Besseling and J. Lyklema, *J. Phys. Chem.*, 1994, **98**, 11610.
36. N. A. M. Besseling and J. Lyklema, *Pure Appl. Chem.*, 1995, **67**, 881.
37. P. Harder, M. Grunze, R. Dahint, G. M. Whitesides and P. E. Laibinis, *J. Phys. Chem. B*, 1998, **102**, 426.
38. R. L. C. Wang, H. J. Kreuzer and M. Grunze, *J. Phys. Chem. B*, 1997, **101**, 9767.
39. S. Herrwerth, W. Eck, S. Reinhardt and M. Grunze, *J. Am. Chem. Soc.*, 2003, **125**, 9359.
40. A. J. Pertsin and M. Grunze, *Langmuir*, 2000, **16**, 8829.
41. H. I. Kim, J. G. Kushmerick, J. E. Houston and B. C. Bunker, *Langmuir*, 2003, **19**, 9271.
42. D. Schwendel, T. Hayashi, R. Dahint, A. Pertsin, M. Grunze, R. Steitz and F. Schreiber, *Langmuir*, 2003, **19**, 2284.
43. J.-M. Zheng, W.-C. Chin, E. Khijniak, E. Khijniak Jr. and G. H. Pollack, *Adv. Colloid Interface Sci.*, 2006, **127**, 19.
44. M. James, T. A. Darwish, S. Ciampi, S. O. Sylvester, Z. Zhang, A. Ng, J. J. Gooding and T. L. Hanley, *Soft Matter*, 2011, **7**, 5309.
45. S. P. Jarvis, G. Kaggwa and K. Sheikh, Liquid Structure at the Solid-Liquid Interface, in *Nanoscale Liquid Interfaces – Wetting, Patterning and Force Microscopy at the Molecular Scale*, ed. T. Ondarçuhu and J.-P. Aimé, Pan Stanford Publishing Pte Ltd., Singapore, 2013, pp. 1–27.
46. S. Sheikh, D. Y. Yang, C. Blaszykowski and M. Thompson, *Chem. Commun.*, 2012, **48**, 1305.
47. N. M. Pawlowska, H. Fritzsche, C. Blaszykowski, S. Sheikh, M. Vezvaie and M. Thompson, *Langmuir*, 2014, **30**, 1199.

48. S. Sheikh, C. Blaszykowski, R. Nolan, D. Thompson and M. Thompson, *J. Colloid Interface Sci.*, 2015, **437**, 197.
49. K. Fedorov, C. Blaszykowski, S. Sheikh, A. Reheman, A. Romaschin, H. Ni and M. Thompson, *Langmuir*, 2014, **30**, 3217.
50. K. Fedorov, A. Jankowski, S. Sheikh, C. Blaszykowski, A. Reheman, A. Romaschin, H. Ni and M. Thompson, *J. Mater. Chem. B*, 2015, **3**, 8623.
51. D. F. Williams, *Biomaterials*, 2014, **35**, 10009.
52. L. Tang and W. Hu, *Expert Rev. Med. Devices*, 2005, **2**, 493.
53. C. Blaszykowski, S. Sheikh and M. Thompson, *Trends Biotechnol.*, 2014, **32**, 61, and references therein.
54. T. Ballet, L. Boulange, Y. Bréchet, F. Bruckert and M. Weidenhaupt, *Bull. Pol. Acad. Sci.: Tech. Sci.*, 2010, **58**, 303.
55. P. Thevenot, W. Hu and L. Tang, *Curr. Top. Med. Chem.*, 2008, **8**, 270.
56. B. Sivaraman and R. A. Latour, *Biomaterials*, 2010, **31**, 1036.
57. B. Sivaraman and R. A. Latour, *Langmuir*, 2012, **28**, 2745.
58. B. Sivaraman and R. A. Latour, *Biomaterials*, 2010, **31**, 832.
59. A. G. Harvey, E. W. Hill and A. Bayat, *Expert Rev. Med. Devices*, 2013, **10**, 257.
60. P. Koegler, A. Clayton, H. Thissen, G. N. C. Santos and P. Kingshott, *Adv. Drug Delivery Rev.*, 2012, **64**, 1820.
61. B. Kasemo and J. Gold, *Adv. Dent. Res.*, 1999, **13**, 8.
62. E. A. Vogler, *J. Biomater. Sci. Polym. Ed.*, 1999, **10**, 1015.
63. P. Vadgama, *Annu. Rep. Prog. Chem., Sect. C: Phys. Chem.*, 2005, **101**, 14.

Subject Index